HADRONIC PRESS MONOGRAPH IN THEORETICAL PHYSICS

ISOTOPIC GENERALIZATIONS OF GALILEI'S AND EINSTEIN'S RELATIVITIES

VOLUME II: CLASSICAL FORMULATIONS

Ruggero Maria Santilli

- 1991 -

HADRONIC PRESS , INC.
35246 US 19 No. # 131
Palm Harbor, FL 34684 USA

ISOTOPIC GENERALIZATIONS OF GALILEI'S AND EINSTEIN'S RELATIVITIES

VOLUME II: CLASSICAL FORMULATIONS

Ruggero Maria Santilli

- 1991 -

HADRONIC PRESS , INC.
35246 US 19 No. # 131
Palm Harbor, FL 34684 USA

First printing: November 14, 1991
Second printing: May 12, 1992

Library of Congress Cataloging-in-Publication Data
(Revised for vol. 2)

Santilli, Ruggero Maria, 1935-
Isotopic generalizations of Galilei's and Einstein's relativities.

(Hadronic Press monographs in theoretical physics)
Includes bibliographical references and indexes.
Contents: v. 1. Mathematical foundations -- v. 2. Classical isotopies.
1. Relativity (Physics)--Mathematics. 2. Gravitatic
--Mathematics. 3. Mathematical physics. I. Title.
II. Series.
QC173.55.S36 1992 530.1'1'0151 92-18676
ISBN 0-911767-55-X (v. 1)
ISBN 0-911767-55-X (v. 2)

dedicated to

Professor **A. JANNUSSIS**
Department of Physics
University of Patras, Patras, Greece

because he is a leader of True Scientists
pursuing **novel** *physical knowledge,without which*
physical research has no thrill of discovery

ISOTOPIC GENERALIZATIONS OF GALILEI'S AND EINSTEIN'S RELATIVITIES
Volume II: Classical Formulations

TABLE OF CONTENTS

CHAPTER IV: ISOTOPIC GENERALIZATIONS OF EINSTEIN'S SPECIAL RELATIVITY

CHAPTER V: ISOTOPIC GENERALIZATIONS OF EINSTEIN'S GRAVITATION

CHAPTER VI: MUTUAL COMPATIBILITY OF THE ISOTOPIC RELATIVITIES

CHAPTER VII: EXPERIMENTAL VERIFICATION OF THE ISOTOPIC RELATIVITIES

PREFACE

OF VOLUMES I AND II

Physics is a science that will never admit *final theories* . No matter how autoritative, the generalization of fundamental physical theories is only a matter of time.

Physics is also a *quantitative science,* that is, requiring mathematically rigorous, quantitative formulations of predictions suitable for direct experimental verifications.

Finally, physics is a science with an absolute standard of values: the *experimental verification.* No matter how plausible a new theory is, it remains conjectural until verified in laboratories. By the same token, no matter how fundamental and authoritative an existing theory is, its validity remains conjectural for all physical conditions under which it has not been directly tested.

Along these lines, the sooner the scientific process is initiated with the submission of possible generalizations of existing theories and their critical examination by independent researchers, the better for the advancement of physical knowledge.

The author has spent his research life in studying possible classical and operator generalizations of *Galilei's relativity, Einstein's special relativity* and *Einsten's general relativity* (or *Einstein's gravitation* , for short). The studies were conducted along the following main lines:

1) Identifation of the *"physical conditions of unequivocal validity"* of conventional relativities;
2) Identification of broader physical conditions under which possible generalized relativities may be physically relevant;
3) Identification of the generalized mathematical tools needed for

a quantitative representation of the broader physical conditions considered;

4) Construction of the generalized relativities, including the identification of their mutual compatibility, implications and quantitative predictions; and, last but not least,
5) Formulation of specific experimental proposals for the verification or disproof of the new relativities.

In particular, this author has studied the above problem, in an evident preliminary way: a) for each of the Galilean, special and general profiles; b) for both classical and operator formulations; and c) in regard to the intrinsic compatibility of the emerging generalizations of Galilei's, the special and the general relativities, first, independently at the classical and operator level and, then, for the identification of a map from the classical into the operator formulations.

After an introductory chapter, Volume I is devoted to the review of the novel mathematical structures needed for a quantitative treatment of the broader physical conditions considered.

Volume II is devoted to: the construction of the classical generalized relativities; the study of their mutual compatibility; the identification of their most important implications; and the proposal of experiments for their verification or disproof.

The scope of these monographs is to identify the status of the studies in the field at this writing (Fall 1991), so that the interested researcher can appraise the new relativities, and participate in their mathematical-theoretical development or experimental verification.

The understanding of these volumes requires a mind open to the possibility that Galilei's relativity, Einstein's special relativity and Einstein's gravitation are not final theories, but only beautiful foundations for expected more general relativities for more complex physical conditions in the Universe.

Ruggero Maria Santilli
Box 1577, Palm Harbor, FL 34682 USA
Fall 1991

CHAPTER III: ISOTOPIC GENERALIZATIONS OF GALILEI'S RELATIVITY

III.1: STATEMENT OF THE PROBLEM

Galilei's relativity in its contemporary presentation (see, e.g., Levy-Leblond (1971), or Sudarshan and Mukunda (1974)) is a body of methodological formulations which, despite their technical advances, preserves in a remarkable way the original conception by Galilei (1638), Newton (1687) and other Founding Fathers of representing:

A) particles which can be effectively approximated as being point-like,

(B) while moving in the inhomogeneous and anisotropic vacuum (empty space), under action-at-a-distance potential interactions;

C) under the (contemporary) conditions that:

C-1) All possible speeds v are much smaller than the speed of light in vacuum c_o;
C-2) All quantum mechanical effects are ignorable (the action A is much bigger than the quantum of energy $\hbar$; and
C-3) All gravitational effects are absent, in the sense that all spaces are assumed to be flat.

In this chapter we shall study an infinite family of generalizations of Galilei's relativity, under the name of *Galilei-isotopic relativities*, or *isogalilean relativities* for short, for the form-invariant description of:

$\hat{A}$) extended and deformable particles which cannot be effectively approximated as being point-like;

$\hat{B}$) while moving within a generally inhomogeneous and anisotropic physical medium, under conventional potential interactions, as well as the contact, nonlinear, nonlocal and nonhamiltonian interactions with the medium itself;

$\hat{C}$) under the additional conditions, as in the conventional case, that:
$\hat{C}$-1) All speeds v are much smaller than the speed of light c_o;
$\hat{C}$-2) All quantum mechanical effects are ignorable; and
$\hat{C}$-3) All spaces are flat.

The Galilei-isotopic relativities were originally submitted in Santilli (1978a), and then formally presented in Chapter 6 of the monograph Santilli (1982a). Specific studies on the isotopic generalization of the rotational symmetry were done in Santilli (1985b), and illustrative examples were presented in Jannussis, Mijatovic and Veljanoski (1991).

All these studies admitted nonlinear and nonhamiltonian internal forces, but of local-differential character. The extension of the research to include nonlocal-integral interactions were conducted in Santilli (1988a), which is the main reference of this Chapter. The operator formulation of the isogalilean relativities was submitted in Santilli (1989a). In this chapter we can evidently present only a summary of the classical profile of this research. The study of the original literature is therefore recommended for a full technical knowledge of the novel relativities.

The objective of this chapter is evidently multifold and consists in the identification, first, of the symmetries of interior systems (II.1.1) of

N particles with nonnull masses m_a, a = 1, 2, ..., N, with nonlinear, nonlocal, nonhamiltonian and nonnewtonian[1] interactions, i.e.,

$$(\dot{a}^\mu) = \begin{pmatrix} \dot{r}^{ia} \\ \dot{p}_{ia} \end{pmatrix} = (\Gamma^\mu(t, a, \dot{a}, ..))$$

$$= \begin{pmatrix} p_{ia}/m_a \\ F^{SA}_{ia}(r) + F^{NSA}_{ia}(t, r, p, \dot{p},..) + \int_\sigma d\sigma\, \mathfrak{F}^{NSA}_{ia}(t, r, p, \dot{p},..) \end{pmatrix}, \tag{1.1}$$

$$i = 1, 2, 3\ (= x, y, z), \quad a = 1, 2, ..., N, \quad \mu = 1, 2, ...\ 6N.$$

when studied as a whole and considered as isolated from the rest of the universe, namely, under the condition of validity of the conventional, total, Galilean conservation laws.

As we shall see, the Lie-isotopic theory outlined in the preceding chapter will allow us to: 1) construct an infinite family of isotopes $\hat{G}(3.1)$ of the conventional Galilei symmetry G(3.1) capable of providing the form invariance of systems (1.1); 2) represent the internal nonlinear, nonlocal, nonhamiltonian and nonnewtonian interactions via their embedding in the isunit of the theory; while 3) guaranteeing the validity of the total conventional Galilean conservation laws because, by construction, the isosymmetry $\hat{G}(3.1)$ preserves the generators and parameters of G(3.1).

In these introductory words we would like to present a few comments to minimize possible misrepresentations due to extended use of concepts which, while fully consistent within the exterior dynamical problem, are intrinsically inconsistent when applied to the different physical arena of the interior problem.

Let us recall that the most dominant *mathematical* notion at the foundations of the isotopies of Galilei's relativity is the generalization of the trivial unit I of the symmetry G(3.1) into the infinitely possible isounits $\hat{1}(t, r, p, \dot{p}, ...)$ of the isosymmnmetries. All various mathematical tools of the theory, such as fields, spaces, algebras, etc., are then generalized accordingly.

The most dominant *physical* notion is that of an extended, and therefore deformable particle moving within an inhomogeneous and anisotropic physical medium which is represented precisely by the isounits $\hat{1}$, as we shall see.

The first aspect the reader should therefore keep in mind is that if, for any reason, the particles considered are point-like, all isotopic

[1] We assume the reader is familiar with the analysis of Chapter I.

formulations recover the conventional ones identically via the reduction $\hat{1} \Rightarrow I$. In fact, points can only experience local-differential, potential, action-at-a-distance interactions without collisions.

The second aspect needed to minimize misconceptions is that *space itself remains perfectly homogeneous and isotropic under our isotopies. Our objective is merely that of representing the inhomogenuity (e.g., local changes of density) and anisotropy (e.g., preferred direction due to intrinsic angular momenta) of physical media, as established by clear experimental evidence.* In particular, such an inhomogenuity and anisotropy is inherent in any nontrivial isotopy of the unit, $I \Rightarrow \hat{1}$. As a result, the reduction $\hat{1} \Rightarrow I$ implies the recovering of the exact homogenuity and isotropy of empty space.

A further notion which is foreign to contemporary theories, but fundamental under isotopies, is the *deformability of extended particles.* In fact, all conventional space-time symmetries are symmetries of rigid bodies, as well known. But rigid bodies do not exist in the physical reality. Thus, the moment the actual extended shape of an object is admitted, it is necessary for physical consistency to be able to represent also all its possible deformations.

This is the physical origin of the infinite number of isotopies of each given conventional space-time symmetry. In fact, the Lie-isotopic theory permits the representation of the actual shape of the particle considered, say, an oblate spheroidal ellispoid, as we shall see, via an isounit of type $\hat{1} = \text{diag.}\,(b_1^{-2}, b_2^{-2}, b_3^{-2})$, $b_k > 0$. Additional isotopies $\hat{1} \Rightarrow \hat{1}' \Rightarrow \hat{1}''$, etc., then permit the representation of all infinitely possible deformations of the original shape caused by sufficiently intense external forces and/or collisions, all this at the purely classical and nonrelativistic level of this chapter, as we shall see.

By comparison, conventional Lie theories can represent the shape only after the rather laborious second quantization; they cannot represent the actual shape itself, but only the remnants of such a shape via the form factors; and positively no deformation of shape is admitted, because this would imply the breaking of a pillar of contemporary theories, the symmetry under rotations.

A further origin of misconception is related to inertial versus noninertial systems. As well known, the contemporary Galilei's and Einstein's special relativities are intrinsically and fundamentally linear and local and, therefore, *inertial* theories. Thus, they are applicable, strictly speaking, only to observers in inertial conditions.

But, inertial systems are a philosophical abstraction because they do not exist in the physical reality of our Earthly environment, nor are they attainable in our Solar or Galactic systems.

A primary purpose of our isotopic relativities is precisely that of achieving generalized relativities which are *direcly* applicable (that is, applicable ***without*** any coordinate transformation) to ***actual experimenters*** (that is, experimenters in *noninertial* conditions).

We shall of course recover inertial conditions in a variety of ways (e.g., via the average of the isounit elements to constants, under which the theory returns to be linear), but only under the strict understanding that this is done as a mere first approximation.

In the final analysis, the reader should always keep in mind equations of motion (1.1) and confront them with any temptation to preserve old notions. As an example, inertial frames are customarily used to study the stability of exterior orbits, such as the center-of-mass orbit of Jupiter in the Solar system, in which case all nonhamiltonian and nonlocal forces of systems (1.1) are identically null, and we have the simple equation

$$\dot{p} = F^{SA}(r) = -\frac{\partial V}{\partial r}. \tag{1.2}$$

It is evident that, within the context of such well identified exterior problem, the inertial approximation of our Earthly observers is fully valid.

The systems under study are fundamentally different than the above, such as a high speed space-ship penetrating Jupiter's atmosphere, for which all forces of systems (1.1) apply

$$\dot{p} = F^{SA}(r) + F^{NSA}(t, r, p, \dot{p}, ...) + \int_{\sigma} d\sigma \, \mathcal{F}^{NSA}(t, r, p, \dot{p}, ...), \tag{1.3}$$

The idea of necessarily preserving the inertial character of the observer while the event considered is under extremely noninertial-nonconservative conditions has no physical basis.

When considering the system satellite-Jupiter as a closed isolated system, we will be in a position to regain the inertial character of the generalized relativities. But, again, this must be considered as a mere approximation of an intrinsically nonlinear, nonlocal and, therefore, noninertial setting.

A further reason for possible misrepresentations is that, in the conventional case, one has only one class of reference frame, the ideal inertial ones, and ignores all the others. In the isotopic case, instead, the different physical conditions imply the necessary existence of classes of reference frames which are not expected to be necessarily equivalent.

In different terms, a frame under inertial approximation on Earth is indeed equivalent to a frame under inertial approximation on Jupiter. However, a noninertial frame on Earth should not be expected to be necessarily equivalent to a noninertial frame on Jupiter. One of the objectives of the Lie-isotopic theory is precisely that of identifying the *classes of equivalence of noninertial frames.* For additional comments, see Fig. III.1.1.

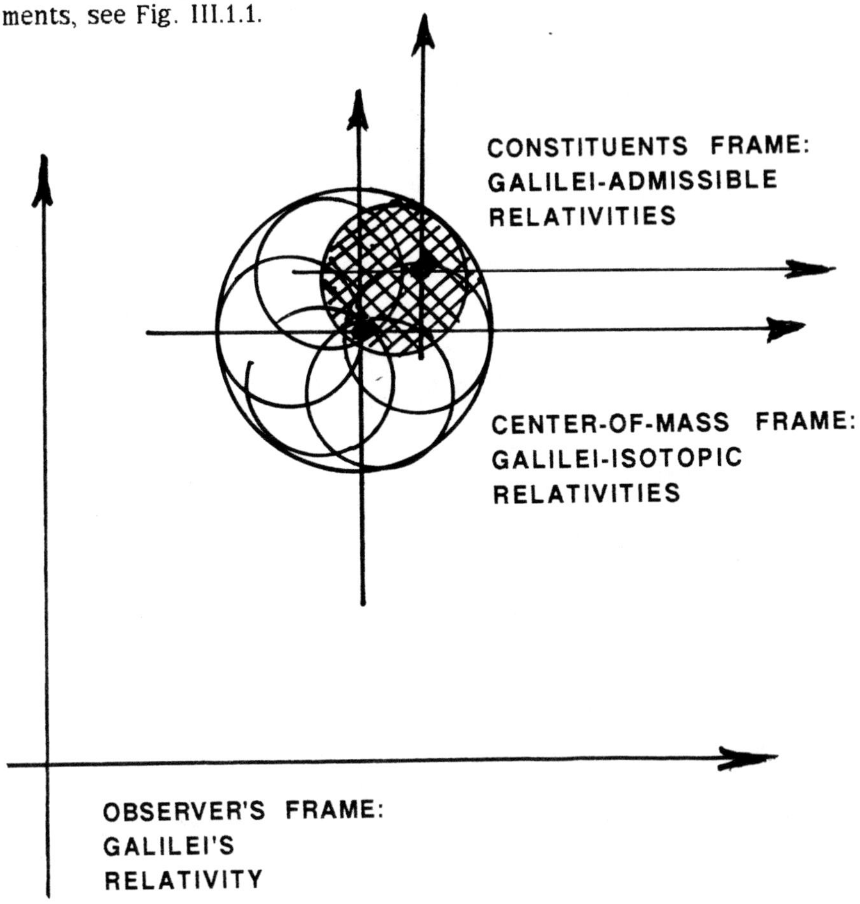

CONSTITUENTS FRAME:
GALILEI-ADMISSIBLE
RELATIVITIES

CENTER-OF-MASS FRAME:
GALILEI-ISOTOPIC
RELATIVITIES

OBSERVER'S FRAME:
GALILEI'S
RELATIVITY

FIGURE III.1.1: A schematic view of the three most important frames and related methodological tools that are recommendable for a comprehensive description of closed nonselfadjoint systems, such as Jupiter. First, we have the external, inertial, observer's frame and related, conventional, Galilei's relativity which describes the center-of-mass trajectory. The reader should be aware that, for such a Galilean setting, Jupiter can only be a structure-less, massive point.

Second, we have the frame at rest with the center-of-mass of Jupiter and the related Galilei-isotopic relativities to be reviewed in Sect. III.6. In this case, Jupiter is indeed represented as an extended structure which verifies all external, conventional, total conservation laws and symmetries; yet it admits nonlinear, nonlocal and nonhamiltonian internal forces. Finally, we have the frame at rest with respect to one individual constituent, while considering the rest of the system as external. In this latter case, we have the broader Lie-admissible methodology of Appendix II.A. The reasons for their emergence as a necessary complement of the Lie-isotopic treatment were indicated in Sect. I.4, and are essentially related to the closed-conservative character of the Lie-isotopic formulations, versus the open-nonconservative character of the Lie-admissible formulations. Since an individual constituent of a closed interior problem is generally in nonconservative conditions, the necessary complementarity of the Lie-admissible algebras follows. The view expressed in this figure can also be obtained from the viewpoiont of the *classes of equivalent frames.* The external, inertial, observer's frame possesses its own class of equivalence, evidently given by all possible inertial frames, as characterized by the *linearity and locality* of the conventional Galilei relativity. In the transition to the representation of Jupiter's structure with a nonlinear, nonlocal and nonhamiltonian interior dynamics, we need an intrinsically *nonlinear and nonlocal* theory to prevent excessive approximations of the type of the perpetual motion in a physical environment. In this latter case, the center-of-mass frame of the system is *noninertial,* because inertial frames do not exist in our physical reality. The Galilei-isotopic relativities then characterize the class of noninertial systems that are equivalent to the center-of-mass frame. It is geometrically possible to show that such (infinite) class *does not* contain the frames of individual constituents, because they generally are in unstable orbits, while the center-of-mass frame of the system is globally stable. In turn, the identification of the class of frames equivalent to (each) constituent's frame can be best done via methods structurally set for nonconservative conditions, such as the Lie-admissible methods.

Finally, we have the issue of preservation or violation of conventional space-time symmetries for systems (1.1). As readily predictable because of extended use, readers may tend to conduct any conceivable effort to salvage conventional symmetries for systems (1.1), such as the rotational symmetry O(3), the Euclidean symmetry E(3), and the Galilei symmetry G(3.1).

It should be stated with clarity that the violation of the symmetries considered for the systems under study has been proved beyond any reasonable (or credible) doubt, and classified into:

> *isotopic, selfadjoint, semicanonical, canonical, and essentially nonselfadjoint breakings (Santilli (1982a), pp. 344-348).*

The technical reasons for the breaking are numerous and of independent nature. First, one should recall Theorems I.3.1 and I.3.2 which prevent the simplistic reduction of nonconservative systems (1.1) to an idealistic collection of stable elementary orbits. Second, one should recall the purely mathematical meaning of idealistic coordinates transformations intended to search for a highly noninertial frame under which conventional symmetries may be preserved, and the other aspects reviewed in Chapter I.

Once the reader recognizes the need to represent systems (1.1) in the experimental r-frame of their detection, numerous, independent reasons for the violation of the conventional O(3), E(3) and G(3.1) symmetries follow.

First these symmetries are strictly *local* in topological character, while systems (1.1) are intrinsically *nonlocal.* Second, the symmetries considered are strictly *linear;* while the systems considered are in the most general possible *nonlinear* conditions. Third, the conventional symmetries are of strictly *inertial* character, while the systems represented are intrinsically *noinnertial.* Fourth, the conventional symmetries necessarily require a *homogeneous and isotropic* medium, while the medium of systems (1.1) is intrinsically *inhomogeneous and anisotropic.* Fifth, the rotational symmetry O(3] and its Euclidfean E(3) and Galilean G(3.1) extensions are strictly and solely a theory for *rigid bodies;* while the objects considered are *deformable* by assumptions. Etc.

Our objectives are those of achieving suitable coverings of the symmetries O(3), E(3) and G(3.1) which:

1) Provide a direct form-invariant description of systems (1.1);

2) Admit the conventional symmetries as particular cases; and, last but not least,

3) Are locally isomorphic to the corresponding conventional symmetries, by coinciding with them at the abstract, coordinate-free level.

III.2: CLOSED NONRELATIVISTIC NONHAMILTONIAN SYSTEMS

It is generally believed that the global stability of a composite system is due to the stability of the individual orbits of its constituents, as it is the case for the planetary and atomic structures. These systems are *closed* in the sense that they verify the ten total, Galilean, conservation laws (when isolated from the rest of the Universe), and are *variationally selfadjoint* (SA), in the sense that the internal forces verify the integrability conditions for the existence of a potential (Helmholtz (1887), Santilli (1978a)).

By no means, these systems exhaust all possible composite systems in our Universe. Another class is provided by *closed nonselfadjoint systems*. These are systems which verify all total, conventional conservation laws (closedness) as it is the case for any isolated system; nevertheless, their internal forces are variationally *nonselfadjoint* (NSA),. i.e., they violate the integrability conditions for the existence of a potential (*loc. cit.*).

The latter systems are considerably more complex than the former. In fact, global stability is achieved, this time, via a collection of orbits each of which is generally unstable. We merely have internal local exchanges of energy, linear momentum and other physical quantities, but in such a way to verify total conservation laws.

An illustration of the latter systems is provided by an individual member of the Solar system, such as Jupiter. As one can see (Figure I.1.1), its global stability is evident. Equally evident is the instability of the orbits of its individual constituents. In fact, direct visual observation via telescopes establishes that Jupiter's total angular momentum is evidently conserved, but its internal structure is characterized by particles with monotonically *nonconserved*[2] angular momenta, and a similar situation occurs for all other quantities.

At a deeper analysis, one can see that Jupiter's interior dynamical equations are nonlinear, nonlocal, nonlagrangian-nonhamiltonian and nonnewtonian, i.e., are precisely of type (III.1.1). The inapplicability of the canonical realizations of Lie's theory and of the symplectic geometry then follows.

---

[2] We here make a distinction between *dissipation*, which implies only the decrease of the energy, angular momentum or other physical quantities, and *nonconservation*, which implies either the decrease or the increase of physical quantities depending on the local physical conditions.

This establishes the physical foundations of these studies: the inapplicability of conventional space-time symmetries and relativities for the interior dynamical problem of closed nonselfadjoint systems, with the consequential need to construct structurally more general symmetries and relativities.

The notion of closed nonselfadjoint systems was introduced, apparently for the first time, in Santilli (1978b), Sect. 3.4, were the classical and operator case of the two bodies was worked out. The notion was then discussed in detail in Santilli (1982a), via the construction of their space-time Lie-isotopic symmetries and conventional conservation laws, and the formulation of their Galilei-isotopic relativities. The systems were then studied from a statistical viewpoint in Fronteau *et al.* (1979) and Tellez-Arenas *et al.* (1979) with rather intriguing implications, e.g., the possibility of introducing a new notion of internal irreversibility which is compatible with a reversible exterior dynamics, exactly as occurring for Jupiter (see the Appendices of Chapter II). Specific classical cases of generalized two- and three-body systems were studied in Jannussis *et al.* (1991) as examples of the Galilei-isotopic relativities, where one can find also generalized Birkhoffian representations.

All the preceding studies were nonlinear, nonhamiltonian and nonnewtonian, but in their local approximation because of the assumption of the symplectic geometry as the background geometry. The extension to nonlocal settings was done in Santilli (1988a) and (1991a).

The predictable (and rather intriguing) connections with Prigogine's statistics (1968) have remained unexplored as of today, and are scheduled for subsequent studies.

The implications of closed nonselfadjoint systems are non-trivial, mathematically and physically.

From a mathematical viewpoint, the systems considered require the construction of covering analytic, algebraic and geometric formulations, besides implying a host of intriguing and fundamental, open mathematical problems (such as the achievement of global topological stability via local instabilities, see Aringazin *et al.* (1990)).

From a physical viewpoint, the implications are equally deep because *closed nonselfadjoint systems require a necessary generalization of conventional relativities at all levels of study, Galilean, relativistic and gravitational, as well as classical and quantum mechanically*, as we shall see.

In this section we shall outline the notions of closed nonselfadjoint systems in their most general possible (nonlocal) formulation, as well as the analytic, algebraic and geometric tools for their treatment.

Let us begin with a representation of closed selfadjoint systems as vector-fields on a manifold. Let $E(r,\delta,\Re)$ be the conventional Euclidean space in three-dimension where $r = (r_k)$, $k = 1, 2, 3$ (= x, y, z) are the physical coordinates of the experimenter, and the metric is given by the familiar form $\delta = \text{diag.}(1,1,1)$ over the reals $\Re$. Introduce in $E(r,\delta,\Re)$ a system of N particles with nonnull masses m_a, $a = 1, 2,..., N$. Let $T^*E(r,\delta,\Re)$ be the cotangent bundle (the conventional phase space) with local chart (coordinates) $a = (a^\mu) = (r,p) = (r_{ka}, p_{ka})$, $\mu = 1, 2,..., 6N$, where the p's are the physical linear momenta, i.e., $p_{ka} = m_a v_{ka}$, $v_{ka} = \dot{r}_{ka} = dr_{ka} / dt$. For simplicity of notation, all indeces of the coordinates and momenta will be treated as subindeces, while the distinction between covariant and contravariant indeces will be kept in $T^*E(r,\delta,\Re)$.

Then, *closed selfadjoint systems* can be defined as the *Hamiltonian vector-field*

$$\dot{a} = (\dot{a}^\mu) = \begin{pmatrix} \dot{r}_{ka} \\ \dot{p}_{ka} \end{pmatrix} = \Xi = (\Xi^\mu(t, a)) = \begin{pmatrix} p_{ka}/m_a \\ f^{SA}(r) \end{pmatrix}, \tag{2.1a}$$

$$\dot{X}_i(t, a) = \frac{\partial X_i}{\partial a^\mu} \dot{a}^\mu + \frac{\partial X_i}{\partial t} \equiv 0, \tag{2.1b}$$

$$k = 1, 2, 3, \qquad a = 1, 2, 3,..., N, \quad \mu = 1, 2,..., 6N$$

where the X's represent the familiar, total, Galilean, conserved quantities

$$\begin{cases} X_1 = H = T(p) + V(r), & (2.2a) \\ (X_2, X_3, X_4) = (P_k) = \sum_a p_{ka}, & (2.2b) \\ (X_5, X_6, X_7) = (M_k) = \sum_a r_{ka} \wedge p_{ka}, & (2.2c) \\ (X_8, X_9, X_{10}) = (G_k) = \sum_a (m_a r_{ka} - t\, p_{ka}), & (2.2d) \end{cases}$$

The selfadjoint character of the forces then ensures the direct applicability of all conventional, canonical, analytic, algebraic and geometric formulations as well known.

The most general possible *closed nonselfadjoint systems* are instead given by the *nonlinear, nonlocal, nonhamiltonian and nonnewtonian vector-fields* on $T^*E(r,\delta,\Re)$

$$\dot{a} = (\dot{a}^\mu) = \begin{pmatrix} \dot{r}_{ka} \\ \dot{p}_{ka} \end{pmatrix} = \Gamma = (\Gamma^\mu(t, a, \dot{a} ...)) =$$

$$= \begin{pmatrix} p_{ka}/m_a \\ F^{SA}{}_{ka}(r) + F^{NSA}{}_{ka}(t, r, p, \dot{p}, ...) + \int_\sigma d\sigma\, \mathcal{F}^{NSA}{}_{ka}(t, r, p, \dot{p}, ...) \end{pmatrix}, \tag{2.3a}$$

$$\dot{X}_i = \frac{\partial X_i}{\partial a^\mu} \dot{a}^\mu + \frac{\partial X_i}{\partial t} \equiv 0, \tag{2.3b}$$

$$i = 1, 2, ..., 10, \quad k = 1, 2, 3, \quad a = 1, 2, ..., N, \quad \mu = 1, 2, ..., 6N,$$

where the X's are exactly the same as in Eq.s (2.2), and the forces $F^{SA}{}_{ka}(t, r, p)$ that are still of potential type, but Galilei-noninvariant, have been incorporated for simplicity in the nonselfadjoint forces.

Systems (2.3) constitute the physical foundations of the studies presented in this monograph. They provide a primitive, classical and nonrelativistic representation of the structure of Jupiter (Santilli (1978a, e), (1982a), (1988a)), as well of a conceivable new structure model of hadrons (Santilli (1978b, d), Myung *et al.* (1982), Mignani *et al.* (1983)).

Note that, while systems (2.1) are unconstrained, Eq.s (2.3) characterize a system with subsidiary constraints, in the sense that conservation laws (2.3b) are now, in general, subsidiary constraints to vector-fields (2.3a).

Finally, while systems (2.1a) are Galilei-invariant, *a necessary condition for the existence of closed nonselfadjoint systems (2.3) is that they are not invariant under the conventional Galilei symmetry.* This is evident on numerous counts outlined in Sect. III.1.

It is easy to see that, under sufficient topological conditions (regularity and analyticity), systems (2.3) are consistent because they are *underdetermined.*

Moreover, it is possible to show that *systems (2.3) admit unconstrained solutions in the nonselfadjoint forces for given Galilean selfadjoint forces.* In fact, by incorporating the nonlocal forces in F^{NSA} for simplicity, subsidiary constraints (2.3b) can be reduced to the following seven conditions in the nonselfadjoint forces (see Santilli (1982a), p. 236 for details)

$$\sum_a F_a^{NSA} = 0, \quad \sum_a p_a \times F_a^{NSA} = 0, \quad \sum_a r_a \wedge F_a^{NSA} = 0, \qquad (2.4)$$

Unconstrained solutions in the nonselfadjoint forces therefore always exist for N > 1, including the case N = 2, as shown in Appendix III.A. The case N = 1 is impossible because one isolated particle is free and cannot experience nonselfadjoint forces.

We now outline the methodological tools for the treatment of closed nonselfadjoint systems.

ANALYTIC FORMULATIONS. A step-by-step generalization of Hamiltonian mechanics under the name of *Birkhoffian mechanics* resulted to be *necessary* for the representation of systems (2.3) (Santilli (1978a), (1982a, (1988a)). In particular, the new mechanics was proven to be *directly universal* for (regular, local and analytic) systems (2.3a), namely, a representation of all systems considered always exists (universality) directly in the a-coordinates of the experimenter (direct universality).

The analytic representation begins with the construction of the following first-order *Pfaffian variational principle in its isotopic form (Sect. II.7)*

$$\delta \hat{A}^\circ = \delta \int_{t_1}^{t_2} dt\, [\, R_\mu(a)\, \dot{a}^\mu - B(t,a)\,] =$$

$$= \delta \int_{t_1}^{t_2} dt\, [\, R^\circ{}_\alpha(a)\, \hat{T}_1{}^\alpha{}_\beta(a)\, \dot{a}^\mu - B(t, a)\,] = 0, \qquad (2.5a)$$

$$R^\circ = (p, 0), \qquad \det \hat{T}_1 \neq 0, \qquad \hat{T}_1 = \hat{T}_1^\dagger, \qquad \mu, \alpha, \beta = 1, 2, ..., 6N, \quad (2.5b)$$

where: the R_μ and B functions are computed from the given equations (2.3a) via one of the several techniques in Santilli (1982a); B(t,a) = B(t,r,p) is called the *Birkhoffian*, because it is generally different than the total energy; and all nonlocal and nonselfadjoint terms are embedded in the isotopic element $\hat{T}_1$ which multiplies the canonical value R°.

Principle (2.5) characterizes a particular form of Birkhoff's equations called the *covariant Hamilton-isotopic equations* (II.7.29), i.e.,

$$\hat{\Omega}^{\circ}_{\mu\nu}(a)\,\dot{a}^{\nu} = \omega_{\mu\alpha}\,\hat{T}_{2}{}^{\alpha}{}_{\nu}(a)\,\dot{a}^{\nu} = \frac{\partial H(t, a)}{\partial a^{\mu}} . \tag{2.6a}$$

$$(\Omega^{\circ}_{\mu\nu}) = \left(\frac{\partial \hat{R}_{\nu}}{\partial x^{\mu}} - \frac{\partial \hat{R}_{\mu}}{\partial x^{\nu}}\right) = \begin{pmatrix} 0_{n\times n} & (T_2)_{n\times n} \\ -(T_2)_{n\times n} & 0_{n\times n} \end{pmatrix} =$$

$$= \begin{bmatrix} 0_{n\times n} & (T_{1\,ij} + p_k \frac{\partial T_1{}^k{}_i}{\partial p_j})_{n\times n} \\ -(T_{1\,ij} + p_k \frac{\partial T_1{}^k{}_i}{\partial p_j})_{n\times n} & 0_{n\times n} \end{bmatrix} . \tag{2.6b}$$

The *contravariant Hamilton-isotopic tensor* has structure (II.7.30), i.e.,

$$(\hat{\Omega}^{\circ\mu\nu}) = (\omega^{\mu\nu}) \times \hat{I}_2 = \hat{I}_2 \times (\omega^{\mu\nu}) = (\omega^{\mu\alpha}\,\hat{I}_{2\alpha}{}^{\nu}) =$$

$$= \begin{bmatrix} 0_{n\times n} & (I_2)_{n\times n} \\ -(I_2)_{n\times n} & 0_{n\times n} \end{bmatrix} , \tag{2.7a)}$$

$$\hat{I}_2 = \hat{T}_2{}^{-1} = \text{diag.}\,(T_2{}^{-1}, T_2{}^{-1}) = \text{diag.}\,(I_2, I_2), \tag{2.7b}$$

$$I_2 = (T_{1\,ij} + p_k \frac{\partial T_1{}^k{}_i}{\partial p_i})^{-1} , \tag{2.7c}$$

The *contravariant Hamilton-isotopic equations* are then given by Eq.s (II.7.32), i.e.,

$$\dot{a}^{\mu} = \hat{\Omega}^{\circ\mu\nu}(a) \frac{\partial H(t, a)}{\partial a^{\nu}} = \omega^{\mu\alpha} \hat{1}_{2\alpha}{}^{\nu}(a) \frac{\partial H(t, a)}{\partial a^{\nu}} , \qquad (2.8)$$

which can be written in the disjoint r- and p-coordinates

$$\dot{r}_i = I_{2\,ij}(r, p) \frac{\partial H(t, r, p)}{\partial p_j}, \qquad (2.9a)$$

$$\dot{p}_i = - I_{2\,ij}(r, p) \frac{\partial H(t, r, p)}{\partial r_j} , \qquad (2.9b)$$

Pfaffian principle (2.5) also implies the following *isotopic generalization of the Hamilton-Jacobi equations*

$$\begin{cases} \dfrac{\partial \hat{A}^{\circ}}{\partial t} + B(t,a) = 0, & (2.10a) \\[2ex] \dfrac{\partial \hat{A}^{\circ}}{\partial r_{ia}} = p_{ia} T_1{}^{ia}{}_{ja}, \qquad \dfrac{\partial \hat{A}^{\circ}}{\partial p_{ia}} = 0, & (2.10b) \end{cases}$$

which have a predictably important role for the operator formulation of systems (2.3), as we hope to show in a subsequent work (see the crucial independence of the isotopic action $\hat{A}^{\circ}$ from the momenta indicated in Sect. II.7).

The rest of the Birkhoffian generalization of Hamiltonian mechanics follows. The reader interested in the studies of these volumes is urged to acquire a technical knowledge of Birkhoffian mechanics because numerous aspects will be tacitly assumed as known during the course of our analysis, some of which are rather insidious.

For instance, the computation of the R_{μ} and B functions from the equations of motion generally yields *nonautonomous representations* with $R_{\mu} = R_{\mu}(t,a)$ and $B = B(t,a)$ even when the system does not dependend explicitly on time. This implies the still more general *nonautonomous Birkhoff's equations* (II.7.11) which violate the conditions to characterize an algebra (Appendix II.A).

Nevertheless, let us recall from Sect. II.7 that nonautonomous representations can be reduced to the *semiautonomous form* with

$R_\mu = R_\mu(a)$, $B = B(t,a)$ via the degrees of freedom of the theory, e.g., the so-called *Birkhoffian gauge transformations* (II.7.12), i.e.,

$$R'_\mu(a) = R_\mu(t,a) + \frac{\partial G(t, a)}{\partial a^\mu}, \tag{2.11a}$$

$$B'(t,a) = B(t,a) + \frac{\partial G(t, a)}{\partial t}. \tag{2.11b}$$

Above all, the reader is expected to be familiar with the techniques of constructing a Birkhoffian representation in which the Birkhoffian characterizes the total energy as the sum $B = H = T(p) + V(r)$ of the kinetic energy $T(p)$ and the potential energy $V(r)$ of all selfadjoint forces (recall that the notion of energy has no mathematical or physical meaning for contact nonselfadjoint forces).

The attentif reader has noted that we have used until now the simplest possible realization of Birkhoffian mechanics, that of Hamilton-isotopic type in which the factorized structure ω is canonical.

Such a structure is amply sufficient for our needs for technical reasons that will be indicated during the course of our analysis. However, the reader should keep in mind that the general treatment of closed nonselfadjoint systems requires the use of the full Birkhoffian-isotopic mechanics as per Definition II.7.1.

Note that *the Birkhoffian-isotopic mechanics is a covering of Hamiltonian mechanics in the sense that: 1) the former mechanics is based on formulations structurally more general than those of the latter; 2) the former mechanics represents physical conditions structurally more general than those of the latter; and 3) the former mechanics admits the latter as a particular case for $R = R^o = (p,0)$ and $T_1 = I$.*

In fact, under values $R = R^o = (p,0)$, $T_1 = I$, the covariant Birkhoff-isotopic tensor assumes the familiar covariant canonical form

$$(\omega_{\mu\nu}) = (\partial_\mu R^o_\nu - \partial_\nu R^o_\mu) = \begin{pmatrix} 0_{3n\times 3n} & -I_{3n\times 3n} \\ I_{3n\times 3n} & 0_{3n\times 3n} \end{pmatrix}, \tag{2.12}$$

with contravariant canonical form

$$(\omega^{\mu\nu}) = (|\omega_{\alpha\beta}|^{-1}) = \begin{pmatrix} 0_{3n\times 3n} & I_{3n\times 3n} \\ -I_{3n\times 3n} & 0_{3n\times 3n} \end{pmatrix}, \tag{2.13}$$

The Hamilton-isotopic equations then recover Hamilton's equations identically

$$\omega_{\mu\nu}\,\dot{a}^{\mu} = \partial_{\mu} H(t,a), \qquad \dot{a}^{\mu} = \omega^{\mu\nu}\,\partial_{\nu} H(t,a), \quad H \equiv B. \tag{2.14}$$

and the same holds for all remaining aspects.

Finally, recall that the various aspects of Birkhoffian mechanics can be constructed via a judicious use of *noncanonical* transformations of the corresponding aspects of Hamiltonian mechanics. In particular, Birkhoff's equations can be constructed via noncanonical transformations of Hamilton's equations and the same occurs for variational principles, Hamilton-Jacobi theory, etc.

As a result, *Birkhoff's equations preserve their form under the most general possible transformations,* trivially, because they already have the most general possible form (Sect. II.8).

These features are important for the subsequent studies of this volume. As an example, we should expect that the isotopic generalizations $\hat{G}(3.1)$ of Galilei's symmetry $G(3.1)$ for systems (2.3) to be studied later on in this chapter, can be constructed via *noncanonical* transformations of $G(3.1)$.

Similarly, the operator image of Birkhoffian mechanics, tentatively submitted in Santilli (1978b) under the name of *hadronic mechanics,* can be expectedly built via nonunitary transformations of conventional quantum mechanics, and then preserve its structure under the most general possible transformations (Santilli (1989)).

ALGEBRAIC FORMULATIONS. The construction of a step-by-step generalization of the conventional formulation of Lie's theory under the name of *Lie-isotopic theory* is necessary for the treatment of closed nonselfadjoint systems, as outlined in Sect. II.6.

The central idea is the generalization of the trivial unit I of current use in both mathematical and physical formulations into the most general possible unit $\hat{1}$, called *isotopic unit* which is nonsingular, and Hermitean, but possesses an otherwise arbitrary dependence on all local variables and quantities

$$\hat{1} = \hat{1}(t, a, \dot{a}, ...). \tag{2.15}$$

The generalization of the unit then implies a corresponding generalization of all major structural aspects of Lie's theory, from the theory of universal enveloping associative algebras, to the theory of Lie algebras, and to the theory of Lie groups.

It is here sufficient to recall that systems (2.3a) under analytic representations (2.8) are characterized by the Lie-isotopic algebras with brackets (II.7.30), i.e.,

$$[A \hat{,} B] = \frac{\partial A}{\partial a^{\mu}} \omega^{\mu\alpha} \hat{1}_{2\alpha}{}^{\nu}(a,..) \frac{\partial B}{\partial a^{\nu}}$$

$$= \frac{\partial A}{\partial r_i} I_{2ij}(t, r, p, ...) \frac{\partial B}{\partial p_j} - \frac{\partial B}{\partial r_i} I_{2ij}(t, r, p, ...) \frac{\partial A}{\partial p_j}, \tag{2.16}$$

where the isounit is now that of the space $T^*\hat{E}(r,\hat{\delta},\hat{\mathfrak{R}})$ (Sect. II.9)

$$\hat{1} = \hat{1}_2 = \text{diag.}\,(I_2, I_2). \tag{2.17}$$

The central problem of this chapter is therefore the study of the *Galilei-isotopic symmetries* $\hat{G}(3.1)$ which, at this preliminary stage, can be characterized by the *Lie-isotopic algebra of Galilean generators* (2.2) with isocommutation rules

$$[X_i \hat{,} X_j] = \hat{C}_{ij}{}^{k}(t, a, \dot{a}, ..) \hat{X}_k, \tag{2.18}$$

and corresponding *Lie-isotopic group*

$$a' = \{ \textstyle\prod_i e_{|\xi}^{w_i \, \omega^{\mu\alpha} \, \hat{1}_{2\alpha}{}^{\nu} \, (\partial X_i/\partial a^{\nu}] \, (\partial/\partial a^{\mu})} \hat{1}_2 \} * a. \tag{2.19}$$

where: the $\hat{C}$'s are the structure functions (Sect. II.6); the w's are the conventional parameters of the Galilei symmetry G(3.1); the exponentiation is in a conventional associative algebra ξ; the transformations are isotopic (Sect. II.3); and we evidently have an infinite number of possible structures $\hat{G}(3.1)$ characterized by an infinite number of possible isounits $\hat{1}_2$.

More specifically, our task is *to identify that particular subclass of general systems (2.3a) which are invariant under the Galilei-isotopic symmetries $\hat{G}(3.1)$.* In fact, the isosymmetry $\hat{G}(3.1)$ will guarantee, on one side, the nonhamiltonian internal structure from the generalized structures (2.18) and (2.19), while, on the other side, ensuring the conservation of the conventional ten total Galilean quantities, from their preservation as generators of the new symmetries.

This objective will be studied in a progressive way, by studying first the isorotational subgroups $\hat{O}(3)$ of $\hat{G}(3.1)$ (Sect. III.3), then the isoeuclidean subgroups $\hat{E}(3)$ (Sect. III.4), and finally passing to the full isosymmetries $\hat{G}(3.1)$ (Sect. III.5).

GEOMETRICAL FORMULATIONS. The necessary additional methodological tools for the study of closed nonselfadjoint systems are evidently of geometrical character. At this first classical and nonrelativistic stage, systems (2.3) require a particular reformulation of the symplectic geometry, submitted under the name of *symplectic-isotopic geometry* or *isosymplectic geometry* for short (Sect. II.9), for the representation of nonlocal interactions (otherwise the conventional symplectic geometry in its exact but general formulation would be sufficient as per the Universality Theorem II.9.1).

Here, let us merely recall for the readers convenience, that the basic geometric quantity is the integrand of principle (2.5) reinterpreted as the *one-isoform* on $T^*\hat{E}(r,\hat{\delta},\hat{\mathfrak{R}})$, Eq.s (II.9.91), i.e.,

$$\hat{\Phi}^\circ_1 = R^\circ_\mu \hat{T}_1{}^\mu{}_\nu \hat{d}x^\nu, \qquad R^\circ = (p, 0), \tag{2.20a}$$

$$\det \hat{T}_1 \neq 0, \qquad \hat{T}_1 = (T_{1\mu}{}^\nu) = (T_1{}^\mu{}_\nu) = T_1{}^\dagger. \tag{2.20b}$$

Its isoexterior derivative on $T^*\hat{E}_2(r,\hat{\mathfrak{R}})$ then produces the two-isoform (II.9.92), i.e.,

$$\hat{\Phi}^\circ_2 = \hat{d}\hat{\Phi}^\circ_1 =$$

$$= \tfrac{1}{2}\delta^{\mu_1\mu_2}{}_{\nu_1\nu_2}\left(\frac{\partial R^\circ_\alpha}{\partial a^\beta} T_1{}^\alpha{}_{\mu_1} T_1{}^\beta{}_{\mu_2} + R_\alpha \frac{\partial T_1{}^\alpha}{\partial a^\beta} T_1{}^\beta{}_\mu \right) \hat{d}a^{\nu_1} \hat{\wedge} \hat{d}a^{\nu_2} =$$

$$= \tfrac{1}{2}\, \omega_{\mu\alpha} \hat{T}_2{}^\alpha{}_\nu \hat{d}a^\mu \hat{\wedge} \hat{d}a^\nu, \tag{2.21}$$

which constitutes the desired geometric counterpart of the Lie-isotopic brackets (2.16).

As a result, the Birkhoffian representation of systems (2.3a) can be "globalized", that is, expressed in a coordinate-free form via their characterization as the *Hamiltonian-isotopic vector-field*

$$\Gamma \rfloor \hat{\Phi}^{\circ}_2 = - dH. \tag{2.31}$$

We assume the reader has learned how to compute the isotopic elements $\hat{T}_2$ from $\hat{T}_1$, and perform the factorization of the canonical form in both the two-isoform (2.21) and in the corresponding isobrackets (2.16).

We also assume the reader is aware that we are dealing with the simplest possible realization of the symplectic-isotopic geometry, that with the canonical factorization. The study of closed nonselfadjoint systems via the general form of the geometry, that of Definition II.9.1 (with the factorized Birkhoffian tensor Ω) is here left as an exercise for the interested reader.

The primary results of the above methodological formulations are therefore the following:

1) In the transition from closed selfadjoint to closed nonselfadjoint systems there is no need to abandon conventional analytic, algebraic and geometric formulations, because
1a) both systems are derivable from a first-order variational principle;
1b) the contravariant algebraic tensor of both systems is Lie; and
c) the covariant geometric tensor of both systems is symplectic.

2) closed nonselfadjoint systems emerge in their Birkhoffian representation when one assumes the most general possible realization of the above structures, while closed selfadjoint systems in their Hamiltonian representation emerge when one assumes the simplest possible (canonical) realization of the same structure. And

3) All distinctions between Birkhoffian and Hamiltonian formulations (and, thus, between closed nonselfadajoint and selfadjoint systems) cease to exist at the abstract, realization-free level. This property is evident within the context of the symplectic geometry, where there is no geometric distinction between the conventional and the isotopic unit and between Hamiltonian and Birkhoffian vector-fields, but it equally holds at the algebraic and analytic levels (see Chap. II, Sect. II.5, II.7 and II.9, in particular).

The above properties are sufficient to anticipate our primary results or, equivalently, to provide advance guidelines for their achievement.

In fact, Properties 1, 2 and 3 above require that the space-time symmetries of closed nonselfadjoint systems must be constructed in such a way to be locally isomorphic to the conventional space-time symmetries, as a necessary condition for their identity at the abstract, realization-free level, in a way compatible with the abstract identity between closed selfadjoint and nonselfadjoint systems and their methodologies.

Specific examples of two-body and three-body closed nonhamiltonian systems are presented in Appendix III.A.

III.3: ROTATIONAL-ISOTOPIC SYMMETRIES

In this section we shall present the infinite family of classical isotopic generalizations $\hat{O}(3)$ of the rotational symmetry O(3), which will be formulated in the infinite family of isotopes $\hat{E}(r,\hat{\delta},\hat{\mathfrak{R}})$ of the conventional Euclidean space $E(r,\delta,\mathfrak{R})$ in three-dimension.

Our objective is, specifically, the study of the isosymmetries $\hat{O}(3)$ of closed nonselfadjoint systems (III.2.3) in their nonlinear, nonhamiltonian and nonnewtonian, as well as nonlocal form.

Isosymmetries $\hat{O}(3)$ were introduced, apparently for the first time, in Santilli (1978a), then expanded in Santilli (1982a), and finally studied in details in Santilli (1985b) in their abstract, and therefore nonlinear and nonlocal version. The classical nonlinear and nonlocal realizations of $\hat{O}(3)$ presented in this section were studied in Santilli (1988a) (1991a, b).

Regrettably, we shall be unable to present, for brevity, the isorepresentation theory of $\hat{O}(3)$, which has been studied within the context of the covering isounitary symmetries $S\hat{U}(2)$ in Santilli ((1989), (1991d)) jointly with other liftings of the conventional theory, e.g., the *iso-Clebsh-Gordon coefficients*, etc.

For the reader's convenience, we shall review first the main results of the abstract formulation, and then pass to its classical realization for closed nonselfadjoint systems. A necessary prerequisite for the understanding of this section is a knowledge of Sect.s II-1 to II-9.

DEFINITION III.3.1 (Santilli (1985a)): The "rotational isotopic groups" $\hat{O}(3)$, or "isorotational groups", are the largest possible isolinear and isolocal groups of isometries of the infinitely possible isotopes $\hat{E}(r,\hat{\delta},\hat{\mathfrak{R}})$ of the three-dimensional euclidean space $E(r,\delta,\mathfrak{R})$, Eq.s (II.3.18), i.e.,

$$E(r,\delta,\mathfrak{R}) \;\Rightarrow\; \hat{E}(r,\hat{\delta},\hat{\mathfrak{R}}], \tag{3.1a}$$

$$\delta = \text{diag.}\, I_{n\times n} \;\Rightarrow\; \hat{\delta} = T(r, \dot{r}, \ddot{r},...)\,\delta, \tag{3.1b}$$

$$\det. T \neq 0, \quad T = T^{\dagger}, \quad \det. \hat{\delta} \neq 0, \quad \hat{\delta}^{\dagger} = \hat{\delta}, \tag{3.1c}$$

$$\mathfrak{R} \Rightarrow \hat{\mathfrak{R}} = \mathfrak{R}\hat{1}, \quad \hat{1} = T^{-1} = \hat{\delta}^{-1} \tag{3.1d}$$

$$(r,r) = r^{i}\,\delta_{ij}\,r^{j} \;\Rightarrow\; (r\,\hat{,}\,r) = (r,\hat{\delta}r)\,\hat{1} = \tag{3.1e}$$

$$= (\hat{\delta}r,r)\,\hat{1} = \hat{1}\,(r,\hat{\delta}r) = [r^{i}\,\hat{\delta}_{ij}(r, \dot{r}, \ddot{r},..)\,r^{j}]\,\hat{1}, \tag{3.1f}$$

characterized by: the right, modular-isotopic transformations

$$r' = \hat{R}(\theta)*r = \hat{R}(\theta)\,\hat{\delta}\,r, \qquad \hat{\delta} = \text{fixed}, \tag{3.2}$$

where the θ's are the conventional Euler's angles, whose elements $\hat{R}(\theta)$ verify the properties

$$\hat{R}*\hat{R}^{t} = \hat{R}^{t}*\hat{R} = \hat{1}, \tag{3.3}$$

or, equivalently, $\hat{R}^{t} = R^{-\hat{1}}$, *and verify the group-isotopic rules*

$$\hat{R}(0) = \hat{1} = \hat{\delta}^{-1}, \tag{3.4a}$$

$$\hat{R}(\theta) * \hat{R}(\theta') = \hat{R}(\theta') * \hat{R}(\theta) = \hat{R}(\theta + \theta'), \quad (3.4b)$$

$$\hat{R}(\theta) * \hat{R}(-\theta) = \hat{1}, \quad (3.4c)$$

Equivalently, the isorotational groups $\hat{O}(3)$ can be defined as the isosymmetries of the infinitely possible deformations of the sphere representable via the particular realization of the isometric

$$\hat{\delta} = \text{diag.}\,(g_{11}, g_{22}, g_{33}), \quad (3.5a)$$

$$r^{\hat{2}} = r_1\, g_{11}\, r_1 + r_2\, g_{22}\, r_2 + r_3\, g_{33}\, r_3 = \text{inv.} \quad (3.5b)$$

Isogroups $\hat{O}(3)$ resulted to be tridimensional simple Lie groups which can be constructed from the sole knowledge of the isometric $\hat{\delta}$ via the generators and parameters of the conventional rotational group O(3).

From Eq. (3.3) it is easy to see that isorotations satisfy the conditions

$$\det(\hat{R}\hat{\delta}) = \pm 1. \quad (3.6)$$

Therefore, $\hat{O}(3)$ is characterized by a continuous semisimple subgroup denoted $S\hat{O}(3)$ for the case $\det(\hat{R}\hat{\delta}) = +1$, and a discrete invariant part for the case $\det(\hat{R}\hat{\delta}) = -1$ representing isoinversions (see below).

Each one of the infinitely many possible $S\hat{O}(3)$ subgroups can be essentially characterized as follows. The abstract, enveloping associative algebra ξ of Sect. II.6 is now realized in the isoform $\hat{\xi}$ characterized by the isounit $\hat{1}$, the conventional generators J_k, k = 1, 2, 3, of SO(3) in their fundamental, 3×3 representation, and all their possible polynomials, resulting in the infinite dimensional basis of the isotopic Poincaré-Birkhoff-Witt Theorem

$$\hat{\xi}(\mathbf{SO}(3)): \quad \hat{1}, \quad J_k, \quad J_i * J_j \quad (i \leq j), \quad J_i * J_j * J_k \quad (i \leq j \leq k), \quad \ldots\ldots\ldots \quad (3.7)$$

The isocommutation rules of the Lie-isotopic algebra $\mathbf{S\hat{O}}(3)$ of $S\hat{O}(3)$ were also studied and shown to be reducible to the form

$$[\hat{J}_i \,\hat{,}\, \hat{J}_j] = \hat{J}_i * \hat{J}_j - \hat{J}_j * \hat{J}_i = \hat{J}_i \hat{\delta} \hat{J}_j - \hat{J}_j \hat{\delta} \hat{J}_i = \epsilon_{ijk} \hat{J}_k, \quad (3.8)$$

under a suitable redefinition $\hat{J}_k$ of the generators J_k (see below), where the tensor ϵ_{ijk} is the conventional totally antisymmetric tensor characterizing the structure constants of **SO**(3).

The Lie-isotopic groups $S\hat{O}(3)$ were obtained via an isoexponentiation of structure (3.7) in $\hat{\xi}$, resulting in the expression

$$S\hat{O}(3): \quad \hat{R}(\theta) = \{e_{|\hat{\xi}}^{J_1\theta_1}\}*\{e_{|\hat{\xi}}^{J_2\theta_2}\}*\{e_{|\hat{\xi}}^{J_3\theta_3}\}, \tag{3.9}$$

which can be rewritten in the conventional associative envelope ξ of **SO**(3)

$$S\hat{O}(3): \quad \hat{R}(\theta) = (\prod_{k=1,2,3} e_{|\xi}^{J_k\hat{\delta}\theta_k})\,\hat{1} = \hat{1}(\prod_{k=1,2,3} e_{|\xi}^{\theta_k\hat{\delta}J_k})$$

$$\stackrel{\text{def}}{=} [S_{\hat{\delta}}(\theta)]\,\hat{1} = \hat{1}\,[S_{\hat{\delta}}^{t}(\theta)]. \tag{3.10}$$

The isorotations can then be written in the simpler form

$$r' = \hat{R}(\theta)*r \equiv S_{\hat{\delta}}(\theta)\,r, \tag{3.11}$$

which is useful for computational convenience. The understanding is that the mathematically correct form remains the isotopic form (to prevent the violation of the linearity condition).

The discrete part is characterized by the *isoinversions*

$$\hat{P}*r = P\,r = -\,r, \tag{3.12}$$

where P characterizes the conventional discrete components of O(3).

The notion of isorotation groups was turned into that of *isorotational symmetries* by noting that, under the conditions of Definition III.3.1, isotransformations (3.4) leave invariant, by construction, the separation in $\hat{E}(\rho,\hat{\delta},\hat{\mathfrak{R}})$ (Theorem II.8.1), i.e.,

$$r^{\hat{2}} = (r'^i\,\hat{\delta}_{ij}\,r'^j)\,\hat{1} \equiv (r^i\,\hat{\delta}_{1ij}\,r^j)\,\hat{1} = r^{\hat{2}} \tag{3.13}$$

owing to the property

$$S_{\hat\delta}^{t}\,\hat\delta\, S_{\hat\delta} \equiv \hat\delta, \tag{3.14}$$

which is identically verified for all possible metrics $\hat\delta$ of the class admitted, plus similar identities for the isoinversions.

The capability for the isorotational symmetries $\hat{O}(3)$ to leave invariant all possible ellipsoidical deformations of the sphere, Eq.s (3.5), then trivially follows from invariance (3.13).

We now outline the classification of all simple isotopes $\hat{O}(3)$ conducted also in Santilli (1985b). To begin, consider the metric (3.5a) of undefined topological structure. By using Eq.s (3.9) or (3.10), it is easy to compute a general isorotation around the third axis

$$\hat\delta = \text{diag.}\,(g_{11}, g_{22}, g_{33}), \tag{3.15a}$$

$$S_{\hat\delta}(\theta) = \begin{bmatrix} \cos[\theta_3(g_{11}g_{22})^{\frac{1}{2}}] & g_{22}(g_{11}g_{22})^{\frac{1}{2}}\sin[q_3(g_{11}g_{22})^{\frac{1}{2}}] & 0 \\ -g_{11}(g_{11}g_{22})^{\frac{1}{2}}\sin[\theta_3(g_{11}g_{22})^{\frac{1}{2}}] & \cos[\theta_3(g_{11}g_{22})^{\frac{1}{2}}] & 0 \\ 0 & 0 & 1 \end{bmatrix} \tag{3.15b}$$

The above notion of abstract isorotational symmetry then leads to the following

LEMMA (III.3.1 (Santilli (loc. cit.): The abstract isotope O(3) of O(3) with a nowhere singular, Hermitean and diagonal isometric (3.5a) of unspecified signature provides a single geometric unification of all possible simple, three-dimensional, Lie groups of Cartan's classification.

This important property provides another illustration of the rather remarkable possibilities of the Lie-isotopic theory. It can be readily seen from the fact that the isosymmetry O(3) in realization (3.15) smoothly interconnects the *compact* realizations $\hat{O}(3) \approx O(3)$ with sig. $\hat\delta$ = (+1,+1,+1), to the *noncompact* realizations $\hat{O}(3) \approx O(2.1)$ with sig. $\hat\delta$ = (−1,−1,+1). The understanding is that Eq.s (3.15) provide the isotopic generalization of the corresponding transformations of O(3) and O(2.1), rather then the conventional tranformations themselves. For additional cases, see Figure III.3.1.

$O_o(3)$: $\hat{\delta} = \delta = \text{dig}\ (+1,+1,+1)$;	$O_o{}^d(3)$: $\hat{\delta} = -\delta = \text{diag}(-1,-1,-1)$
$O_1(3)$: Sig. $\hat{\delta} = (+1,+1,+1)$;	$O_1{}^d(3)$: Sig. $\hat{\delta} = (-1,-1,-1)$
$O_2(3)$: Sig. $\hat{\delta} = (+1,+1,-1)$;	$O_2{}^d(3)$: Sig. $\hat{\delta} = (-1,-1,+1)$
$O_3(3)$: Sig. $\hat{\delta} = (+1,-1,+1)$;	$O_3{}^d(3)$: Sig. $\hat{\delta} = (-1,+1,-1)$
$O_4(3)$: Sig. $\hat{\delta} = (-1,+1,+1)$;	$O_4{}^d(3)$: Sig. $\hat{\delta} = (+1,-1,-1)$
$\hat{O}(3) : \hat{\delta} = \text{diag.}\ (g_{11}, g_{22}, g_{33})$	

FIGURE III.3.1: A classification of all possible isotopes $\hat{O}(3)$ of O(3) submitted in Santilli (1985b). They can be presented via the classification of all possible underlying isoeuclidean spaces $\hat{E}(r,\hat{\delta},\hat{\mathfrak{R}})$ or directly, via the classification of all possible topologies of the isometric $\hat{\delta}$. The first group $O_o(3)$ is the conventional one. The isotopic theory initiates with the isodual $O_o{}^d(3)$ as per Definition II.3.3 which can be formulated only via the use of a bona-fide isounit $\hat{I} = -I$ Then eight classes of isotopes follow, each one conftaining an infinite number of isotopes, grouped into classes connected by isoduality. The classification of the isotopes $\hat{O}(3)$ therefore includes, not only the conventional O(3) and O(2.1), but also two infinite classes of nonlinear and nonlocal realizations of O(3) and O(2.1) interconnected by isotopic duality. The isotope $\hat{O}(3)$ of the last list of the diagram is the abstract isotope of Lemma III.3.1 unifyuing all the preceding realizations.

The possibilities of geometrical unification offered by the Lie-isotopic theory are therefore remarkable, and expressible via the following

CONJECTURE III.3.1: The simple, abstract, n-dimensional isotopes $\hat{G}(n)$ unify in one single algorithm all possible simple, nonexceptional Lie algebras of the same dimension in Cartan's classification.

Lemma III.3.1 proves the conjecture for the case n = 3. In Chapter IV we shall prove it for the case n = 6. The proof for the general case is left to the interested scholar[3].

This completes our studies of the classification of all possible isotopes of O(3). The analysis of this volume is restricted hereon only to the first infinite class of isotopes

def

[3] In studying the conjecture one should also keep in mind the "degrees of freedom" of the isofields expressed by Proposition II.2.1

$$\hat{O}(3) = \hat{O}_1(3), \quad \text{sig.}\,\hat{\delta} = (+1, +1, +1), \quad \hat{\delta} > 0. \qquad (3.16)$$

We shall therefore consider only the isotopes isomorphic to O(3) and ignore all others for brevity. Only the isoduals

$$\hat{O}^d(3) \stackrel{\text{def}}{=} \hat{O}_1{}^d(3)\,, \text{sig.}\,\hat{\delta} = (-1, -1, -1), \quad \hat{\delta} < 0, \qquad (3.17)$$

will be considered later on in this chapter.

The terms "isorotations" or "rotational-isotopic transformations" are therefore referred to in this volume, specifically, to those characterized by positive-definite isometrics $\hat{\delta}$.

By recalling that all nonsingular and Hermitean metrics and isometrics can be diagonalized, all positive-definite isometrics can therefore be written in the diagonal form

$$\hat{\delta} = \text{diag.}\,(b_1{}^2, b_2{}^2, b_3{}^2), \qquad (3.18a)$$

$$b_k = b_k(t, r, p, \dot{p}, \ldots) > 0, \qquad (3.18b)$$

which is assumed hereon as our basic form.

The first physical motivation for the restriction of the isometrics $\hat{\delta}$ to be positive-definite is the following. As well known, *mathematically* we can indeed deform the sphere

$$r^2 = r_1 r_1 + r_2 r_2 + r_3 r_3 > 0, \qquad (3.19)$$

into all infinitely possible compact (ellipsoidical) and nonconpact (hyperboloid) forms

$$r^{\hat{2}} = r_i\, g_{ij}\, r_j \gtrless 0, \qquad (3.20)$$

which then produce the classification of all possible, compact and noncompact isotopes reviewed earlier.

However, on physical grounds, a given sphere can only be deformed into ellipsoids, and there exists no known physical process capable of turning a sphere into a hyperboloid.

Additional reasons are of geometrical nature, and are motivated by the intent (see the remarks at the end of Sect. III.2) of reaching the unification of isotopic and conventional theories at the abstract, realization-free level.

Along these lines, one note that a most salient geometric axiom of the conventional theory of O(3) is the positive-definiteness of its invariant, Eq.s (3.19). In order to achieve an isotopic theory of $\hat{O}(3)$ capable of coinciding with that of O(3) at the abstract level, one must therefore preserve the same axiom of positive-definiteness of the underlying invariant (3.18).

Some of the main properties of isorotations can then be expressed as follows

THEOREM III.3.1 (Santilli (1985b)): The groups of (compact) isometries $\hat{O}(3)$ of all infinitely possible ellipsoidical deformations of the sphere on the isoeuclidean spaces $\hat{E}(r,\delta,\hat{\Re})$, $\hat{\Re} = \Re\hat{I}$, $\hat{I} = \delta^{-1}$, $\delta > 0$, verify the following properties:

1) The groups $\hat{O}(3)$ consist of infinitely many different groups corresponding to the infinitely many possible deformations of the sphere (explicit forms of the isometric δ), Eq. (3.18a);

2) All isosymmetries $\hat{O}(3)$ are locally isomorphic to O(3) under conditions (3.18b) herein assumed; and

3) The groups $\hat{O}(3)$ constitute "isotopic coverings" of O(3) in the sense that:

3.a) The groups $\hat{O}(3)$ are constructed via methods (the Lie-isotopic theory) structurally more general than those of O(3) (the conventional Lie's theory);

3.b) The groups $\hat{O}(3)$ represent physical conditions (deformations of the sphere; inhomogeneous and anisotropic interior physical media; etc.) which are broader than those of the conventional symmetry (perfectly rigid sphere; homogeneous and isotropic space; etc.); and

3.c) All groups $\hat{O}(3)$ recover O(3) identically whenever $\hat{I} = I$ and they can approximate the latter as close as desired for $\hat{I} \approx I$.

A first illustration of the nontriviality of the above results can be expressed via the property (which disproves a rather widespread belief)

COROLLARY III.3.1.1 (Santilli (loc. cit.): The rotational symmetry is

not broken by ellipsoidical deformations $\hat{\delta}$ of the sphere δ, but it is instead exact, provided that it is realized at the higher Lie-isotopic level with isounit $\hat{1} = \hat{\delta}^{-1}$.

The above results essentially confirm the existence of only one abstract rotational symmetry $\bigcirc(3)$, and infinitely many different, but isomorphic realizations. Of these, the conventional symmetry O(3) is obtained when one selects the simplest possible Lie product AB - BA, while the infinitely many isotopes $\hat{O}(3)$ are obtained when one selects our less trivial Lie-isotopic product A*B - B*A.

In different terms, when the conventional rotations do not constitute a symmetry of the system considered, this is not sufficient to imply that the rotational symmetrty is broken, because the symmetry can be reconstructed as exact at the more general isotopic level, by embedding all symmetry breaking terms in the isounit, exactly as done in Definition III.3.1.

We encounter in this way our first example of *reconstruction of an exact space-time symmetry when claimed to be broken.* Later on, we shall encounter the same reconstruction for all remaining connected space-time symmetries, e.g., for the Galilei symmetry studied in Section III.5, and for the Lorent symmetry studied in the next chapter.

The same mechanism of isotopic reconstruction of the exact symmetry is conceivable also for discrete symmetries, although it has not yet been studied in detail at this writing. We are referring to a conceivable isotopic reconstruction of the exact parity under weak interactions (Santilli (1984)), or a conceivable isotopic formulation of the exact time-reversal invariance of the center-of-mass behavior of strongly interacting particles with irreversible interior dynamics (Santilli (1983b)). It should be stressed here that we have merely mentioned *possibilities* of the Lie-isotopic techniques which require predictable additional studies for their resolution.

Note that, for general ellipsoids (3.18) *the "rotational symmetry" is exact, but the "conventional rotations" do not constitute a symmetry any longer.* This and other occurrences will require a suitable generalization of conventional relativities, even though the underlying space-time symmetries are locally isomorphic to the conventional ones (see Sect. III.6).

The latter aspect is rendered necessary by the following property.

COROLLARY III.3.1.2 (Santilli (loc. cit.)): While conventional rotations are linear in $E(r,\delta,\Re)$, isorotations are formally *isolinear and isolocal in $\hat{E}(r,\hat{\delta},\hat{\Re})$, but generally nonlinear and nonlocal in $E(r,\delta,\Re)$, i.e.,*

$$r' = \hat{R}(\theta)*r = \hat{R}(\theta)\,\hat{\delta}(t, r, \dot{r}, \ddot{r},...)\,r \qquad (3.21)$$

A further important result is the isotopic generalization of the conventional *Euler's theorem* on the general displacement of a rigid body with one point fixed (see, e.g., Goldstein (1950)), which we can express via the following

COROLLARY III.3.1.3 (Santilli (loc. cit.)): The general displacement of an elastic body with one fixed point is an isorotation $\hat{O}(3)$ around an axis through the fixed point.

In different terms, isorotations characterize not only a rotation of a given body, but also, jointly, its possible deformations. Thus, while the theory of rotations characterizes *rigid bodies*, as well known, the theory of isorotations characterizes *elastic bodies*. The covering nature of the latter over the former is then evident. For a conceptual anticipation of conceivable applications to elementary particle physics, see Sect. III.7.

This completes our review of the abstract treatment of the isorotational symmetry $\hat{O}(3)$. We are now sufficiently equipped to study the classical realizations of the isorotations under the conditions of: 1) being directly applicable to classical, closed, nonrelativistic, nonselfadjoint systems (III.2.3); 2) permitting the achievement of the conservation of the total angular momentum via the invariance of the systems under isorotations (without any need of subsidiary constraints); and 3) allowing the inclusion of nonlocal internal forces.

The first step toward these objectives is the identification of the physical role of carrier spaces. To begin, let us consider the isoeuclidean spaces $\hat{E}(r,\hat{\delta},\hat{\Re})$ with a positive-definite, diagonal isometric $\hat{\delta}$, Eq.s (3.18).

The phase space of the theory is then the cotangent bundle $T^*\hat{E}(r,\hat{\delta},\hat{\Re})$ in which we introduce N particles with our now familiar local coordinates

$$a = (a^{\mu}) = (r_{ka}, p_{ka}), \qquad (3.22)$$
$$\mu = 1, 2, ..., 6N,\ k = 1, 2, 3\ (= x, y, z),\ a = 1, 2, ..., N$$

The next step is to equip the space with the one-isoforms (Sect. II.9)

$$\hat{\Phi}_1 = R^\circ_\mu \hat{T}_1{}^\mu{}_\nu da^\nu = p_{ia} \hat{\delta}_{ij} dr_{ja}, \tag{3.23a}$$

$$R^\circ = (p, 0), \qquad \hat{T}_1 = (\hat{T}_1{}^\mu{}_\nu) = \text{diag.}\,(\hat{\delta}, \hat{\delta}), \tag{3.23b}$$

which is the fundamental space for the representation of systems (III.2.3a) via Pfaffian variational principles.

To study the isosymmetries of the systems we have to consider the isospace $T^*\hat{E}_2(r,\hat{\delta},\hat{\mathfrak{R}})$ of two-isoforms constructed from one-isoforms (3.23)

$$\hat{\Phi}_2 = \hat{d}\hat{\Phi}_1 = \tfrac{1}{2}\,\omega_{\mu\alpha}\,\hat{T}_2{}^\alpha{}_\nu\,\hat{d}a^\mu \hat{\wedge} \hat{d}a^\nu =$$

$$= \tfrac{1}{2}\,[\,\omega_{\mu\nu}\, b_\mu{}^2\, b_\nu{}^2 + (\,R^\circ_\mu \frac{\partial b_\mu{}^2}{\partial a^\nu}\, b_\nu{}^2 - R^\circ_\nu \frac{\partial b_\nu{}^2}{\partial a^\mu} b_\mu{}^2\,)\,]\,\hat{d}a^\mu \wedge \hat{d}a^\nu, \tag{3.24}$$

where $\omega_{\mu\nu}$ is the canonical symplectic tensor, and the isotopic element $\hat{T}_2$ is generally different than $\hat{T}_1$, with the explicit form (II.9.94)[4]

$$(\,T_2{}^{\mu\nu}) = (\,b^{\mu 2}\, b^{\nu 2} +$$

$$\omega^{\alpha\rho}\,\delta^\nu{}_\sigma\,(R^\circ_\rho \frac{\partial b_\rho{}^2}{\partial a^\sigma}\, b_\sigma{}^2 - R^\circ_\sigma \frac{\partial b_\sigma{}^2}{\partial a^\rho}\, b_\rho{}^2\,)\,) \stackrel{\text{def}}{=} \text{diag}\,(\hat{G}, \hat{G}), \tag{3.25}$$

The isometric and isotopic element $\hat{1}_2$ of $T^*\hat{E}_2(r,\hat{\delta},\hat{\mathfrak{R}})$ can then be written

---

[4] We introduce here an important notation which deserves a clarification. Recall from Sect. II.7 that we have two isospaces with generally different isometrics, the isospace $T^*\hat{E}_1(r,\hat{\delta},\hat{\mathfrak{R}})$ for one-isoforms (i.e., for the integrands of the variational principles), and the different isospaces $T^*\hat{E}_2(r,\hat{\delta},\hat{\mathfrak{R}}) = T^*\hat{E}(r,\hat{G},\hat{\mathfrak{R}})$ for the two-isoforms (i.e., for the characterization of symmetries and conserved quantities), with interconnecting rules given precisely by Eq.s (3.15). From hereon, when dealing with the *"characteristics b-functions"* of the interior medium, we shall tacitly refer to $T^*\hat{E}_1(r,\hat{\delta},\hat{\mathfrak{R}})$ while, when dealing with the *"characteristics B-functions"*, we shall refer to $T^*\hat{E}_1(r,\hat{\delta},\hat{\mathfrak{R}}) = T^*\hat{E}(r,\hat{G},\hat{\mathfrak{R}})$. Evidently, when the b's are constants we have $b_k \equiv B_k$, and $T^*\hat{E}_1(r,\hat{\delta},\hat{\mathfrak{R}}) \equiv T^*\hat{E}_2(r,\hat{G},\hat{\mathfrak{R}})$.

$$T^*\hat{E}_2(r,\delta,\mathfrak{R}) \equiv T^*\hat{E}(r,\hat{G},\mathfrak{R}) : \quad r = [r_{ia}\hat{G}_{ij}(t, r, p, \dot{p},..)\, r_{ja}\,]\,\hat{1}_2, \tag{3.26a}$$

$$\hat{1}_2 = (\hat{T}_2)^{-1} = \mathrm{diag.}\ \hat{G}^{-1}, \hat{G}^{-1}), \tag{3.26b}$$

Thus, the actual invariant of the isorotational theory under study is invariant (3.26a).

By recalling the interplay between geometry and algebras of Sect. II.9, the Lie-isotopic brackets of the theory are given by

$$[A \hat{,} B] = \frac{\partial A}{\partial a^\mu}\ \omega^{\mu\alpha}\,\hat{1}_{2\alpha}{}^{\nu}\ \frac{\partial B}{\partial a^\nu} \tag{3.27}$$

$$= \frac{\partial A}{\partial r_{ia}}\,\hat{G}_{ij}{}^{-2}(t, r, p,..)\,\frac{\partial B}{\partial p_{ja}} - \frac{\partial B}{\partial r_{kia}}\,\hat{G}_{ij}{}^{-2}(t, r, p, ..)\,\frac{\partial A}{\partial p_{ja}} .$$

Our objective is evidently that of reviewing the theory of isorotations $\hat{O}(3)$ via Lie-isotopic brackets (3.27). For clarity, we shall proceed in stages, and begin with the study first of the case of constant isometrics

$$\hat{\delta} = \mathrm{diag.}\ (b_1{}^2, b_2{}^2, b_3{}^2), \tag{3.28a}$$

$$b_k = \mathrm{constants} > 0, \tag{3.28b}$$

for which $\hat{T}_1 \equiv \hat{T}_2$, and the Lie-isotopic brackets (3.27) assume the simpler form

$$[A \hat{,} B] = \frac{\partial A}{\partial r_{ka}}\, b_k{}^{-2}\, \frac{\partial B}{\partial p_{ka}} - \frac{\partial B}{\partial r_{ka}}\, b_k{}^{-2}\, \frac{\partial A}{\partial p_{ka}}, \tag{3.29}$$

To identify the Lie-isotopic algebra $\mathbf{S\hat{O}}(3)$ characterized by brackets (3.29), let us compute first the *fundamental isotopic commutation rules* which are readily given by

$$(\,[a^\mu \hat{,}\ a^\nu]\,) = \begin{pmatrix} [r_i \hat{,}\ r_j] & [r_i \hat{,}\ p_j] \\ [p_i \hat{,}\ r_j] & [p_i \hat{,}\ p_j] \end{pmatrix} = (\hat{\Omega}^{\circ\mu\nu}) = \begin{pmatrix} 0 & \hat{\delta}^{-1} \\ -\hat{\delta}^{-1} & 0 \end{pmatrix}.$$

Next, we introduce the generators of the Lie-isotopic algebra **$\hat{SO}(3)$** of $\hat{SO}(3)$ which, by central assumption of Lie-isotopies (Sect. II.6), are given by the *conventional generators* of O(3), i.e., by the components of the angular momentum

$$J_k = \epsilon_{kij} r_i p_j, \tag{3.31}$$

The above quantities are called the components of the *Birkhoffian angular momentum* to emphasize the fact that they characterize a generalized notion because they are no longer defined on $T^*E_2(r,\delta,\mathfrak{R})$, but on $T^*\hat{E}_2 r,\hat{\delta},\hat{\mathfrak{R}})$. As a result, while the magnitude of the Hamiltonian angular momentum is given by the familiar expression

$$J^2 = J_k J_k, \tag{3.32}$$

the magnitude of the Birkhoffian angular momentum is instead given by

$$J^{\hat{2}} = J * J = J_i \hat{\delta}_{ij} J_j = J_k b_k^2 J_k. \tag{3.33}$$

Note that the interpretation of components (3.31) as isoscalars in $\hat{\mathfrak{R}}$ would imply the expressions (Sect. II.2)

$$\hat{J}_k = \hat{\epsilon}_{kij} * \hat{r}_i * \hat{p}_j = (\epsilon_{kij} r_i p_j) \hat{1} = J_k \hat{1}, \tag{3.34}$$

called the *trivial isotopy* (Sect. II.8) because it does not provide a generalized invariance, as the reader is encouraged to verify.

Also, the reader should keep in mind that we are dealing with the classical realization of **$\hat{SO}(3)$**, rather than its matrix realization as in Santilli (1985b). This implies that the generators of the isosymmetries must be ordinary functions, while quantities (3.34) are matrices.

To compute the isocommutation rules of $\hat{SO}(3)$, we first compute the isotopic liftings of the commutation rules between the angular momentum, and the local variables, resulting in the expressions

$$[J_k \hat{,} r_i] = \epsilon_{kij} b_i^2 r_j, \tag{3.35a}$$

$$[J_k \hat{,} p_i] = \epsilon_{kij} b_i^2 p_j, \tag{3.35b}$$

(where there is evidently no summation on the i-index).

The desired isocommutation rules of (compact) isorotational algebra $\mathbf{S\hat{O}}(3)$ are then given by (Santilli (1988a)),

$$\mathbf{S\hat{O}}(3): \qquad [J_i \hat{,} J_j] = C_{ij}^{\ k} J_k = \epsilon_{ijk} b_k^{-2} J_k. \tag{3.36}$$

which, under the redefinition

$$\hat{J}_1 = b_2 b_3 J_1, \qquad \hat{J}_2 = b_1 b_3 J_2, \qquad \hat{J}_3 = b_1 b_2 J_3, \tag{3.37}$$

can be written

$$[\hat{J}_i \hat{,} \hat{J}_j] = \epsilon_{ijk} \hat{J}_k, \tag{3.38}$$

This confirms the existence of a classical realization of the isocommutation rules of $S\hat{O}(3)$ possessing the same structure constants of SO(3). In turn, this confirms the local isomorphism between all possible isotopes $S\hat{O}(3)$ and SO(3) in accordance with Theorem III.3.1.

The isocenter of the enveloping algebra (Sect. II.6) is given by the isounit, which is the zero-order isocasimir, $C^{(o)} = \hat{1}$, and magnitude (3.33) of the Birkhoffian angular momentum, $C^{(2)} = J^{\hat{2}}$, as expected. In fact,

$$[J^{\hat{2}} \hat{,} J_i] = [J_k b_k^2 J_k \hat{,} J_i] = 2\epsilon_{kij} J_k J_j \equiv 0. \tag{3.39}$$

Note that the isosquare of J has the particular geometrical significance

$$J^{\hat{2}} = (\det \hat{\delta})\, J^2, \tag{3.40}$$

with intriguing implications in particle physics we hope to indicate at some future time.

Note also that $J^2 = J_k J_k$ *is not* an isocasimir of Lie-isotopic algebra (3.36) or of (3.38), as the reader can verify. This occurrence is important inasmuch as it confirms the correctness of selection (3.33) for the Birkhoffian angular momentum.

The occurrence also indicates that expression (3.38) of the isocommutation rules has a primary *mathematical* significance, inasmuch as it is formally identical to the conventional commutation rules. However, the isocommutation rules of direct *physical* significance are those in the physical angular moments J, i.e., rules (3.36).

The desired classical realization of the Lie-isotopic group structure is readily given via isoexpenentiations of Eq.s (3.36) resulting in the forms (Santilli (*loc. cit.*))

$$\hat{SO}(3): \quad \hat{R}(\theta) = [\prod_{k=1,2,3} e|_{\xi}^{\theta_k \omega^{\mu\alpha} \hat{1}_{2\alpha}{}^{\nu} (\partial J_k/\partial A^{\nu})(\partial/a^{\mu})} \} \hat{1}$$

$$\stackrel{\text{def}}{=} S_{\hat{\delta}}(\theta)\, \hat{1}, \tag{3.41}$$

where the exponentials are expanded in the conventional associative envelope ξ.

Note the true realization of the notion of isotopic lifting of a Lie symmetry, i.e., the preservation of the original generators and parameters of the symmetry, and the isotopic generalization of the *structure* of the Lie group via the liftings $I \Rightarrow \hat{1}_2$.

The computation of examples is straightforward. For instance a (classical) *isorotation around the third axis* is given by (Santilli (*loc. cit.*))

$$r' = \hat{R}(\theta_3)*r = S_{\hat{\delta}}(\theta_3)\, r \tag{3.42}$$

$$= \begin{pmatrix} r'_1 & & r_1\cos(\theta_3 b_1 b_2) - r_2 \dfrac{b_2}{b_1}\sin(\theta_3 b_1 b_2) \\ r'_2 & = & r_1 \dfrac{b_1}{b_2}\sin(\theta_3 b_1 b_2) + r_2\cos(\theta_3 b_1 b_2) \\ r'_3 & & r_3 \end{pmatrix}.$$

The proof of the invariance of isoseparation (3.28) under the above transformation is an instructive exercise for the reader interested in acquiring a knowledge of isotopic techniques. The computation of other examples can be readily done via Eq.s (3.41).

Note that the convergence of series (3.41) into finite transformations of type (3.42) is reduced to the convergence of the original series prior to the lifting, plus sufficient continuity and regularity conditions on the isounit.

Note also the appearance of the isotopic elements b_k directly in the angles of isorotation. This occurrence is useful for the reconstruction of the exact rotational symmetry according to the rule

$$\theta_3|_{Ham.} = \theta_3|_{Birk.} b_1 b_2, \tag{3.43}$$

which has important applications in particle physics (Sect. III.7)

In different terms, *the deformation experienced by the body considered, and represented by the b-quantities, is compensated by the isorotation in such a way that the combination of the deformation and isorotation equals the angle of rigid rotation* In this way, the exact rotational symmetry of a rigid body, the l.h.s. of Eq.s (3.43), is decomposed into the product of an isorotation and the b-quantities.

This is the realization in Birkhoffian mechanics of the property that all distinctions between conventional and isotopic symmetries cease to exist at the abstract, realization/free level.

We now pass to the application of the general theory of *isoinvariance* outlined in Sect. II.8, to the isorotations of closed nonselfadjoint systems. For this purpose, we have to verify first that the J's are indeed the generators of isorotations.

Consider an infinitesimal isorotation $\delta\theta$ around a fixed axis with unit isovector $n = (n_1, n_2, n_3)$ in $\hat{E}(r,\hat{\delta},\hat{\mathfrak{R}})$, i.e.,

$$r_k \Rightarrow r'_k + \delta\theta\, \epsilon_{kij}\, n_i\, r_j, \tag{3.44a}$$

$$p_k \Rightarrow p_k + \delta\theta\, e_{kij}\, n_i\, p_j. \tag{3.44b}$$

The isoexponentiation of the above quantities yields the relations

$$\{e|_\xi^{-\delta\theta\, n*J}\}\, r_k \approx r_k - \delta\theta\, [n*J \hat{,} r_k] = r_k + \delta\theta\, \epsilon_{kij}\, n_i\, r_j, \tag{3.45a}$$

$$\{e|_\xi^{-\delta\theta\, n*J}\}\, p_k \approx p_k - \delta\theta\, [n*J \hat{,} p_k] = p_k + \delta\theta\, \epsilon_{kij}\, n_i\, p_j, \tag{3.45b}$$

where the * product is evidently that in $\hat{E}(r,\hat{\delta},\hat{\mathfrak{R}})$. This confirms that the conventional components of the angular momentum are indeed the generators of isorotations.

The notion of *isorotational symmetry* is then given by a simple isotopy of the conventional one (Definition II.8.3). In fact a Birkhoffian B(r, p) is invariant under an isorotation around the n-axis iff it verifies the invariance property

$$B(r, p) = B(r + \delta\theta\, n\hat{\wedge}J, p + \delta\theta\, n\hat{\wedge}J)$$

$$= \{ e_{|\xi}^{-\delta\theta\, n\hat{\wedge}J} \} B(r, p), \qquad (3.46)$$

where $\hat{\wedge}$ is the vector product computed in $T^*\hat{E}(r,\hat{\delta},\hat{\mathfrak{R}})$, which can hold iff

$$[J_k \hat{,} B] = 0, \quad k = 1, 2, 3, \qquad (3.47)$$

For a more rigorous and general presentation, see Theorems II.8.2 and II.8.3.

We reach in this way the rather simple conclusion that *a Birkhoffian vector-field is invariant under isorotations when properly written in $T^*\hat{E}(r,\hat{\delta},\hat{\mathfrak{R}})$*, i.e., when all operations of contraction, power, etc., are properly made with the isometric $\hat{\delta}$.

Explicitly, the achievement of the $\hat{O}(3)$ invariance requires, first, the construction of a Hamilton-isotopic representation characterized by the tensor $\omega^{\mu\alpha}\, \hat{1}_{2\alpha}{}^{\nu}$, and then the restriction of the admissible Hamiltonians to those forms on $T^*\hat{E}_2(r,\hat{\delta},\hat{\mathfrak{R}})$ which are invariant under isorotation, .e.g., of the type

$$B = H = T(p) + V(r) = \frac{p_{ia}\, \hat{\delta}_{ij}\, p_{ja}}{2m_a} + V(r), \qquad (3.48a)$$

$$r = |r_{ia}\, \hat{\delta}_{ij} r_{ja}|^{\frac{1}{2}}. \qquad (3.48b)$$

Finally, note from Theorem II.8.3, that *conditions (3.47) are necessary and sufficient for the complete invariance of nonlinear, nonlocal, nonhamiltonian and nonnewtonian systems (III.2.3a) represented via the Hamilton-isotopic equations.*

We now pass to a review of isometrics with a nontrivial functional dependence, namely, for general Lie-isotopic brackets (3.27) for $\hat{G}$ =

Diag. (B_1^2, B_2^2, B_3^2).

It is easy to see that the isocommutation rules remain structurally unchanged under the generalization herein considered, with the only replacement of the b- with the B-quantities, e.g.,

$$[J_i \hat{,} r_j] = \epsilon_{ijk} B_j^2 r_k, \quad [J_i \hat{,} p_j] = \epsilon_{ijk} B_j^2 p_k. \tag{3.49}$$

The *general isocommutation rules of the isorotational algebras* **SÔ**(3) are then given by

$$\mathbf{S\hat{O}}(3): \quad [J_i \hat{,} J_j] = C_{ij}^{\ k}(r, p, ...) J_k = \epsilon_{ijk} B_k^{\ 2}(r, p, ...) J_k, \tag{3.50}$$

and provides a first illustration of the *structure functions* of the Lie-isotopic theory (Sect. II.6). The reformulation of the above algebra to reach the same structure constants of the conventional symmetry, as in Eq.s (3.38), is here left as an instructive exercise for the interested reader.

As one can see, under the condition of positive-definiteness of the isometric G, all infinitely possible isotopes SÔ(3) remain isomorphic to SO(3), by therefore preserving the semisimple and connected properties of SO(3).

The study of the *global isocasimir invariants*, that is, the isocasimirs valid everywhere in $T^*\hat{E}_2(r,\hat{\delta},\hat{\mathfrak{R}})$, under a nontrivial functional dependence of the isometric, is involved on technical grounds, inasmuch as it requires a deeper knowledge of the Birkhoffian realization of universal enveloping isoasssociative algebras and related neutral elements (see the remarks at the end of Sect. II.6)..

We shall therefore content ourselves with the identification of the *local isocasimirs*, that is, isocasimirs valid in a (star shaped) neighborhood of a point $\bar{a}$ of the local variables a.

It is easy to see that, in this local sense, the isocasimirs of realization (3.50) persist, i.e., are given by

$$C^{(0)} = \hat{1}_2 \big|_{\bar{a}}, \quad C^{(2)} = \hat{J^2} = (J \hat{G} J) \big|_{\bar{a}} \tag{3.51}$$

A simple example of a global isocasimir is given when

$$B_1 = B_2 = B_3 \equiv B(p), \quad \hat{T}_2 = B^2 I, \quad \hat{1}_2 = B^{-2} I, \tag{3.52}$$

in which case the magnitude of the angular momentum

$$\hat{J}^2 = J \hat{G} J = B^2(p)\ J^2, \tag{3.53}$$

is indeed a neutral element of the Lie-isotopic envelope, as the reader can verify.

Additional examples will be given when considering specific physical problems. The case of brackets (3.27) for nondiagonal isometrics will be studied at some later time.

We close this section with comments on the physical significance of our isoeuclidean spaces $\hat{E}(r,\hat{G},\hat{\mathfrak{R}})$. The first point the reader should keep in mind is that *the experimenter observing interior dynamical problems is not in $\hat{E}(r,\hat{G},\hat{\mathfrak{R}})$, but in the physical space $E(r,\delta,\mathfrak{R})$. Therefore, the spaces $\hat{E}(r,\hat{G},\hat{\mathfrak{R}})$ essentially provide a geometrization of interior physical media, with the understanding that actual measures remain done in the physical space $E(r,\delta,\mathfrak{R})$.*

This dichotomy of geometrical space $\hat{E}(r,\hat{\delta},\hat{\mathfrak{R}})$ versus physical space $E(r,\delta,\mathfrak{R})$ will appear clear with further studies and applications of the isotopic relativities. At this point, the following introductory comments are in order.

The first meaning of the isoeuclidean space $\hat{E}(r,\hat{G},\hat{\mathfrak{R}})$ is that of providing a geometrization of the physical characteristics of the interior physical media via the B-functions, called *characteristics B-functions*. Recall that contact nonhamiltonian interior forces have no effect in the total energy, but only on the local internal exchanges of energy. These features suggest a geometrization of the interior media which is such to provide no global effect, exactly along the notion of closed nonhamiltonian systems.

Moreoever, one of the central objectives of any acceptable theory for the interior problem is that of representing the inhomogeneous and anisotropic character of the interior physical media. Our isoeuclidean spaces achieve these objectives in their entirety. In fact, they represent the *inhomogenuity* of the interior media via different values of the elements of the isometrics as well as their dependence on the locally varying density $\mu(r)$, temperature $\tau(r)$, possible index of refraction $n(r)$, etc.

$$B_1(t, r, p, \dot{p}, \mu, \tau, ...) \neq B_2(t, r, p, \dot{p}, \mu, \tau, n, ...) \neq B_3(t, r, p, \dot{p}, \mu, \tau, n, ...). \tag{3.54}$$

The *anisotropy* can be represented, e.g., via a factorization of Finsler's type

$$B_k = F(r)\, \tilde{B}_b, \qquad (3.55)$$

where r represents a preferred direction in the media, such as that of the intrinsic angular momentum.

Of particular importance is the possibility of averaging the characteristics B-quantities to constants, via any appropriate averaging procedure

$$<|\, B_k(t, r, p, \mu, \tau, n, \ldots) \,|> \; = \; b_k \; = \; \text{constants} > 0. \qquad (3.56)$$

In fact, the characterization of the physical media with the *B-functions* is necessary when describing a specific interior trajectory. However, the averaging of the B-quantities to constants, Eq.s (3.56), is sufficient for a global study of interior dynamics, e.g., when studying Jupiter's structure from an outside observer, or studying the global behavior of light propagating within inhomogeneous and anisotropic atmospheres (Chapter IV).

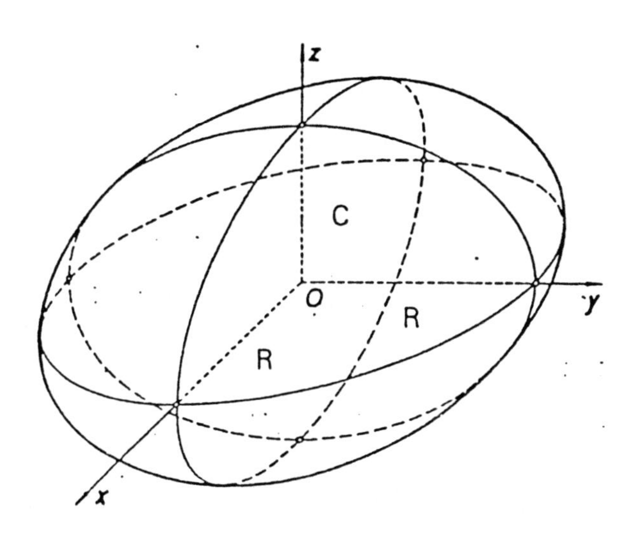

FIGURE III.3.2: An illustration of the possibilities of the theory of isorotations in classical and particle physics: the direct representation of the actual shape of a given charge distribution (an oblate spheroidal ellipsoid in the figure), as well as of all its infinitely possible deformations under sufficiently intense external fields, collisions or other causes. Note that the isorotational theory can first represent the actual shape of the considered charge distribution (say, an oblate spheroidal ellipsoid as in the figure), and then its possible deformations, all this already at a classical level (Santilli (1985b),(1988a)). By comparison, contemporary theories can indeed reach a representation of the extended character of particles via the form factors, but, as well known: 1) the representation is achieved only after the rather complex second quantization; 2) the form factors provide only a remnant of the actual shape and cannot characterize the actual shape itself (at any rate, the representation of non-spherical shapes would imply the breaking of the conventional rotational symmetry); and 3) the representation cannot possibly characterize the deformations of said charge distributions (because, again, the conventional rotational symmetry refers strictly to rigid bodies and it is broken under the deformations of the physical reality). The advantage of the isorotations over the conventional rotations is then evident. In short, *the theory of isorotations is a theory of extended, elastic and deformable bodies.* This concept alone is sufficient to require a suitable isotopy of the conventional relativities which is best illustrated by the simplest possible case of a free particle obeying the Galilei-isotopic relativity (Sect. III.7).

As a result, the realization of the isorotational symmetry with characteristic B-*constants*, Eq.s (3.36) and (3.40), is fully sufficient for the global study of closed nonselfadjoint systems, while the more general form with nontrivial characteristic B-*functions* is primarily needed for the local internal behavior.

In turn, this illustrates the sufficiency of the isocasimir (3.33) for the global behavior, as well as provides a physical reason for the lack of existence of global isocasimirs with nontrivial B-quantities.

A further application of our isospaces is the representation of the actual shape of the particle considered. Recall that one of the geometrical meanings of the invariant (3.19) of O(3) is the perfect sphere. Therefore, one of the geometrical meanings of the isoinvariants (3.18) is the representations of the actual, generally nonspherical shape of the particle considered, e.g., an oblate spheroidal ellispoid, jointly with all its infinitely possible deformations (see Figure III.3.2 for more details).

The epistemological implications of the averaging process (3.56) should be known since these introductory words. Recall from Corollary III.3.1.2, that the isotransformations are generally nonlinear and nonlocal in the physical space $E(r,\delta,\Re)$. This is however the case for characteristics B-functions with a nonlinear and nonlocal dependence in their variables. When these B-functions are averaged to constants, isotransformations (3.21) reacquire their conventional linear and local character, with the understanding that the isorotations themeselves remain generalized, Eq.s (3.42).

In conclusion, the averaging (3.56) of the characteristics B-functions to constants for global studies of closed nonhamiltonian systems implies the full recovering of the linearity and locality of the isotransformations and, thus, the preservation of the conventional inertial approximation of the observer.

III.4: EUCLIDEAN-ISOTOPIC SYMMETRIES

In the preceding section we have introduced the largest possible nonlinear and nonlocal, classical realization of the isorotational symmetries $\hat{O}(3)$ on the Euclidean-isotopic spaces $T^*\hat{E}_2(r,\hat{\delta},\hat{\mathfrak{R}}) = T^*\hat{E}(r,\hat{G},\hat{\mathfrak{R}})$. In this section we shall generalize the results to present the infinitely possible isotopic generalizations of the Euclidean group $E(3)$.

For this purpose, consider again system (III.2.3) of N particles in isospace $T^*\hat{E}(r,\hat{G},\hat{\mathfrak{R}})$, but this time considered with respect to the coordinate differences r_{ab} among generic particles a and b, i.e.,

$$T^*\hat{E}_2(r,\hat{\delta},\hat{\mathfrak{R}}): \quad (r_{ab})^{\hat{2}} = (r_{abi}\,\hat{G}_{ij}\,r_{abj})\,\hat{1}, \tag{4.1a}$$

$$r_{ab} = r_a - r_b, \quad \hat{\mathfrak{R}} = \mathfrak{R}\hat{1}_2, \quad \hat{1}_2 = \text{diag.}(\hat{G}^{-1}, \hat{G}^{-1}) > 0, \tag{4.1b}$$

$$i, j = 1, 2, 3, \quad a, b = 1, 2, \ldots, N,$$

where we have kept the notations of the preceding section and the isometric $\hat{G}$ has the most general possible nonlinear and nonlocal dependence on all needed local quantities, such as: coordinates and their derivaties r, $\dot{r}$, $\ddot{r}$, ...; density μ of the medium in which motion occurs; temperature τ; index of refraction n; etc.

Since $\hat{G}$ is nonsingular and Hermitean, it can always be diagonalized to the form

$$\hat{G} = \text{diag.}\,(B_1^2, B_2^2, B_3^2), \quad B_k = B_k(t, r, \dot{r}, \ddot{r}, \mu, \tau, n, ..) > 0. \tag{4.2}$$

DEFINITION: The "Euclidean-isotopic symmetries" $\hat{E}(3)$, also called "isoeuclidean symmetries", are the largest possible nonlinear and nonlocal Lie-isotopic groups of isometries of isoseparations (4.1a) on $T^*\hat{E}(r,\hat{G},\hat{\mathfrak{R}})$.

As well known, the conventional Euclidean group E(3) is given by the semidirect product

$$E(3) = O(3) \otimes T(3), \tag{4.3}$$

where: O(3) is the conventional group of (linear, local) rotations in $T^*E(r,\delta,\mathfrak{R})$,

$$O(3): \quad r' = R(\theta)\, r, \quad p' = R(\theta)\, p, \qquad R^t R = R R^t = I; \tag{4.4}$$

T(3) is the conventional group of (linear, local) translations

$$T(3): \; r' = r + r^\circ, \quad p' = p, \tag{4.5}$$

and the parameters are given by (the ordered set)

$$w = (w_k) = (\theta, r^\circ), \quad k = 1, 2, ..., 6, \tag{4.6}$$

namely, they are the conventional Euler's angles $\theta = (\theta_i)$ and the conventional translation constants $r^\circ = (r^\circ_i)$.

By introducing the (ordered set of) generators

$$E(3): \qquad X = (X_k) = (J, P), \quad k = 1,2,..6, \tag{4.7a}$$

$$J_k = \Sigma_a \epsilon_{kij}\, r_{ia}\, p_{ja}, \quad P_k = \Sigma_a\, p_{ka}, \tag{4.7b}$$

the conventional Poisson brackets among functions A, B on $T^*E(r,\delta,\mathfrak{R})$

$$[A\, , B] = \frac{\partial A}{\partial r_{ka}} \frac{\partial B}{\partial p_{ka}} - \frac{\partial A}{\partial r_{la}} \frac{\partial B}{\partial p_{ka}} \quad - \tag{4.8}$$

yield the familiar commutation rules of the Euclidean algebra

$$\mathbf{E}(3): \quad [J_i, J_j] = \epsilon_{ijk} J_k, \qquad [J_i, P_j] = \epsilon_{ijk} P_k, \qquad [P_i, P_j] = 0, \tag{4.9}$$

and the connected component of the Euclidean group

$$E(3): \; a' = \{\prod_k e^{[w_k \omega^{\mu\nu} (\partial_\nu X_k)(\partial_\mu)}_{|\xi}\} a, \tag{4.10}$$

To study the structure of the infinite family of possible isotopes $\hat{E}(3)$ of $E(3)$, we begin by recalling that the transformation theory on $T^*\hat{E}(r,\hat{G},\hat{\Re})$ is isotopic (Sect. II.3), i.e., it is formally linear and local in $T^*\hat{E}(r,\hat{G},\hat{\Re})$, but generally nonlinear and nonlocal in $T^*E(r,\delta,\Re)$. Also, the isometrics $\hat{G}$ enter directly into the exponential structure of the Lie-isotopic group and this ensures the intrinsic nonlinearity as well as nonlocality of the theory.

Finally, we recall that the Lie-isotopic brackets in $T^*\hat{E}(r,\hat{G},\hat{\Re})$ are given by Eq.s (III.3.27) for diagonal elements (4.2)

$$[A \,\hat{,}\, B] = \frac{\partial A}{\partial r_k} B_k^{-2} \frac{\partial B}{\partial p_k} - \frac{\partial A}{\partial p_k} B_k^{-2} \frac{\partial B}{\partial r_k}. \tag{4.11}$$

The following property then holds.

THEOREM III.4.1 (Santilli (1988a): The isoeuclidean symmetries $\hat{E}(3)$ of isoseparation (4.1) on $\hat{E}(r,\hat{G},\hat{\Re})$ possess the structure

$$\hat{E}(3) \; : \; \hat{O}(3) \otimes \hat{T}(3), \tag{4.12}$$

where $\hat{O}(3)$ is the isorotational symmetries of Sect. III.3,

$$\hat{O}(3): \quad r'_a = \hat{R}(\theta) * r_a = \hat{R}(\theta)\, \hat{G}(r,\dot{r},\ddot{r},\mu,\tau,\ldots)\, r_a, \tag{4.13a}$$

$$\hat{R}^{\hat{t}} * \hat{R} = \hat{R} * \hat{R}^{\hat{t}} = \hat{1} = \hat{G}^{-1}, \tag{4.13b}$$

$$r_a = (r_{ia}), \quad i = 1, 2, 3 \;\; (=x, y, z), \quad a = 1, 2, \ldots, N,$$

and $\hat{T}(3)$ is the largest possible group of nonlinear and nonlocal isotranslations

$$\hat{T}(3): \quad r'_{ia} = r_{ia} + r_i^{\circ}\, \hat{B}_i^{-2}(t, r, \dot{r}, \ddot{r}, \mu, \tau, n....), \tag{4.14a}$$

$$p'_{ia} = p_{ia}, \tag{4.14b}$$

with Birkhoffian realization in $T^*\hat{E}(r,\hat{G},\hat{\Re})$ characterized by:

1) The same set of parameters $(w_k) = (\theta_i, r^{\circ}_i)$ of the conventional symmetry E(3);

2) the same set of generators $(X_k) = (J, P)$ of the conventional symmetry E(3),

3) the isocommutations rules of the Lie-isotopic algebras

$$\hat{E}(3): \quad [J_i \,\hat{,}\, J_j] = \epsilon_{ijk}\, B_k^{-2}\, J_k, \tag{4.15a}$$

$$[J_i \,\hat{,}\, P_j] = \epsilon_{ijk}\, B_j^{-2}\, P_k, \tag{4.15b}$$

$$[P_i \,\hat{,}\, P_j] = 0. \tag{4.15c}$$

4) the Lie-isotopic group structure

$$\hat{E}(3): \quad \hat{e}(w) * a = \{\hat{R}(\theta), \hat{T}(r^{\circ})\} * a =$$

$$= \{ [\prod_k e^{w_k \omega^{\mu\sigma} \hat{1}_{2\sigma}{}^{\nu} (\partial_\nu X_k)(\partial_\mu)}|_{\xi}\,]\, \hat{1}_2\} * a, \tag{4.16}$$

5) the realization of the $\hat{B}$-functions

$$\hat{B}_i^{-2}(t,r,p,....) = B_i^{-2} + r^{\circ}_j\, [B_i^{-2} \,\hat{,}\, P_j]\, /2! + r^{\circ}_j\, r^{\circ}_k\, [[B_i^{-2} \,\hat{,}\, P_j] \,\hat{,}\, P_k] \,/\, 3! + ... \tag{4.17}$$

and the local isocasimir invariants given by [5]

---

[5] It is appropriate to recall here that the realization of the isoeuclidean symmetry under consideration is classical, that is, via *functions* in $T^*\hat{E}(r,\hat{G},\hat{\Re})$. As a result, the generators and, therefore, the isocasimir invariants, must necessarily be functions, while the isounit $\hat{1}_2$ of Eq.s (4.18) is a metrix of functions. Thus, only the *functional*

$$\hat{C}^{(o)} = \hat{1}_2, \qquad \hat{C}^{(1)} = (P*P)\,\hat{1}_2, \qquad \hat{C}^{(2)} = (J*P)\,\hat{1}_2, \tag{4.18}$$

Finally, all the above infinitely possible Euclidean-isotopic symmetries $\hat{E}(3)$ result to be locally isomorphic to the conventional Euclidean symmetry E(3) under the sole conditions of sufficient smoothness, nonsingularity and positive-definiteness of the isometrics $\hat{G}$.

PROOF. By central assumption, the generators and parameters remain unchanged under isotopy, and only the Lie structure is generalized in an axioms preserving way. Under these conditions, isocommutator rules (4.15) follow from the Lie-isotopic brackets (4.11), while isotopic group structure (4.16) follows from the exponentiation of the above brackets according to the rules of Sect. II.6. Similarly, *local* isocasimirs (4.18) follow from their isocommutativity with all generators (see the comments on the locality of the isocasimirs of Sect. III.3).

The isorotational symmetries $\hat{O}(3)$ were studied in the preceding section. Therefore, we have to study here only the isotranslations. They can be written

$$\hat{T}(3): \qquad r'_{ia} = \hat{T}(r^\circ) * r_{ia} = \tag{4.19}$$

$$= \{ \prod_j e_{|\xi}^{\, r^\circ_j\, \omega^{\mu\sigma}\, \hat{1}_{2\sigma}{}^{\nu}\, (\partial_\nu P_j)\,(\partial_\mu)} \}\, \hat{1}_2 * r^a{}_i =$$

$$= r_{ia} + r^\circ_j\,[r_{ia} \,\hat{,}\, P_j] / 2! + r^\circ_j r^\circ_k\,[[r_{ia} \,\hat{,}\, P_j] \,\hat{,}\, P_k] / 3! + \ldots\ldots$$

and

$$p'_{ia} = \hat{T}(r^\circ) * p_{ia} = \tag{4.20}$$

$$\{ \prod_a \exp [\, r^\circ_j\, \omega^{\mu\sigma} \times \hat{1}_{2\sigma}{}^{\nu}\, (\partial_\nu P_j)\,(\partial_\mu)\}\, p_{ia}$$

$$= p_{ia} + r^\circ_j\,[p_{ia} \,\hat{,}\, P_j] / 2! + r^\circ_j r^\circ_k\,[[r_{ia} \,\hat{,}\, P_j] \hat{,}\, P_k] / 3! + \ldots\ldots$$

from whch Eq.s (4.14) and (4.17) follow.

---

factors of the isounit in Eq.s (4.17) are the correct isocasimirs of the isoeuclidean symmetries.

We finally remain with the proof of the local isomorphisms $\hat{E}(3) \approx E(3)$. The local isomorphisms $\hat{O}(3) \approx O(3)$ have been proved in the preceding section and shown to hold for all positive-definite isometrics $\hat{G}$. The local ismorphisms $\hat{T}(3) \approx T(3)$ then trivially follow from the preservation of the original Abelian character under isotopies. QED.

The application of the isoeuclidean symmetry to closed nonhamiltonian systems (II.2.3) is a particular case of Theorems II.8.2 and II.8.3 here left to the interested reader.

For future needs, let us consider the global treatment of systems (II.2.3) under average (III.3.56) of the charactereristic B-functions to b-constants, and let us introduce the redefinitions

$$\hat{J}_k = \sum_a \epsilon_{kij} \hat{r}_{ia} \hat{p}_{ja}, \qquad \hat{P}_k = \sum_a \hat{p}_{ka}, \tag{4.21a}$$

$$\hat{r}_{ia} = r_{ia} b_i, \qquad \hat{p}_{ia} = p_{ia} b_i \quad \text{(no sum)}. \tag{4.21b}$$

Then, isocommutation rules (4.15) become

$$\hat{\mathbf{E}}(3): [\hat{J}_i \hat{,} \hat{J}_j] = \epsilon_{ijk} \hat{J}_k, \ [\hat{J}_i \hat{,} \hat{P}_j] = \epsilon_{ijk} P, \ [\hat{P}_i \hat{,} \hat{P}_j] = 0, \tag{4.22}$$

namely, the structure constants of $\hat{\mathbf{E}}(3)$ and $\mathbf{E}(3)$ coincide, thus illustrating again the local isomorphisms $\hat{E}(3) \approx E(3)$.

The study of the classification of all possible isotopes $\hat{E}(3)$ under the relaxation of the positive-definiteness of the isotopic element $\hat{T}_2$ is left to the interested reader.

Intriguingly, this section constitutes a preliminary step for the Lie-isotopic generalizations $\hat{P}(3.1)$ of the Poincaré symmetry $P(3.1)$ to be considered in Chapter IV.

As a matter of fact, Theorem III.4.1 can be easily modified to produce the infinite family of Lie-isotopic generalizations $\hat{P}(2.1)$ of the Poincaré symmetry $P(2.1)$ in (2+1-space-time dimensions, via the mere assumption

$$\hat{G} = T\eta, \ \eta = \text{diag.}\,(+1, +1, -1), \tag{4.23}$$

namely, that the isometrics $\hat{G}$ are the isotopes of the (2+1)-dimensional Minkowski metric η, rather than of the Euclidean metric δ.

These relativistic aspects will be studied in details in Chapter IV.

III.5: GALILEI-ISOTOPIC SYMMETRIES

As well known, the conventional *Galilei symmetry* G(3.1) (see, e.g., Levy-Leblond (1971) or Sudarshan and Mukunda (1974)) can be defined as the largest Lie group of *linear and local* transformations leaving invariant the separations

$$t_a - t_b = \text{inv.},$$

$$(r_{ia} - r_{ib})\, \delta_{ij}\, (r_{ja} - r_{jb}) = \text{inv. at } t_a = t_b, \tag{5.1}$$

$$i, j = 1, 2, 3 \ (= x, y, z), \qquad a = 1, 2, \ldots, N$$

in $\Re_t \times T^*E(r,\delta,\Re)$, where $\Re_t$ represents time, $E(r,\delta,\Re)$ is the conventional Euclidean space, and T^*E its cotangent bundle (phase space), with metric $\delta = \text{diag.}\,(1,1,1)$ over the reals $\Re$.

The explicit form of the *Galilei transformations* is given by the familiar expressions

$$\begin{cases} t' = t + t^\circ, & \text{translations in time} \quad (5.2a) \\ r'_{ia} = r_{ia} + r^\circ_i, & \text{translations in space} \quad (5.2b) \\ r'_{ia} = r_{ia} + t^\circ v^\circ_i, & \text{Galilei boosts} \quad (5.2c) \\ r'_a = R(\theta)\, r_a, & \text{rotations.} \quad (5.2d) \end{cases}$$

A classical realization of G(3.1) (for the case of all non-null masses, $m_a \neq 0$, $a = 1, 2, \ldots, N$, herein assumed) is characterized by the (ordered sets of) parameters

$$w = (w_k) = (\theta_i, v^\circ_i, r^\circ_i, t^\circ), \quad k = 1,2,\ldots 10, \quad i = 1, 2, 3 \tag{5.3}$$

and generators

$$X = (X_k) = (J_i, G_i, P_i, H), \tag{5.4a}$$

$$J_i = \textstyle\sum_a \epsilon_{ilm}\, r_{la}\, p_{ma}, \qquad P_i = \sum_a p_{ia}, \tag{5.4b}$$

$$G_i = \textstyle\sum_a (m_a r_{ia} - t p_{ia}), \qquad H = p_{ia}\, p_{ia}/2m_a + V(r_{ab}), \tag{5c}$$

$$r_{ab} = r_a - r_b, \quad i = 1, 2, 3, \quad k = 1, 2, \ldots, 10, \quad a, b = 1, 2, \ldots, N,$$

with canonical realization of the Lie algebra **G**(3.1) via the conventional Poisson brackets

$$\mathbf{G}(3.1): \qquad [J_i, J_j] = \epsilon_{ijk} J_k, \quad [J_i, P_j] = \epsilon_{ijk} P_k, \tag{5.5a}$$

$$[J_i, G_j] = \epsilon_{ijk} G_k, \quad [J_i, H] = 0, \tag{5.5b}$$

$$[G_i, P_j] = \delta_{ij} M, \qquad [G_i, H] = P_i, \tag{5.5c}$$

$$[P_i, P_j] = [G_i, G_j] = [P_i, H] = 0, \tag{5.5d}$$

$$M = \sum_a m_a, \tag{5.5e}$$

Casimir invariants

$$C^{(o)} = I, \qquad C^{(1)} = P^2 - 2MH, \quad C^{(2)} = (MJ - G \wedge P)^2, \tag{5.6}$$

and canonical realization of the group structure G(3.1)

$$G(3.1): \quad a' = g(w)\, a = \{ \prod_k \exp [w_k\, \omega^{\mu\nu} (\partial_\nu X_k)(\partial_\mu)]\}\, a \tag{5.7}$$

$$\partial_\mu = \partial / \partial a^\mu, \quad a = (a^\mu) = (r_{ia}, p_{ia}), \ \mu = 1,2,...,6N,$$

where $\omega^{\mu\nu}$ is the canonical Lie tensor (III.2.13).

The main lines of the the construction of the infinite family of Lie-isotopic generalizations $\hat{G}(3.1)$ of the Galilei symmetry G(3.1), under the name of the Galilei-isotopic symmetries, were submitted in the original proposal for the Lie-isotopic theory (Santilli (1978a,c)). A step-by-step generalization of classical Hamiltonian mechanics , under the name of *Birkhoffian mechanics*, was subsequently constructed, and the Galilei-isotopic symmetries were formally proposed in Chapter 6 of Santilli (1982a) for their most general possible, nonlinear, nonhamiltonian and nonnewtonian, but local-differential realization.

A comprehensive study for the nonlocal classical realization was conducted in Santilli (1988a) and (1991b), which constitute the basis of this presentation.

Examples of closed two- and three-body nonhamiltonian systems invariant under the Galilei-isotopic symmetry were studied by Jannussis, Mijatovic and Veljanoski (1991) (as reviewed in Appendix

III.A).

As now familiar, the starting basis is the infinite number of isotopes $\hat{E}(r,\hat{\delta},\hat{\mathfrak{R}})$ of the Euclidean space $E(r,\delta,\mathfrak{R})$ which is, in turn, extended to the isocotangent bundle $T^*\hat{E}(r,\hat{\delta},\hat{\mathfrak{R}})$. A nonhamiltonian system of N particles, Eq.s (III.2.3a), is then considered in such isospace with the familiar local coordinates $a = (a^\mu) = (r_{ka}, p_{ka})$, $\mu = 1, 2, ..., 6N$, $k = 1, 2, 3$ (= x, y, z), and $a = 1, 2, ..., N$.

The system is then represented via the Birkhoffian variational principle (II.7.23) which essentially introduces one-isoforms on $T^*\hat{E}_1(r,\hat{\delta},\hat{\mathfrak{R}})$

$$\hat{\Phi}_1 = \theta_1 \times \hat{T}_1 = R^\circ{}_\mu \hat{T}_1{}^{\mu\nu}\, \hat{d}a^\nu, \tag{5.8a}$$

$$R^\circ = (p, 0), \quad \hat{T}_1 = \text{diag.}\ (\hat{\delta}, \hat{\delta}), \tag{5.8b}$$

$$\hat{\delta} = \text{diag.}\ (b_1{}^2, b_2{}^2, b_3{}^2), \ b_k = b_k(t, r, p, \dot{p}, ...) > 0. \tag{5.8c}$$

The isospaces for the characterization of the symmetries of the systems are $T^*\hat{E}_2(\hat{\delta},\hat{\mathfrak{R}}) = T^*\hat{E}(r,\hat{G},\hat{\mathfrak{R}})$ characterized by the two-isoform (II.9.92) which, in their Hamiltonian-isotopic expression, can be written in the local chart a

$$\hat{\Omega}^\circ{}_2 = [\tfrac{1}{2}\, \hat{T}_{2\mu}{}^\alpha(a...)\, \omega_{\alpha\nu}]\, \hat{d}a^\mu \hat{\wedge} \hat{d}a^\nu = d[\hat{T}_1(a,..)\times R^\circ]_\mu\, da^\mu, \tag{5.9a}$$

$$\hat{T}_2 = \text{Diag.}\ (\hat{G}, \hat{G}), \quad \hat{G} = \text{diag.}\ (B_1{}^2, B_2{}^2, B_3{}^2), \ B_k > 0, \tag{5.9b}$$

where the methods to construct the isometrics $\hat{G}$ from $\hat{\delta}$ are assumed to be known (see Sect. II.9 and the outline of Sect. III.2).

The Lie algebra brackets characterized by two-forms (5.9) are given by the now familiar expression in $T^*\hat{E}(r,\hat{G},\hat{\mathfrak{R}})$

$$[A \,\hat{,}\, B] = \frac{\partial A}{\partial a^\mu}\, \omega^{\mu\alpha}\, \hat{1}_{2\alpha}{}^\nu \frac{\partial B}{\partial a^\nu}, \tag{5.10a}$$

$$\hat{1}_2 = \text{diag.}\ (\hat{G}^{-1}, \hat{G}^{-1}), \tag{5.10b}$$

where $\omega^{\mu\alpha}$ is the familiar canonical Lie tensor.

We are now equipped to introduce the following

DEFINITION III.5.1 (loc. cit.): The "general (nonlinear, nonlocal, classical) Galilei-isotopic symmetries", or "general isogalilean symmetries" $\hat{G}(3.1)$ are given by the Lie-isotopic groups of the most general possible transformations on $\hat{\Re}_t \times T^*\hat{E}(r,\hat{G},\hat{\Re})$

$$t_a - t_b = \text{inv.}, \tag{5.11a}$$

$$(r_{ka} - r_{kb})\, B_k^2(t, r, p, ...)\, (r_{ka} - r_{kb}) = \text{inv.} \quad \text{at } t_a = t_b, \tag{5.11b}$$

$$t_a, t_b \in \hat{\Re}_t, \qquad r_a, r_b \in T^*\hat{E}(r,\hat{G},\hat{\Re}) \tag{5.11c}$$

where $\hat{\Re}_t$ is an isotopic lifting of the conventional field $\Re_t$, here called "isotime field", with explicit structure

$$\hat{\Re}_t = \Re\, \hat{1}_t, \quad \hat{1}_t = B_4^{-2}(t, r, p, ...), \qquad B_4 > 0, \tag{5.12}$$

$T^*\hat{E}(r,\hat{G},\hat{\Re})$ is the isocotangent bundle for sisosymplectic two-isoforms with isometrics $\hat{G}$ (5.9b), and the four functions B_1, B_2, B_3 and B_4 besides being independent and positive-definite, are arbitrary nonlinear and nonlocal (e.g., integral) functions on all possible, or otherwise needed local variables and quantities. The "restricted isogalilean symmetries" occur when the characteristic B-quantities are constants different than one.

The reason for the additional lifting $\Re_t \Rightarrow \hat{\Re}_t$ in the trasition from the isoeuclidean symmetries $\hat{E}(3)$ of the preceding section to the isogalilean symmetries $\hat{G}(3.1)$ of this section, will be evident shortly, although their ultimate meaning will appear in the study of the nonrelativistic limit of the Poincaré-isotopic symmetries to be studied in the next chapter.

At this point we merely recall from Sect. II.2 that the use of the isotime field does not affect the physical time. In fact, isofields possess the conventional sum,

$$\hat{t}_1 + \hat{t}_2 = (t_1 + t_2)\, \hat{1}_2, \tag{5.13}$$

but have the isotopic multiplication

$$\hat{t}_1 * \hat{t}_2 = (t_1 t_2)\, \hat{1}_t. \tag{5.14}$$

As a result, the multiplication of the isotime by any quantity, say, A, is conventional, $\hat{t}*A \equiv tA$, which justifies the setting of the measurement theory with respect to the ordinary time.

THEOREM III.5.1 (Santilli (1978a), (1982a), (1988a)): The general nonlinear and nonlocal, classical realization of the Galilei-isotopic (or isogalilean) symmetries $\hat{G}(3.1)$ on $\hat{\Re}_t \times T^*\hat{E}(r,\hat{G},\hat{\Re})$ as per Definition III.5.1, can be written

$$\begin{cases} t' = t + t^\circ \tilde{B}_4^{-2}, & \text{iso-time translations} \quad (5.15a) \\ r_i' = r_i + r^\circ_i \tilde{B}_i^{-2}, & \text{iso-space translations} \quad (5.15b) \\ r_i' = r_i + t^\circ v^\circ_i \tilde{B}_i^{-2}, & \text{iso-Galilei boosts} \quad (5.15c) \\ r' = \hat{R}(\theta) * r, & \text{isorotations,} \quad (5.15d) \end{cases}$$

where the $\tilde{B}$-functions are generally nonlinear and nonlocal in all possible local variables and quantities to be identified shortly. Moreover, the Galilei-isotopic symmetries $\hat{G}(3.1)$ are characterized by the Lie-isotopic brackets (5.10) underlying the exact symplectic-isotopic two-forms, with explicit expression

$$[A\hat{,} B] = \frac{\partial A}{\partial r_{ka}} B_k^{-2} \frac{\partial B}{\partial p_{ka}} - \frac{\partial A}{\partial p_{ka}} B_k^{-2} \frac{\partial B}{\partial r_{ka}}, \qquad (5.16)$$

and possess the following structure:
1) the conventional parameters (5.3), i.e.

$$w = (w_k) = (\theta_i, r^\circ_i, v^\circ_i, t^\circ), \quad k = 1,2,...10, \qquad (5.17)$$

and the conventional generators (5.4), but now defined on isospace $\hat{\Re}_t \times T^*\hat{E}(r,G,\hat{\Re})$, i.e.

$$J_i = \Sigma_a \epsilon_{ijk} r_{ja} p_{ka}, \qquad P_i = \Sigma_a p_{ia}, \qquad (5.18a)$$

$$G_i = \sum_a (m_a r_{ia} - t p_{ia}), \tag{5.18b}$$

$$H = p_{ka} B_k^2 p_{ka} / 2m_a + V(r_{ab}), \tag{5.18c}$$

$$r_{ab} = |r_a - r_b|^{\hat{2}} = \{(r_{ka} - r_{kb}) B_k^2 (r_{ka} - r_{kb})\}^{\frac{1}{2}}, \tag{5.18d}$$

2) the Lie-isotopic algebra

$$\hat{\mathbf{G}}(3.1): \ [J_i \hat{,} J_j] = \epsilon_{ijk} B_k^{-2} J_k, \qquad [J_i \hat{,} P_j] = \epsilon_{ijk} B_j^{-2} P_k, \tag{5.19a}$$

$$[J_i \hat{,} G_j] = \epsilon_{ijk} B_j^{-2} G_k, \qquad [J_i \hat{,} B] = 0, \tag{5.19b}$$

$$[G_i \hat{,} P_j] = \delta_{ij} M B_j^{-2}, \qquad [G_i \hat{,} B] = 0, \tag{5.19c}$$

$$[P_i \hat{,} P_j] = [G_i \hat{,} G_j] = [P_i \hat{,} B] = 0, \tag{5.19d}$$

3) the Lie-isotopic group

$$\hat{G}(3.1): \quad r' = \{ [\prod_k e_{|\xi}^{w_k \omega^{\mu\sigma} \times I_{2\sigma}^{\nu} (\partial_\nu X_k)(\partial_\mu)}] \hat{I}_2 \} * r, \tag{5.20}$$

4) the local isocasimir invariants

$$\hat{C}^{(o)} = \hat{I}_2, \quad \hat{C}^{(1)} = (P\hat{G}P - MH) \hat{I}_2, \tag{5.21a}$$

$$\hat{C}^{(2)} = (MJ - G \wedge P)^{\hat{2}} = \{ (MJ - G \wedge P) \hat{G} (MJ - G \wedge P)] \hat{I}2, \tag{5.21b}$$

5) the explicit expressions of the $\tilde{B}_i$ functions

$$\tilde{B}_i^{-2}(r^\circ) = B_i^{-2} + r^\circ_j [B_i^{-2} \hat{,} P_j] / 2! + r^\circ_m r^\circ_n [[B_i^{-2} \hat{,} P_m] \hat{,} P_n] / 3! + \ldots \tag{5.22a}$$

$$\tilde{B}_i^{-2}(v^\circ) = B_i^{-2} + v^\circ_j [B_i^{-2} \hat{,} G_j] / 2! + v^\circ_m v^\circ_n [[B_i^{-2} \hat{,} G_m] \hat{,} G_n] + \ldots$$

(5.22b)

while $\tilde{B}_4^{-2}(t^\circ)$ is the solution of the algebraic equation

$$r(t + t^\circ \hat{\tilde{B}}_4^{-2}) = \{e_{|\xi}^{t^\circ \omega^{\mu\sigma} \times \hat{1}_2^{\sigma\nu} (\partial_\mu H)(\partial_\nu)}\}\, r. \tag{5.23}$$

The infinite family of Galilei-isotopic symmetries so constructed result to be all locally isomorphic to the conventional Galilei symmetry under the conditions of sufficient smoothness, nonsingularity and positive-definitness of the isounits. Finally, all isosymmetries $\hat{G}(3.1)$ can approximate the conventional symmetry $G(3.1)$ as close as desired whenever the isounits approach the conventional unit, and they all admit the conventional symmetry as a particular case by construction.

PROOF. As now familiar, the Lie-isotopic theory preserves, by central condition, the parameters and generators of the conventional symmetries, and this illustrates property 1). The Lie-isotopic algebra $\hat{\mathbf{G}}(3.1)$ can then be readily computed via the use of brackets (5.16), and this proves property 2). The exponentiation to the Lie-isotopic group $\hat{G}(3.1)$ (property 3) then follows uniquely via the use of the general theory. The validity of the local isocasimirs (5.21) also follows via the use of the same isocommutators (Property 4)). The application of such exponentiations to the local coordinates then yields the explicit forms (5.15) with explicit form (5.22) of the $\tilde{B}$-functions. Finally, the isounit of the time isofield, $\tilde{B}_4^{-2}(t^\circ)$ is provided by the solution of Eq. (5.23). The local isomorphism $\hat{G}(3.1) \approx G(3.1)$ trivially follows from the isocommutation rules. Q.E.D.

Note that, for $\hat{1}_2 = I$ and $\tilde{B}_4 = 1$, one recovers the conventional Galilei symmetry G(3.1) identically, because in this case $\tilde{B}_k = \hat{B}_4 = 1$, thus recovering the conventional, canonical representation of linear and local Galilei's transformations, including the Galilean translations in time $t' = t + t^\circ$. However, for $\hat{1}_2 \neq I$, Eq.s (5.23) cannot evidently hold any longer for $t' = t + t^\circ$. The lifting to form (5.15a) then follows. Explicit examples will be worked out later on.

The preceding results evidently include those for the Euclidean-isotopic symmetries $\hat{E}(3)$, as well as of the isorotational symmetries of the preceding sections.

It is an instructive exercise for the interested reader to prove that the infinite family of isosymmetries $\hat{G}(3.1)$ so constructed do indeed verify the conditions of Definition III.5.1 and, in particular, do constitute isosymmetries of invariants (5.11).

COROLLARY III.5.1A: In the particular case of constant isometrics $\hat{\delta}$, we have

$$\hat{\mathfrak{R}}_t \times T^*\hat{E}_1(r,\hat{\delta},\hat{\mathfrak{R}}) \equiv \hat{R}_t \times T^*\hat{R}_2(r,G,\hat{\mathfrak{R}}) \equiv \mathfrak{R}_t \times T^*\hat{E}(r,\hat{\delta},\hat{\mathfrak{R}}), \qquad (5.24a)$$

$$g = G = \text{diag.}\,(b_1^2, b_2^2, b_3^2), \quad b_k = \text{const.} > 0; \qquad (5.24b)$$

$$\hat{1}_t = b_4^{-2} = \text{const.} > 0, \qquad (5.24c)$$

the $\tilde{B}$-quantities coincide with the diagonal elements of the isounits,

$$\tilde{B}_i^{-2}(r^\circ) \equiv \tilde{B}_i^{-2}(v^\circ) \equiv B_i^{-2} \equiv b_i^{-2}, \quad \tilde{B}_4^{-2}(t^\circ) \equiv b_4^{-2}. \qquad (5.25)$$

and the general isogalilean transformations (5.15) become linear and local, i.e., they assume the simplified form

$$t' = t + t^\circ b_4^{-2}, \qquad (5.26a)$$

$$r'_i = r_i + r^\circ_i\, b_i^{-2}, \qquad (5.26b)$$

$$r'_i = r_i + t^\circ\, v^\circ_i\, b_i^{-2}, \qquad (5.26c)$$

$$r' = \hat{R}(\theta) * r. \qquad (5.26d)$$

which are the "restricted isogalilean transformations" of Definition III.1.

The above properties, whose proof is trivial, have important implications from a relativity viewpoint. In fact, they imply that the Galilei-isotopic symmetries can indeed preserve inertial frames, but, of course, in their linear particularization, e.g., following averaging of the characteristic B-functions of the interior medium of type (III.3.56).

The problem of the isocasimirs for the global case requires a study of the isoscalar extensions of the Galilei-isotopic symmetries, isoassociative envelopes in classical realization and their neutral elements. As such, this study will be conducted at some later time.

The application of the Galilei-isotopic symmetries $\hat{G}(3.1)$ to the characterization of closed nonselfadjoint systems can now be formulated. In fact, Theorems II.8.2 and II.8.3 readily yield the following

LEMMA III.5.1 (Santilli (1988a)): Necessary and sufficient conditions for the isoinvariance of closed nonselfadjoint systems (II.3.23) under the isogalilean symmetries $\hat{G}(3.1)$ are that they can be consistently written in isospace $\hat{R}_t \times T^*\hat{E}(r,\hat{G},\hat{\Re})$ and admit the representation in terms of the symplectic-isotopic or, equivalently, Lie-isotopic representation

$$\omega_{\mu\sigma}\, T_2{}^{\sigma}{}_{\nu} \frac{da^{\nu}}{dt} = \omega_{\mu\sigma} T_2{}^{\sigma}{}_{\nu} \Gamma^{\nu} = \frac{\partial H}{\partial a^{\mu}}, \tag{5.27a}$$

$$\frac{da^{\mu}}{dt} = \Gamma^{\mu} = \omega^{\mu\sigma} \hat{1}_{2\sigma}{}^{\mu}(a) \frac{\partial H}{\partial a^{\nu}}, \tag{5.27b}$$

$$H = p_{ia}\, \hat{G}_{ij}(t, r, p, ...)\, p_{ja}/2m_a + V(r_{ab}), \tag{5.27c}$$

$$r_{ab} = \{(r_{ia} - r_{ib})\, \hat{G}_{ij}(r, p, ...)\, (r_{ja} - r_{jb})\} \tag{5.27d}$$

in which case all total quantities (5.18) are not subsidiary constraints, but first integrals of the equations of motion.

It should be understood that the imposition of the Galilei-isotopic invariance *restricts* closed nonselfadjoint systems, from the general class (III.2.3) with subsidiary constraints, to that particular subclass in which the total quantities are automatically conserved in virtue of the isosymmetry $\hat{G}(3.1)$.

This provides the nonlocal extensions of the nonlinear, nonhamiltonian and nonnewtonian results of Santilli (1982a).

The classification of all possible isotopes $\hat{G}(3.1)$ of $G(3.1)$ via the relaxation of the positive-definiteness of the isounits $\hat{1}_2$ and $\hat{1}_t$ is a rather intriguing problem, which we are forced, for brevity, to leave to the interested reader.

A notion particularly intriguing for the study of the discrete symmetries and other aspects is the following one derived from Definition II.3.3.

DEFINITION III.5.2 (Santilli (1988a), (1991a)): The "isoduals" $\hat{G}^d(3.1)$ of the Galilei-isotopic symmetries $\hat{G}(3.1)$ on isospaces

$$\hat{\mathfrak{R}}_t \times T^*E(r,\hat{G},\hat{\mathfrak{R}}), \qquad \hat{\mathfrak{R}}_t = \mathfrak{R}_t \hat{1}_t, \qquad \hat{\mathfrak{R}} = \mathfrak{R}\hat{1}_2, \tag{5.28a}$$

$$\hat{1}_t > 0, \qquad \hat{1}_2 = \text{diag.}(\hat{G}^{-1}, \hat{G}^{-1}), \quad \hat{G} > 0, \tag{5.28b}$$

are given by the isosymmetries on the isodual spaces

$$T^*\hat{E}'(r,\hat{G}^d,\hat{\mathfrak{R}}^d), \qquad \hat{\mathfrak{R}}_t{}^d = \mathfrak{R}_t \hat{1}_t{}^d, \qquad \hat{\mathfrak{R}}^d = \mathfrak{R}\hat{1}_2{}^d, \tag{5.29a}$$

$$\hat{1}_t{}^d = -\hat{1}_t, \qquad \hat{1}_2{}^d = \text{diag.}(\hat{G}^{d-1}, \hat{G}^{d-1}), \qquad \hat{G}^d = -\hat{G}. \tag{5.29b}$$

A number of comments are now in order. First, it should be noted that the notion of duality *necessarily* requires the Lie-isotopic theory, evidently because it cannot be defined without a generalized notion of unit.

Second, it is important to see that *isoduality does not represent inversions (or iso inversions)*. For this purpose, let us consider the isoinversions

$$r' = Pr = \hat{P}*r = \hat{P}\,T\,r - r, \tag{5.30}$$

where P is the conventional inversion element and $\hat{P} = P\,\hat{1}$ its isotopic image. It is easy to see that the above inversions coincide with their isodual. In fact, by introducing the isodual $T^d = -T$, we have

$$\hat{P}^d *^d r = \hat{P}\,\hat{1}^d\,T^d\,r \equiv P\,\hat{1}\,T\,r \equiv Pr = -r \tag{5.31}$$

Inversions can indeed be representred via an isotopic lifting but of type different than that occurring in isoduality, and given by the particular value $T^{\frac{1}{2}} = -1$, under which we have

$$r^{\hat{2}} = r^t\,\hat{\delta}\,r = r^t\,T\,\delta\,r = r^t\,T^{\frac{1}{2}t}\,\delta\,T^{\frac{1}{2}}\,r = r'^{\,2} =$$

$$= r'^t\,\delta\,r' \equiv r^t\,\delta\,r = r^2 \quad \text{for } T^{\frac{1}{2}} = -1, \ r' = -r. \tag{5.32}$$

Third, the following property (which is introduced here apparently for the first time) is rather simple at the isoalilean level, but has intriguing implications at the isorelativistic and isogravitational levels of the subsequent chapters.

PROPOSITION III.5.1: All systems which are invariant under the isogalilean symmetries are isodual invariant.

PROOF. Hamilton-isotopic equations (5.27a) can be written

$$\hat{G}_{ik}\,\dot{r}_k - \frac{\partial H}{\partial p_k} = 0, \qquad \hat{G}_{ki}\,\dot{p}_k + \frac{\partial H}{\partial r_k} = 0. \tag{5.33}$$

Then, for isogalilean invariant Hamiltyonians (5.27c), we have

$$H^d = p_{ia}\,\hat{G}^d{}_{ij}\,p_{ja}\,/\,2m_a + V(r^d{}_{ab}), \tag{5.34a}$$

$$r^d{}_{ab} = \{\,(r_{ia} - r_{ib})\,\hat{G}^d{}_{ij}\,(r_{ja} - r_{jb})\,\}^{\frac{1}{2}}, \tag{5.34c}$$

$$\hat{G}^d = -\,\hat{G}, \tag{5.34c}$$

under which

$$\hat{G}^d{}_{ki}\,\dot{r}_i - \partial H^d\,/\,\partial p = -\,(\,\hat{G}_{ki}\,r_i - \partial H\,/\,\partial p_k) = 0 \tag{.5.35a}$$

$$\hat{G}^d{}_{ki}\,\dot{p}_i \equiv \partial H^d\,/\,\partial r = -\,\hat{G}_{ki}\,\dot{p}_i + \partial H\,/\,\partial r_k) = 0 \tag{5.35b}$$

which, when combioned with the isodual invariance of isoinversions, Eq.s (5.31), prove the proposition. QED.

In different terms, *the invariance under isoduality appears to be a basic law of nature independent from their invariance under inversions*.

We close this section with a few remarks on the problem of the *explicit construction of the isogalilean symmetries for a given Galilei-noninvariant system*.

As well known, in the conventional canonical treatment of mechanics, the Galilei symmetry G(3.1) is preassigned. Physical systems are then restricted to those which are G(3.1)-invariant. This evidently results in severe limitations in the class of systems admitted, which are essentially given by the local and Hamiltonian systems of closed selfadjoint type.

In the covering Birkhoffian mechanics, the situation is reversed. In fact, one considers, first, the equations of motion as provided by the experimental evidence, and then searches for their space-time symmetries.

The Birkhoffian realization of the Lie-isotopic theory has been conceived also for the explicit construction of the isotopic covering $\hat{G}(3.1)$ of G(3.1) from given, G(3.1)-noninvariant equations of motion, with the consequential, substantial broadening of the class of admitted systems, while preserving the conventional class as particular case.

The rules for the explicit construction of the covering $\hat{G}(3.1)$ symmetries from given equations of motion are rather simple. In fact, one merely has to write the system considered in the symplectic-isotopic form (5.28) on $\hat{\mathfrak{R}}_t \times T^*\hat{E}_2(r,G,\hat{\mathfrak{R}})$. This provides the fundamental isounit $\hat{1}_2$ which characterizes the Lie-isotopic structure (5.21) of $\hat{G}(3.1)$. The rest of the isosymmetry (5.21) is characterized by the *conventional parameters* w_k and by the *conventional generators* X_k (only properly written in $\hat{\mathfrak{R}}_t \times T^*\hat{E}_2(r,\hat{G},\hat{\mathfrak{R}})$).

Note finally that, under a sufficient smoothness of the isounit, the existence and convergence of the infinite expansions (5.21) is guaranteed by those of the conventional structure (5.7). The reader should, however, not expect easily summable series (see the examples of sums into *transcendental functions* of the original proposal in Santilli (1978a)).

The first (and perhaps most important) examples of $\hat{G}(3.1)$-invariant systems are the two-body and three-body, closed nonselfadjoint systems studied in Appendix A. Additional explicit examples will be given in Section III.7, when studying the notion of particle characterized by the $\hat{G}(3.1)$ isosymmetry.

III.6: ISOTOPIC LIFTINGS OF GALILEI'S RELATIVITIES

As well known, the *Galilei relativity* (see, e.g., Levy-Leblond (1971) or Sudarshan and Mukunda (1974)) is a description of physical systems via their form-invariance under the Galilei's symmetry $G(3.1) = [O_\theta(3) \otimes T_{r°}(3)] \times [T_{v°}(3) \times T_{t°}(1)]$, or, equivalently, under the celebrated *Galilei's transformations*

$$\begin{cases} t' = t + t^\circ, & \text{translations in time,} \quad (6.1a) \\ r'_i = r_i + r^\circ_i, & \text{translations in space,} \quad (6.1b) \\ r'_i = r_i + t^\circ v^\circ_i, & \text{Galilei boosts,} \quad (6.1c) \\ r' = R(\theta)\, r\,, & \text{rotations,} \quad (6.1d) \\ r' = P\, r = -\, r, & \text{inversions.} \quad (6.1e) \end{cases}$$

The relativity is verified in our physical reality only for a rather small class of Newtonian systems, called *closed selfadjoint systems.* These are systems (such as our planetary system) which verify the conventional total Galilean conservation laws, and admit internal forces which are local (differential), potential and selfadjoint.

For all remaining Newtonian systems, Galilei's relativity is violated according to a number of mechanisms indicated in Sect. III.1. In the final analysis, the limitations of Galilei's relativity are inherent in its mathematical structure. In fact,

1) The *linear* character of Galilei's transformations is at variance with the generally *nonlinear* structure of the systems of the physical reality of the interior dynamical problem, as established by incontrovertible evidence;

2) The *local* (differential) character of Galilei's relativity is at variance with the generally *nonlocal* (integral) nature of the systems in our Earthly environment; and

3) The strictly *Hamiltonian* (canonical) structure of Galilei's relativity is at variance with the generally *nonhamiltonian* character of physical systems of our reality.

An infinite family of Lie-isotopic generalizations of the Galilei symmetry, under the name of *Galilei-isotopic symmetries* $\hat{G}(3.1)$ has been presented in the preceding section to represent a broader class of systems. In particular, we have shown that:

A) The Galilei-isotopic symmetries characterize *closed non-selfadjoint systems* (III.2.3). These are systems (such as Jupiter) which verify the conventional, total, Galilean conservation laws, while admittin the additional class of nonlocal, nonhamiltonian and nonnewtonian internal forces.

B) The Galilei-isotopic symmetries possess the structure

$$\hat{G}(3.1) = [\hat{O}_\theta(3) \otimes \hat{T}_{r^\circ}(3)] \times [\hat{T}_{v^\circ}(3) \times \hat{T}_{t^\circ}(1)]\,, \quad (6.2)$$

and result to be locally isomorphic to the conventional symmetry G(3.1) under the positive-definiteness of the underlying isounits, by admitting the latter as a particular case. In this sense, $\hat{G}(3.1)$ provides an infinite family of *Lie-isotopic coverings* of G(3.1).

C) All symmetries $\hat{G}(3.1)$ can be explicitly constructed via the Lie-isotopic theory, that is, via the use of the same parameters and generators (conserved quantities) of the conventional symmetry, but via the most general possible, axiom-preserving realizations of Lie algebras and Lie groups. In this way, an infinite number of symmetries $\hat{G}(3.1)$ can be constructed for each given Hamiltonian H = T + V (i.e., for given potential-selfadjoint forces), as characterized by an infinite number of possible isometrics for the interior space, which represent the infinitely possible interior physical media.

The isogalilean transformations are defined in the isotopic generalizations $\hat{\mathfrak{R}}_t \times T^*\hat{E}(r,\hat{G},\hat{\mathfrak{R}})$ of the conventional space $\mathfrak{R}_t \times T^*E(r,\delta,\mathfrak{R})$ of Galilei's relativity, and are explicitly given by (Sect. III.5)

$$\begin{cases} t' = t + t^\circ \,\hat{B}_4^{-2}(t, r, p, ..), & \text{isotime translations,} & (6.3a) \\ r'_i = r_i + r^\circ_i \,\hat{B}_i^{-2}(t, r, p, ..), & \text{isospace translations,} & (6.3b) \\ r'_i = r_i + t^\circ v^\circ_i \,\hat{B}_i^{-2}(t, r, p, ...), & \text{isogalilei boosts,} & (6.3c) \\ r' = \hat{R}(\theta) * r = \hat{R}(\theta)\, G(t, r, p, ...)\, r, & \text{isorotations,} & (6.3d) \\ r' = \hat{P} * r = \hat{P}\, T(t, r, p, ...)\, r, & \text{isoinversions,} & (6.3e) \\ T \Rightarrow T^d = -T, \quad \hat{1} \Rightarrow \hat{1}^d = -\hat{1}, & \text{isoduality,} & (6.3f) \end{cases}$$

where the $\hat{B}$'s are generally nonlinear as well as nonlocal functions of all variables, they vary from system to system, and they can be explicitly computed via the Lie-isotopic techniques for each system.

The reader should keep in mind, not only the nontrivial difference in the functional dependence of isotransformations (6.3) with transformations (6.1), but also the appearance of the additional isodual transformations (6.3f) which are not introducved for the conventional Galilean relativity owing to the restriction of the unit to the trivial form I = diag. (1, 1,).

It is easy to see that transformations (5.3) leave invariant the following isoseparations in $\hat{\Re}_t \times T^* \hat{E}(r, \hat{G}, \hat{\Re})$

$$t_a - t_b = \text{inv.}, \quad (r_{ia} - r_{ib})\, \hat{G}_{ij}(t, r, p, ..)\, (r_{ja} - r_{jb}) = \text{inv.}, \tag{6.4}$$

$$i, j = 1, 2, 3\ (= x, y, z), \quad a, b, = 1, 2, ..., N,$$

the latter one applying for $t_a = t_b$.

In this section we shall study the relativities characterized by the Galilei-isotopic symmetries $\hat{G}(3.1)$ under the name of *Galilei-isotopic relativities* or *isogalilean relativities* for short. The generalized relativities were first submitted in Santilli (1978a), and then studied in details for the nonlinear, nonhamiltonian and nonnewtonian but local case in Santilli (1982a) (see Chapter 6, particularly Definition 6.3.9, p. 243 and ff.). The extension to nonlocal systems was reached in Santilli (1988a).

DEFINITION III.6.1: The "general, nonlinear and nonlocal, Galilei-isotopic relativities", or "general isogalilean relatiuvities" for short, are given by the form-invariant description of physical systems characterized by the infinite family of Galilei-isotopic symmetries $\hat{G}(3.1)$ on $\hat{\Re}_t \times T^*\hat{E}(r, \hat{G}, \hat{\Re})$, $\hat{\Re} = \Re\, \hat{I}_2 = \text{diag.}\,(\hat{G}^{-1}, \hat{G}^{-1})$,, $\hat{\Re}_t = \Re\, \hat{I}_t$, $\hat{G} > 0$, $\hat{I}_2 > 0$, $\hat{I}_t > 0$, and their isodual $G^d(3.1)$, with corresponding, infinite family of general isogalilean transformations (5.3). The "restricted (linear and local) isogalilean relativities" occur for diagonal isometrics with constant diagonal elements other than one.

The reader should be aware of the *uniqueness of transformations* (6.3) for each given isometric $\hat{G}$ (up to the degrees of freedom of the Lie-isotopic theory which are broader than the conventional ones, and include, e.g., the *Birkhoffian gauge transformations*, see Sect. II.7).

The restriction $\hat{G}(3.1) \approx G(3.1)$ in the above definition should also be kept in mind. This is due to the fact that, if such restriction is lifted (i.e., if the isounits are not necessarily positive-definite), isosymmetries $\hat{G}(3.1)$ still formally exist, but they do not qualify for the characterization of covering relativities. See in this respect the classification of all possible compact and noncompact isotopes $\hat{O}(3)$ of O(3) of Sect. III.3.

Finally, one should keep in mind that all our isotopic formulations have been constructed in such a way to coincide with the original formulations at the abstract, realization-free level. We should therefore expect that the Galilei-isotopic relativities coincide, by construction, with the *conventional relativity* at the coordinate-free level.

The form-invariant description of physical systems characterized by the Galilei-isotopic relativities has been studied in the preceding section. We therefore remain here with the problems of:

a) identifying the physical laws characterized by the isogalilean relativities;

b) prove their form-invariance under the isogalilean symmetries $\hat{G}(3.1)$; and their idodual $G^d(3.1)$ and

c) prove their abstract equivalence to the conventional physical laws.

The above results are expected from the very structure of our isotopies. Consider the historical *Galilei's boosts*

$$r'_i = r_i + t^\circ v^\circ_i, \quad p'_i = p_i + mv^\circ_i, \tag{6.5}$$

which, as well known, apply for the simple case of a particle with constant speed, under the (often tacit) assumption that motion occurs in vacuum.

Suppose now that the particle considered is extended and penetrates within a physical medium at a given instant of time t. Then, the Galilei transfortmations are evidently inapplicable, e.g., because of their linearity, locality and Hamiltonian character, while the particle experiences a drag force that is nonlinear, nonlocal and nonhamiltonian.

Our generalized transformations

$$r'_i = r_i + t^\circ v^\circ_i \hat{B}_i^{-2}(t, r, p, ...), \tag{6.6a}$$

$$p'_i = p_i + mv^\circ_i \hat{B}_i^{-2}(t, r, p, \tag{6.6b}$$

are then applicable to represent the *deviations* from the original uniform motion. In particular, Eq.s (6.6) can represent a (monotonic) increase or decrease of speed depending on the sign of the v°-parameter (since the $\hat{B}^{-2}$-terms are always positive definite). In the former case we have the usual drag force caused by motion within the physical medium. In the latter case we have instead a particle penetrating a highly turbulent medium which causes an increase of its speed.

The physical law characterized by the isogalilean boosts is therefore the representation of an arbitrary rectilinear motion, when all forces derivable from a potential are null, and the local, monotonic changes (increase or decrease) of the speed are due to the contact nonpotential forces originating from the physical medium.

The important point is that, in the transition from the linear, local and Hamiltonian transformations (6.5) to their nonlinear, nonlocal and nonhamiltonian generalizations (6.6) the geometric axioms are preserved, as established by the local isomorphisms $\hat{T}(v^\circ) \approx T(v^\circ)$.

In different terms, recall that our isotopic liftings leave unchanged (by central assumption) the original generators, that is, they leave unchanged the existing potential forces. Then, the isotopic liftings $G(3.1) \Rightarrow \hat{G}(3.1)$ characterize the transition from the original, $G(3.1)$-invariant system of particles moving in vacuum, to the same system moving within physical media. The infinite number of isotopes $\hat{G}(3.1)$ is needed for physical consistency in order to represent the infinite variety of different physical media in which the original system can be immersed.

This identifies property a) above, namely, the generalized physical laws representing rectilinear motion within physical media.

We now pass to the study of properties b) and c), namely, the form-invariance of the generalized physical laws (6.6) under $\hat{G}(3.1)$ and their axiomatic equivalence to the Galilean laws (6.5).

For this purpose, we note that the laws of the uniform motion in vacuum is geometrically expressed by the structures

$$T(v^\circ)\, r_i = r_i - t^\circ v^\circ{}_i, \quad T(v^\circ)\, p_i = p_i - m v^\circ{}_i, \tag{6.7a}$$

$$T(v^\circ) = e_{|\xi}^{\,v^\circ{}_j\, \omega^{\mu\nu}\, (\partial_\nu G_j)\, (\partial_\mu)}, \tag{6.7b}$$

namely, *the structure of the the Galilean law of uniform motion is provided by the right modular (associative) action of the finite Galilei boosts $T(v^\circ)$ on the coordinates and momenta.*

But isotopic laws (6.6) are geometrically expressed by

$$\hat{T}(v^\circ) * r_i = r_i - t^\circ v^\circ_i \hat{B}_i^{-2}, \quad \hat{T}(v^\circ) * p_i = p_i - m v^\circ_i B_i^{-2}, \tag{6.8a}$$

(9a)

$$\hat{T}(v^\circ) = \{ [e_{|\xi}^{v^\circ_j \omega^{\mu\sigma} \hat{1}_2^{\sigma\nu} (\partial_\nu G_j)(\partial_\mu)}] \hat{1}_2 \}. \tag{6.8b}$$

Thus, *the structure of the variable motion within a physical medium is characterized by the modular-isotopic (associative-isotopic) action of the finite isogalilei boosts on coordinates and momenta.*

But the modular action $T(v^\circ)\, r$ coincides with the modular-isotopic action $\hat{T}(v^\circ)*r$ at the abstract, realization-free level by construction. This shows that *the abstract axioms underlying the Galilean uniform motion are preserved by our covering Galilei-isotopic relativities,* and proves property c) above.

The proof of property b) is trivial and merely follows from the composition law of Lie-isotopic groups

$$\hat{T}(r^\circ) * \hat{T}(r'^\circ) = \hat{T}(r'^\circ) * \hat{T}(r^\circ) = \hat{T}(r^\circ + r'^\circ), \tag{6.9}$$

Note the unity of physical and mathematical thought between the generalized and conventional relativities. In fact, we can introduce only one abstract law of rectilinear motion, say, $\mathbf{T(\upsilon^0)r}$, with infinitely many different, but locally isomorphic realizations $\hat{T}(v^\circ)*r$ representing the infinitely many nonuniform motions within different physical media, and only one canonical realization $T(v^\circ)\, r$, representing uniform motion in vacuum.

The invariance of the physical laws under isoduality $\hat{G}(3.1) \Rightarrow \hat{G}^d(3.1)$ then follows from Proposition III.5.1 and the property that the mapping $T \Rightarrow T^d = -T$ implies $B^2 \Rightarrow -B^2$, thus resulting to be equivalent to the inversions of the Galilean values $(t^\circ, r^\circ) \Rightarrow (-t^\circ, -r^\circ)$.

The extension of the above results to other physical laws is straightforward, and is here left to the interested reader for brevity.

Note that, despite their nonlinearity, *all Galilei-isotopic transformations (6.3) locally coincide with the conventional transformations (6.1),* i.e., at a given, fixed value $\bar{t}, \bar{r}, \bar{p}, \ldots$ of the local variables, we have

$$t^\circ \hat{B}_i^{-2}(\bar{t}, \bar{r}, \bar{p}, \ldots) \equiv t'^\circ = \text{cost.}, \qquad r^\circ \hat{B}_i^{-2}(\bar{t}, \bar{r}, \bar{p}, \ldots) \equiv r'^\circ, \tag{6.10a}$$

$$v^{\circ}\, \tilde{\hat{B}}_i^{-2}(t, \bar{r}, \bar{p}, ...) \equiv v'^{\circ} = \text{const.}, \qquad \hat{R}(\theta)_{|\bar{r}, \bar{p}, ...} \equiv R(\theta'). \tag{6.10b}$$

A similar situation must then hold for the physical laws, namely, we can state that *the physical laws characterized by our Galilei-isotopic relativities are locally equivalent to the conventional Galilean laws.*

We have reached in this way the most important physical result of the analysis of this volume until now, which can be expressed as follows:

THEOREM III.6.1 (Santilli (1988a)): All infinitely possible isogalilean relativities on $\hat{R}_t \times T^*\hat{E}(r,\hat{G},\hat{\Re})$ coincide with the conventional Galilei relativity on $\Re_t \times T^*E(r,\delta,\Re)$ at the abstract, realization-free level, that is, not only all infinitely possible isogalilean symmetries $\hat{G}(3.1)$ coincide with the conventional Galilei symmetry $G(3.1)$, but also the infinite class of isogalilean transformations (6.3a, b, c, d, e) coincide with the conventional Galilei transformations (6.1a, b, c, d), and the same holds for the related physical laws.

The above properties illustrate the ultimate physical and mathematical unity of the isogalilean relativities with the conventional one, exactly as anticipated earlier from the abstract unit of the underlying methodological tools illustrated in Chapter II.

Such unity, however, will appear in its full light only at the gravitational level of Chapter V where we shall show that the axiomatic unity between Galilei's boosts and their isotopic extensions is a particular case of a much broader geometric unity within the context of the Riemannian-isotopic geometry of Sect. II.11.

In particular, in Chapter V we shall show that *the transition from the Galilean, exterior, uniform motion in vacuum to its isotopic extensions within physical media, does not imply a change in geodesic motion, but only the transition from geodesics within conventional spaces, to geodesics within isospaces (Sect. II.12).*

Despite such mathematical and physical unity, the physical differences between the Galilei-isotopic relativities and the conventional one are nontrivial. To begin the illustration of this point, let us recall that *Galilei's relativity establishes the equivalence of all inertial frames,* as well known.

On the contrary, the Galilei-isotopic relativities establish *equivalence subclasses of noninertial frames, those with respect to the center-of-mass frame of the system,* each class being characterized by each relativity (i.e., by each physical medium). The understanding is that different systems imply different subclasses of isotopically equivalent frames.

To put it differently, physical events can occur in the Universe according to a multiple infinity of noninertial conditions. The Galilei-isotopic relativities essentially indicate that all these noninertial frames cannot be reduced to one single Lie-isotopic class of equivalence, but require their classification into subclasses of frames, isotopically equivalent to the observer's frame at rest with the interior problem considered.

But the Galilei-isotopic relativities are coverings of the conventional one. This means that the conventional inertial aspects are not lost, but fully included and actually generalized in the broader Galilei-isotopic setting. This concept can be made more clear via the use of the Corollary III.6.1.a under which we have the following

COROLLARY III.6.1.a (loc. cit.): The isogalilean relativities admit an infinite subclass of linear and local, generalized relativities on $\Re_t \times T^*\hat{E}(r,\hat{\delta},\hat{R})$ for $\hat{\delta}$ = constant > 0, $\hat{\delta} \neq I$, called "restricted isogalilean relativities", which are nontrivially different than the conventional Galilei's relativity.

In fact, under the assumption of the Corollary, the general Galilei-isotopic transformations (6.3) assume the particular, manifestly linear and local form

$$\begin{cases} t' = t + t^\circ b_4^{-2}, & \text{isotime translations,} \quad (6.11a) \\ r'_i = r_i + r^\circ_i\, b_i^{-2}, & \text{isospace translations} \quad (6.11b) \\ r'_i = r_i + t^\circ\, v^\circ_i\, b_i^{-2}, & \text{isogalilei boosts,} \quad (6.11c) \\ r' = \hat{R}(\theta) * r = \hat{R}(\theta)\, g\, r, & \text{isorotations,} \quad (6.11d) \\ r' = \hat{P} * r = -r, & \text{isoinversions,} \quad (6.11d) \\ \hat{\delta} \Rightarrow \hat{\delta}^d = -\hat{\delta}, & \text{isoduality,} \quad (6.11f) \end{cases}$$

$$\hat{\delta} = \text{diag.}\,(b_1^2\,,\, b_2^2,\, b_3^2), \quad b_k = \text{cost.} > 0. \qquad (6.11g)$$

$$\hat{\delta}^d = \text{diag.}\,(-\,b_1^2, (b_2^2, -\,b_3^2) \qquad (6.11h)$$

The lack of equivalence between the restricted isogalilean

relativities and the conventional one is soon illustrated by the fact that the formers characterize *deformable bodies,* while the latter characterizes *rigid bodies,* as pointed out in Sect. III.3. For further comments, see Figure III.6.1.

In this chapter we shall consider the following two different types of applications of the isogalilean relativities:

I) Characterization of a generalized notion of particles which is studied in the next section;

II) Constructions of systems invariant under the isogalilean symmetries, also studied in the next section; and

III) Characterization of closed-isolated systems of generalized particles, studied in Appendix III.A;

the above studies being a mere rudimentary ground for our intended operator formulations.

A direct way for the construction of the isogalilean relativities was submitted in Santilli (1982a), pp. 246-247, and it may be advantageous to outline it here because instructive.

Consider a system (III.2.3a), for simplicity, in its local, but nonlinear and nonhamiltonian form,

$$\dot{r} = p, \qquad \dot{p} = F^{NSA}(r, p, ..), \tag{6.12}$$

which violates Galilei's symmetry in any of the mechanisms indicated in Sect. III.1. As recalled in Sect. I.3, a rather natural tendency when facing these systems is that of transforming them into a form which recovers the G(3.1) invariance.

Under sufficient topological conditions (locality, analyticity and regularity in a star-shaped neighborhoods of the local variables), the *Lie-Koening theorem* (see, e.g., Santilli (1982a), Hill (1967) and others) ensures that a transformation

$$a = (r,p) \Rightarrow a^* = (r^*, p^*) = a^*(a), \tag{6.13}$$

capable of reducing system (6.12) to a Hamiltonian form always exists. In particular, the transformed system can indeed be of the "free" form

$$\dot{r}^* = p^*, \qquad p^* = 0. \tag{6.14}$$

under which the Galilei's symmetry exactly holds, i.e.

$$G(3.1) = a^{*\prime} = \{e_{|\xi}^{\ w_k \omega^{\mu\nu} (\partial_\nu X^*_k)(\partial_\mu)}\}. \tag{6.15a}$$

$$\dot{r}^{*\prime} = p^{*\prime}, \quad \dot{p}^{*\prime} = 0. \tag{6.15b}$$

Systems: closed. Forces: local, SA. Space: $\Re_t \times T^*E(r,\delta,\Re)$ $\delta = I.$ Frames: Inertial. Methods: Lie's theory. Relativity: Convent. Galilei's relativity.	Systems: closed. Forces: local NSA. Space: $\Re_t \times T^*\hat{E}(r,g,\hat{\Re})$ $\hat{\delta} = \text{const} > 0.$ Frames: Inertial. Methods: Lie-isotopic theory. Relativity: Restricted isogalilean relativities.	Systems: closed. Forces: nonloc. NSA. Space: $\hat{\Re}_t \times T^*\hat{E}_2(r,G,\hat{\Re})$ $\hat{G} = \hat{G}(r,p,..) > 0.$ Frames: Noninertial. Methods: Lie-isotopic theory. Relativity: General isogalilean relativities.

FIGURE III.61: A classification of physical systems, with their carrier spaces, observer's frames (assumed at rest with respect to the center-of-mass of the system), and related methodology. The first column depicts the conventional linear-local-inertial-Hamiltonian setting; the second column depicts the first nontrivial isotopic generalization, that of linear-local-inertial-isotopic type; and the third column depicts the most general possible nonlinear, nonlocal, nonhamiltonian *and noninertial* setting. The first two columns have equivalent inertial characterizations, because they are both defined on inertial frames. However, the first column treats rigid bodies, while the second represents deformable bodies. The third column represents instead the most general possible conditions of extended-deformable bodies in regard to both the existing forces and the observer frames.

But the original systems is nonlinear and nonhamiltonian by assumption. Transformations (6.13) are, therefore necessarily nonlinear and noncanonical. As a result, the new frame a^* is in noninertial conditions, and it is not realizable in an actual laboratory. System (6.14), therefore, has a purely mathematical meaning, as stressed in Sect. I.3.

The re-transformation of the conventional Galilean symmetry (6.15) in the mathematical a^*-frame back to the original a-frame of the experimenter, yields precisely the covering Galilei-isotopic symmetry $\hat{G}(3.1)$ according to the rules

$$w_k \omega^{\mu\nu}(\partial_\nu X^*(t, a^*)) (\partial_\mu) \equiv w_k \; \Omega^{\mu\nu}(a) (\partial_\nu X_k(t, a)) (\partial_\mu), \tag{6.16a}$$

$$\Omega^{\mu\nu}(a) = \frac{\partial a^\alpha}{\partial a^{*\mu}} \omega^{\alpha\beta} \frac{\partial a^\beta}{\partial a^{*\nu}}, \quad X^*_k(t, a^*) \equiv X_k(t, a), \tag{6.16b}$$

and the same evidently holds for the physical laws.

In conclusion, another method for the construction of the covering relativities (besides the main method offered by the Lie-isotopic techniques) is the use of the Lie-Koening theorem, when applicable, for the reduction of nonhamiltonian systems to a Hamiltonian form, and then their retransformation back to the physical variables of the experimenter.

It appears recommendable to indicate at this point some of the implications of the isogalilean relativities in particle physics because they may provide the reader with a view of the *novel* implications expected from all covering relativities. It should be stressed that the following possibilities are merely speculative at this writing and in need of detailed operator studies currently available only in a preliminary form.

As illustrated in Appendix III.A, one of the most visible implications of the isogalilean relativities is the characterization of a new class of bound systems. In fact, as we shall see, the isogalilean invariance ultimately implies closed nonhamiltonian systems with contact internal forces.

To put it diffrently, the conventional Galilean symmetry can be visualized with the familiar Kepler's system with a central nucleus constituted by a body heavier than the remaining ones, and the peripheral bodies moving in vacuum without collisions.

On the contrary, the isogalilean symmetries can be visualized with aggregates of bodies in mutual physical contact, whose nucleus, called *isonucleus* (Santilli (1988a)), can have an arbitrary (nonnull) mass bigger, equal or smaller than the peripheral ones.

In the transition to nuclear physics, these results imply the familiar visualization of the Galilean symmetry with the atomic structure and the visualization of the operator isogalilean symmetries with the nuclear structure (Santilli (1989)). In fact, unlike the atomic structure, nuclei are composed of nucleons in conditions of mutual contact (and actually of mutual penetration for about 10^{-3} units of their volume). This results precisely in bound states as aggregates of particles under contact interactions whose centers are not the Keplerian one but our isonucleui.

Still in turn, this indicates that the historical chain of successive approximation of the nuclear forces via the additional of a considerable number of potentials in the Hamiltonian has been halted, because the nuclear structure appear to have precisely a closed nonhamiltonian structure as ultimately represented by our isogalilean symmetries (Santilli (1989)).

Note the expected variation of the nonpotential/nonhamiltonian terms from nucleus to nucleus, thus implying different isounits for different nuclei and consequently different isogalilean symmetries.

Some of the most important examples of isogalilean systems are the two-body and three-body, closed nonselfadjoint systems outlined in Appendix III.A. The two-body case implies the simplest conceivable isotopy of Galilei's relativity, that characterized by the *scalar isotopy* of the canonical tensor

$$\omega^{\mu\nu} \Rightarrow \hat{\Omega}^{\circ\mu\nu} = b^2\, \omega^{\mu\nu}, \quad b = \text{const.} > 0. \tag{6.17}$$

Despite this simplicity, the isotopy implies the appearance of acceleration-dependent forces which, in turn, imply rather profound modifications of the conventional two-body bound states exhibiting a sort of "mass renormalization"already at the classical-nonrelatuivistic level (Appendix III.A).

In turn, the operator formulation of such a generalized bound state possesses such departures from the conventional quantum mechanical two-body bound state, to permit a quantitative representation of *all* the intrinsic characteristics of the π^0 particle as a *"compressed positronium"*, i.e., as a generalized state of one ordinary electron and one ordinary positron isotopically bounded one inside the other at mutual distances smaller than the size of their wavepackets (see the calculations of Sect. 5 in Santilli (1978b), and the recent reviews Santilli

(1990), (1991e).

Evidently the model offers realistic possibilities of *identifying the π^o constituents with ordinary particles freely produced in the spontaneous decays*, as previously established for the nuclear and atomic cases.

By comparison, such a model is fundamentally inconsistent within the context of the conventional two-body, quantum mechanical systems, as the interested reader is encouraged to verify (see the various reasons identified in Santilli (1974), (1978b))

For the three-body closed nonselfadjoint system in stable configurations (i.e., the straight line and triangular configurations of Appendix III.A), we have an isotopy more general than the preceding one of the type

$$\omega^{\mu\nu} \Rightarrow \hat{\Omega}^{\circ\mu\nu}(a) \qquad (6.18)$$

See Jannussis, Mijatovic and Veljanoski (1991).

In this case too the physical implications are nontrivial. In fact, the isogalilean relativities permit a central nucleus with mass much smaller than that of the peripheral constituents, thus allowing fundamentaklly new structure models of the tritium as well as of hadrons. (Santilli (1989)).

As an illustration, such a structure (once implemented with our covering isorotational symmetry $S\hat{U}(2)$ for the spin) offers realistic hopes of achieving a quantitatively consistent representation of Rutherford's (1920) historical conception of the neutron as a *"compressed hydrogen atom"*, that is, as an electron totally compressed (say, in a supernova explosion) in the center of the proton, thus acquiring our configuration of isonucleus (see Santilli (1990) and (1991e) for preliminary operator studies).

Still in turn, if such a model is proved consistent in due time, it could allow the identification of one of the neutron's constituents (e.g., the d-quark) with Rutherford's electron (although in the modified form indicated in the next section).

The ultimate possibility is therefore the elimination of the now vexing and unattained confinement of the hadronic constituents, and its replacement with the free tunneling of the constituents, as historically established for atoms and nuclei.

But, to avoid major inconsistencies, these possibilities require a *generalized notion of constituent, that is, a generalized relativity valid in the interior problem only*. We reach in this way again the main line of research of these volumes regarding the dichotomy:

conventional relativities for the exterior problem and generalized relativities for the interior dynamics.

A particularly intriguing example of Gisogalilean systems is given by the three-body systems in *Jannussis' configuration* (1991), Eq.s (III.A.15), with genuine subsidiary constraints, which we regret to be unable to study at this time for brevity.

We are also unable to study closed nonselfadjoint systems with a large number of constituents N. In fact, for a sufficiently high value of N, the systems acquire intriguing statistical aspects (e.g., an internal irreversibility compatible with the center-of-mass revisibility), with a number of expected connections with *Prigogine's statistics* (1968).

In the final analysis, $\hat{G}(3.1)$-invariant systems with a large number of constituents constitute our *nonrelativistic closed nonhamiltonian structure model of Jupiter,* as pointed out in the introductory words of Sect. I.1.

Moreover, we have intriguing connections with Mach's principle originating in the interior, acceleration-dependent forces (prior any Riemannian structure), as suggested by Assis (1990) and Graneau (1990).

We cannot close this section without a few comments regarding the virtual complete restriction of the physical literature of this century to inertial frames.

On epistemological grounds, we stressed in Sect. III.1 that *inertial frames are a philosophical abstraction, because they do not exist in our Earthly environment, nor can they be attained in our Planetary or Galactic systems.* The Galilei-isotopic relativities are therefore intended to identify the equivalence class of each *actual frame,* that is, of each *noninertial frame.*

While inertial frames are now part of the history of physics and, as such, they must be preserved in any future development, one should be truly aware of the rather serious implications for any restriction of physical treatments to inertial frames *only.* In fact, such a restriction implies the necessary linearity (as well as locality and potential character) of the theory. A restriction of this type implies:

♣ *classically,* the implicit acceptance of the perpetual motion in a physical environment, and

♥ *operationally,* the impossibility of resolving fundamental open problems of contemporary physics.

It should be brought to the attention of the interested researcher that, quite likely, the restriction of all physical descriptions to inertial frames, and therefore to linear, local and potential theories, may well be responsible for the primary unsolved problems of contemporary physics, such as:

a) The problematic aspects of the controlled fusion, which may be due to lack of suitable representation of the highly noninertial conditions at the instant of initiation of the fusion process); or

b) The impossibility of achieving a full quark confinement until now, which may be due to our inability to treat the highly noninertial, acceleration-dependent processes expected in the hadronic structure; or

c) The problematic aspects in achieving a true grand unification of all interactions, which may be due precisely to the intrinsically noninertial character of the strong interactions as compared to the inertial structure of the electromagnetic interactions;and others.

III.7: NONRELATIVISTIC ISOPARTICLES AND ISOQUARKS

The most effective way of appraising the possible physical relevance of the covering isogalilean relativities is by identifying their implications in the characterization of the notion of particle.

As well known, the conventional notion of *classical nonrelativistic particle* is a representation of the Galilei group $G(3.1) = [O_\theta(3) \otimes T_r \circ(3)) \times (T_v \circ(3) \times T_t \circ(1)]$ on $\Re_t \times T^*E(r,\delta,\Re)$ and, as such, it is characterized by conventional units, the scalar unit 1 for the time field $\Re_t$, and the six-dimensional unit matrix I for the phase space.

By recalling that the Galilei symmetry holds only for interactions which are of local (differential) and potential (selfadjoint) type, the notion essentially characterizes the historical Galilei's concept of *"massive point"* moving in vacuum under action-at-a-distance interactions.

In particular, the intrinsic characteristics of the particle (mass, spin, charge, etc.) are immutable, classically and quantum mechanically, because points are immutable geometrical objects.

This perennial character of the intrinsic characteristics of particles was challenged in the first half of this century, particularly in nuclear physics. As an example, the total magnetic moments of nuclei have remained essentially unresolved, despite over half a century of research. One can therefore read in Blatt and Weiskopf (1952), p. 31, the possibility that

"*the intrinsic magnetic moment of a nucleon is different when it is in close proximity to another nucleon*".

In different terms, the early (but not the contemporary) studies in nuclear physics admitted the possibility that the value of the magnetic moments of nucleons changes in the transition from the physical conditions under which they have been measured until now (long range electromagnetic interactions only), to the *different* physical conditions when members of a nuclear structure. Similar doubts were also expressed for the spin (see, e.g., (*loc. cit.*), p. 254).

A conjecture was then submitted by Santilli (1978b) according to which massive physical particles can experience an alteration of their intrinsic characteristics, called *mutation,* when experiencing physical conditions broader than those permitted by the Galilean symmetry, such as:

a) composite particles with an extended charge distribution experiencing a deformation of their shape under sufficiently intense external, potential interactions; or

b) extended charge distributions experiencing a deformation of their shape due to contact, nonhamiltonian-nonselfadjoint interactions (as coinceivable in collisions); or

c) particles totally immersed in an external physical medium, such as the core of a star.

In all these cases, *the deformation of the shape of the particle constitutes the first physical origin for a necessary mutation of the intrinsic magnetic moment,* as well established in classical and atomic physics. The mutation of all the remaining intrinsic characteristics, under sufficient physical conditions, can then be inferred from a number of arguments, e.g., of relativistic character.

However, *the primary physical origin of mutation was identified (loc. cit.) as being due precisely to the (operator version of) the contact, nonlinear, nonlocal and nonhamiltonian interactions studied in this volume,* as expected for a proton in the core of a star, or to a lesser extent, in a nuclear structure.

More generally, it was pointed out in (*loc. cit.*) that *a necessary condition for the mutation of elementary particles is the existence of interactions which violate the Galilei (and Lorentz) symmetry.* The deformation of the charge distribution is the simplest possible mechanism of breaking conventional space-time symmetries, because they are well known to be applicable only for rigid bodies.

The contact, nonlinear, nonlocal and nonhamiltonian interactions under study here characterize the most general possible violation of the Galilei (and Lorentz) symmetry at all their structural levels, e.g., inertial, local, canonical, etc. It is hoped that, in this way, the reader begins to see the implications of the interactions herein considered.

The above results were reached in Santilli (*loc. cit.*), Sect. 4.19, via the addition of a (variationally) nonselfadjoint coupling to the conventional Dirac's equation. In fact, these interactions are notoriously velocity-dependent, as it must be the case for all drag forces, whether Newtonian or field theoretical. In turn, the addition of a velocity-dependent coupling to Dirac's equation implies the necessary alteration of the conventional gamma matrices. The mutation, in general, of the intrinsic magnetic momentum, spin and other characteristics is then a necessary consequence.

According to these results, we can visualize a hierarchy of different physical conditions, of increasing complexity and methodlogical needs, such as:

A) The *atomic structure,* in which no mutation is possible because of the large mutual distances among the constituents;

B) The *nuclear structure,* in which small mutations are conceivable because available experimental data on the volumes of nuclei and of individual nucleons establish that nucleons, when members of a nuclear structure, are not only in contact, but actually in conditions of mutual penetration of about 10^{-3} parts of their charge volume. In turn, such mutual penetrations are expected to characterize an additional (small) term in the nuclear force, precisely of the short range, nonlinear, nonlocal and nonhamiltonian type studied in these notes;

C) The *hadronic structure,* in which case we expect a

proportionately higher mutation because, as indicated earlier, the size of all hadrons is approximately the same and coincides with the size of the wavepackets of all known massive particles. The hadronic constituents, to be massive physical particles, are therefore expected to be in conditions of total mutual penetration, resulting precisely in the short range nonlocal and nonhamiltonian interactions under consideration here. Still in turn, these interactions are expected to require a generalized notion of particle as hadronic constituent;

D) The *core of stars*, where a proportionately higher mutation is expected because, in addition to the hadronic conditions of total mutual penetration, we have their compression; and

E) The *gravitational collapse*, where we expect the most extreme possible mutations because we have the most extreme conceivable physical conditions of particles in the Universe, consisting not only of total mutual penetration, and their compression, but also the condensation of an extremely large number of particles in an extremely small region of space.

The studies on the Lie-isotopic liftings of the Galilei relativity outlined in the preceding sections were conducted for the purpose of permitting a quantitative study of the conjecture of mutation of the intrinsic characteristics of particles suitable for experimental verifications.

DEFINITION III.7.1 (Santilli (1988a)): A nonrelativistic isoparticle is an isorepresentation of one of the infinitely possible isogalilean symmetries $\hat{G}(3.1)$ on isospace $\hat{\Re}_t \times T^*\hat{E}(r,G,\hat{\Re})$

$$\hat{G}(3.1): \qquad a' = \hat{g}(w) * a = \hat{g}(w)\, \hat{T}_2 a$$

$$= \{[e^{w_k\, \omega^{\mu\sigma}\, \hat{I}_2^{\ \sigma\nu}\, (\partial_\nu X_k)\, (\partial_\mu)}_{|\xi}]\, \hat{I}_2\} * a, \tag{7.1}$$

$$a = (r,p), \quad \hat{I}_2 = \hat{T}_2^{\ -1} > 0.$$

Equivalently, a nonrelativistic isoparticle can be defined as the generalization of the conventional notion of particle induced by the isotopic liftings of the units

$$I_t = 1 \in \hat{\Re}_t \;\Rightarrow\; \hat{I}_t \in \hat{\Re}_t, \tag{7.2a}$$

$$I \in T^*E(r,\delta,\mathfrak{R}) \Rightarrow \hat{I}_2 = \text{diag}\,(\hat{G}^{-1}, \hat{G}^{-1}) \in T^*\hat{E}(r,\hat{G},\mathfrak{R}). \tag{7.2b}$$

$$\delta = \text{diag.}\,(1,1,1) \Rightarrow G = \text{diag.}\,(B_1{}^{-1}, B_2{}^{-2}, B_3{}^{-2}) > 0. \tag{7.2c}$$

A first central advanceof the notion of isoparticles over that of conventional particles is the possibility to represent the actual shape of the particle considered and, consequentially, of all its infinitely possible deformations, at the primitive classical Galilean level. As a result, when a particle is realistically represented, it can possess an infinite number of different intrinsic characteristics depending on the infinitely possible local conditions, e.g., as per classification A–E above.

On operator grounds, the ideal conditions of applicability of the Galilei-isotopic relativities are given by the hadronic structure, evidently because of the open historical legacy of its nonlocal character by Fermi (1949), Bogoliubov (1963) and other Founders of particle physics.

The ideal isoparticles are therefore expected to be the quarks, resulting in the following

DEFINITION III.7.2 (Santilli (1988a) Mignani and Santilli (1991)): The "nonrelativistic isoquarks", hereon denoted with the symbols $\hat{u}$, $\hat{d}$, etc., are ordinary quarks under short range nonlocal and nonhamiltonian interactions due to mutual wave overlappings as characterizable by the isorepresentations of the Lie-isotopic symmetries $\hat{G}(3.1) \times S\hat{U}(3)$, where $\hat{G}(3.1)$ represents the space-time structure, and $S\hat{U}(3)$ represents the isotopic-unitary lifting of SU(3).(Mignani (1984)).

The above definition is introduced to finally initiate a quantitative study of the open historical legacy of the nonlocality of the hadronic structure, and also in the hope of resolving at least some of the now vexing problems of contemporary hadron physics.

Needless to say, a long chain of studies is needed for a quantitative, mathematical, theoretical and experimental appraisal of the above possibilities. The fundamental step is, and will remain, the primitive Newtonian setting which is the arena of our direct intuitions.

In this section we shall initiate this proces by presenting a few classical nonrelativistic examples of isoparticles. As we shall show in subsequent works, the operator formulation is merely consequential, and actually enhances the classical mutations of this volume.

A technical knowledge of the preceding analysis is necessary for a true understanding of this section.

FREE ISOPARTICLE. In this case, with reference to systems (III.2.3a) and their representation (III.2.9), $N = 1$, all selfadjoint and nonselfadjoint forces are null, the isometrics $\hat{\delta}$ and $\hat{G}$ must evidently coincide and be constants,

$$\hat{\delta} = \hat{G} = \text{diag.}\,(b_1^2, b_2^2, b_3^2), \qquad b_i = \text{constants} > 0. \tag{7.3}$$

Hamilton-isotopic equations (III.2.9) then describe the free particle

$$\dot{r}_i = b_i^{-2}\, \partial H / \partial p_i = p_i/m = v_i, \tag{7.4a}$$

$$\dot{p}_i = - b_i^{-2}\, \partial H / \partial r_i = 0. \tag{7.4b}$$

namely, *the Galilei-isotopic equations of motion are identical to those of the conventional Galilei's relativity.*

Despite that, the use of the Galilei-isotopic relativities is not trivial, because it permits the direct representation of:

1) the extended character of the particle;

2) the actual shape of the particle considered; and

3) an infinite class of possible deformations of the original shape (see below);

all the above already at our primitive, classical, nonrelativistic level.

By comparison, if one insists in preserving the conventional Galilei relativity:

1') the Galilean particle is strictly a massive point, and its extended character can be represented only after the rather complex process of *second quantization;*

2') the second quantization does not represent the actual shape of a particle, say, an oblate spheroidal ellipsoid as per capability 2) above, but provides only the remnants of the actual shape; and, last but not least,

3') possible deformations of extended particles are strictly excluded, as well known, for numerous reasons, e.g., because they imply the breaking of the conventional rotational symmetry.

As an illustration, there are reasons to suspect that the charge distribution of the proton is not perfectly spherical, but characterized instead by a deformation of the sphere of the oblate type computed by Nishioka and Santilli (1991)

$$\delta = \text{diag.}\ (1,1,1) \quad \Rightarrow \quad \hat{\delta} = \hat{G} = (\text{diag.}\ (1,\ 1,\ 0.6), \tag{7.5}$$

(where the third axis is assumed to be that of the intrinsic angular momentum), which permits an interpretation of the anomalous magnetic moment of the particle.

Oblate spheroidal ellipsoid (7.5) can be directly and exactly represented by our Galilei-isotopic relativities, already at the classical nonrelativistic level of this treatment via the value of our isometric

$$b_1{}^2 = 1, \quad b_2{}^2 = 1, \quad b_3{}^2 = 0.6. \tag{7.6}$$

It is evident that such an actual, direct and immediate representation of the shape of the proton is impossible within the context of the conventional Galilei's relativity.

We shall indicate in subsequent studies that, in the transition to the operator version of the theory, the representational capabilities are enhanced because of the appearance of additional degrees of freedom besides that offered by the isounit of the enveloping algebra (e.g., that of the isotopy of the Hilbert space, see Santilli (1980), Myung *et al.* (1982) and Mignani *et al.* (1983)).

The above case illustrates the simplest conceivable (and perhaps most fundamental) mutation of a Galilean particle. In fact, the original particle has the perfectly spherical shape expressed by the underlying metric δ = diag. (1,1,1), while our Galilei-isotopic particle can acquire any one of the infinitely many ellipsoidical deformations of the original sphere expressed by the isometrics $\hat{\delta}$.

The case also illustrates a first use of our isoeuclidean spaces $\hat{E}(r,\hat{\delta},\hat{\mathfrak{R}})$ for the characterization of shape only without any force.

In particular, it should be indicated here that this is a sort of limiting case because the notion of isoparticle generally requires nontrivial interactions. With the terms "free" isoparticle we therefore refer to a conventionally free particle which however represented in an isospace, thus acquiring nongalilean characteristics.

In conclusion, the mutation of shape under consideration at this primitive Newtonian level is intrinsically contained in the lifting of the underlying metric $\delta \Rightarrow \hat{\delta}$, with consequential liftings of fields, metric spaces, space-time symmetries, etc. Equivalently, it can be geometrically expressed by the symplectic isotopy $\omega_2 \Rightarrow \hat{\Omega}^\circ_2 = \omega_2 \times T_2$ and it is algebraically/group theoretically characterized by the Lie-isotopy $\omega^{\mu\nu} \Rightarrow \Omega^{\circ\mu\nu} = \omega^{\mu\sigma} \hat{1}_2{}^{\sigma\nu}$, $\hat{1}_2 = T_2^{-1} > 0$.

ISOPARTICLE UNDER EXTERNAL SELFADJOINT INTERACTIONS. The simplest generalization of the preceding case is the extended-deformable particle under conventional, external, *potential* interactions. In this case, again with reference to system (III.2.3a) in representation (III.2.9), $N = 1$, $V \neq 0$, $F^{SA} \neq 0$, $F^{NSA} = \mathcal{F}^{NSA} = 0$, and the b-quantities of Eq.s (7.3) can still be assumed to be independent of the local coordinates in first approximation, although they can be dependent on the local strength of V, local pressure and other quantities. The equations of motion are then given by

$$\dot{r}_i = b'^{-2}_i \, \partial H / \partial p_i = p_i / m = v_i, \tag{7.7a}$$

$$\dot{p}_i = - b'^{-2}_i \, \partial H / \partial r_i = -(\partial V/\partial r) \, r_i/r, \tag{7.7b}$$

where one should assume that the deformation of shape $\hat{\delta} \Rightarrow \hat{\delta}'$ is volume preserving

$$\hat{\delta} \Rightarrow \hat{\delta}', \quad \det \hat{\delta} \equiv \det. \hat{\delta}'. \tag{7.8}$$

Equations of motion (7.7) also coincide with the conventional Galilean equations when

$$r = (r_i \, b_i^2 r_i)^{\frac{1}{2}} = \overline{r} = (\overline{r}_i \delta_{ij} \overline{r}_j)^{\frac{1}{2}}, \tag{7.9a}$$

$$\overline{r}_i = r_i \, b_i \ \text{(no sum)}, \tag{7.9b}$$

namely, when the distance r in our geometrical space $\hat{E}(r,\hat{\delta},\hat{\mathfrak{R}})$ coincides with the distance $\overline{r}$ in our physical space $E(\overline{r},\delta,\mathfrak{R})$.

Note also the general rule from Eq.s (7.9) of considerable value in practical applications that *the isotopic contraction of a conventional vector can always be written as a conventional contraction of a generalized vector.*

Again, the transition from the Galilei to our Galilei-isotopic relativities is not trivial. In fact, it first allows the direct representation of the actual shape of the particle, as in the free case. In addition, *the isogalilean relativities can represent the deformations of the original shape caused by the external force.*

In fact, starting from an extended particle with the shape represented by the isometric $\hat{\delta}$, we have to expect from simple mechanical considerations that the application of the external force F^{SA} causes a deformation of the shape into the isometric $\hat{\delta}'$.

Needless to say, one may argue that such deformation could be small for given conditions. The point is that perfectly rigid bodies do not exist in the physical reality. The *amount* of deformations for given conditions is evidently an open scientific question, but its *existence* is out of any scientific doubt.

In conclusion, the rotational and Galilei symmetries characterize a *theory of rigid bodies*, as well known. Our isorotations and isogalilean symmetries characterize instead a *theory of deformable bodies* (Figure III.3.2) without violating the abstract O(3) and G(3.1] symmetries, but by realizing them instead in their most general possible form. This completes our consideration for an isoparticle under conventional, external *potential* forces.

ISOPARTICLE UNDER EXTERNAL NONSELFADJOINT INTERACTIONS. The next example is that of an extended-deformable isoparticle under, this time, *nonpotential* external fields caused by motion within a physical medium. Note that this class of interactions is strictly excluded by the conventional Galilei relativity, but it is rather natural for our covering isogalilean relativities.

In this case, N = 1, the selfadjoint interactions can be assumed to be null (V = 0) for simplicity, but we have nontrivial nonselfadjoint interactions represented via our Hamilton-isotopic, Lie-isotopic and symplectic-isotopic methods.

A first simple case in one space-dimension is given by a particle moving within a resistive medium under a quadratic damping force

$$m\ddot{r} + \gamma \dot{r}^2 = 0, \qquad (7.10)$$

with the Birkhoffian representation (Santilli (1982a))

$$\hat{R}^{\circ} = (p, 0), \qquad T = \text{diag.}\ (\hat{\delta}, \hat{\delta}), \tag{7.11a}$$

$$H = p^{\hat{2}} / 2m = p\hat{\delta}p / 2m, \quad \hat{\delta} = e^{2\gamma r/m}, \tag{7.11b}$$

which provides a first approximation of systems such as a satellite penetrating Jupiter's atmosphere or, along similar conceptual grounds, a proton moving within the core of a star.

The above case illustrates a second use of our isospaces, namely, that for the characterization of nonpotential forces of the interior dynanmics. In fact, the lifting $E(r,\delta,\mathfrak{R}) \Rightarrow \hat{E}(r,\hat{\delta},\hat{\mathfrak{R}})$, $\delta = I \Rightarrow \hat{\delta} = \delta \exp 2\gamma r/m > 0$, essentially represents the local, but nonlinear and nonselfadjoint resistive force $F^{NSA} = -\gamma \dot{r}^2$.

While such resistive forces imply an evident breaking of the conventional Galilei symmetry, our techniques permit its exact restoration at the broader isotopic level because the emerging isometric $\hat{\delta}$ is positive-definite.

Additional uses of our isospaces will be indicated later on when dealing with specific applications,

The interested reader can readily enlarge the above example to three dimensions, e.g., for motion along the third axis

$$\hat{R}^{\circ} = (p, 0), \qquad T = \text{diag.}\ \hat{\delta}, \hat{\delta}), \tag{7.12a}$$

$$H = p^{\hat{2}} / 2m = p_i\ \hat{\delta}_{ij}\ p_j / 2m, \quad \hat{\delta} = \text{diag.}\ (b_1{}^2, b_2{}^2, b_3{}^2)\ e^{2\gamma n \times r/m} \tag{7.12b}$$

where $n \times r$ represents the direction of motion, with a deformation of shape, this time, due to contact interactions.

Along similar lines, one can have an isoparticle subject to selfadjoint ($V \neq 0$) and nonselfadjoint ($\hat{\delta} = \hat{\delta}(r)$) interactions. In this case, a simple example is given by the quadratically damped oscillator

$$\ddot{r} + r + \tfrac{1}{2}\gamma\dot{r}^2 + \tfrac{1}{2}\gamma r^2 = 0, \qquad m = k = 1, \tag{7.13}$$

with Birkhoffian representation (Santilli (*loc. cit.*))

$$\hat{R}^{\circ} = (p, 0), \quad T = \text{diag}\ (\hat{\delta}, \hat{\delta}) = \text{diag.}\ (e^{\gamma r}, e^{\gamma r}), \tag{7.14a}$$

$$H = \tfrac{1}{2}\, p^{\hat{2}} + \tfrac{1}{2}\, r^{\hat{2}} = \tfrac{1}{2} p e^{\gamma r} p + \tfrac{1}{2} r e^{\gamma r} r. \tag{7.14}$$

An illustration of nonlinear, nonlocal and nonhamiltonian internal forces is provided by the equations of motion

$$\ddot{r} + \gamma \dot{r}^2 \int_\sigma d\sigma\, \mathcal{F}(r) = 0, \qquad m = 1, \qquad (7.15)$$

where $\dot{r}$ and $\ddot{r}$ are referred to the behaviour of the center-of-mass under a point-like approximation of the particle, and the integral represents the correction in the trajectory due to the extended shape.

System (7.15) can be represented via Hamilton-isotopic equations in terms of the following quantities

$$R^\circ = (p, 0), \qquad T = \mathrm{diag}\,(\hat{G}, \hat{G}), \qquad (7.16a)$$

$$H = \tfrac{1}{2}\, \hat{p^2} = \tfrac{1}{2}\, p\, \hat{G}\, p, \qquad \hat{G} = \hat{\delta} \exp \{\gamma\, r \int_\sigma d\sigma\, \mathcal{F}(r)\, \}. \qquad (7.16b)$$

Another example is given by the systems characterized by

$$R^\circ = (p, 0), \quad T = \mathrm{diag.}\, (\hat{G},\hat{G}), \qquad H = \tfrac{1}{2}\, \hat{p^2} = \tfrac{1}{2}\, p\hat{G}p, \qquad (7.17)$$

and the following isometric

$$\hat{G} = \exp [\, \gamma r + k\int_\sigma d\sigma\, \mathcal{F}(t, r, p,..)]. \qquad (7.15)$$

where, again, one can see two separate representations, one for the damping of the center-of-mass, and the corrections in the trajectory due to the shape of the particle.

Numerous additional examples can be worked out by the interested reader in any desired combination of selfadjoint and nonselfadjoint forces, the latter being local-differential or nonlocal-integral., as desired.

In all the above cases, the isogalilean relativities permit the explicit construction of the generalized invariance $\hat{G}(3.1)$ via the computation of the Lie-isotopic tensor (III.2.7), and its use in the exponential characterization of $\hat{G}(3.1)$ as per Sect. III.5, all in a way which reconstructs the exact Galilei symmetry in isospaces $\hat{\mathfrak{R}}_t \times T^*\hat{E}(r,\hat{G},\hat{\mathfrak{R}})$, while the conventional symmetry is manifestly broken in $R_t \times T^*E(r,\delta,\mathfrak{R})$.

A first understanding is that, to have a full $\hat{G}(3.1)$-invariant model, examples (7.10)-(7.15) have to be interpreted as a two-body system with relative coordinate r, otherwise one has only the full $\hat{O}(3)$-invariance.

Another understanding, stressed earlier during the course of our analysis, is that examples (7.10)-(7.15) represent an elementary or composite isoparticle within an external physical medium. As a result, the total energy is generally *nonconserved* by assumption. In this case, the conserved Birkhoffian merely represents a first integral of the equations of motion, and not a physical quantity (see the examples in Santilli (1982a).

The reader should therefore be aware of the fundamental distinction between the open-nonconservative models here considered, and the closed nonhamiltonian systems studied in Appendix III.A.

This completes our examples of Galilei-isotopic symmetries for *one* elementary or composite isoparticle under the most general, possible external interactions.

The attentive reader has noted that the notion of "isoparticle" studied in this section is a particular case of the notion of "genoparticle" of Appendix II.

III.8: ISODUAL ISOGALILEAN RELATIVITIES

In this section we study in more detail the new space-time and space-time symmetries identified by our isotopies, which we have called *isodual space-time* , and *isodual isosymmetries* , respectively (Sect.s II.3 and III.5).

These novel notions originate from the property that, for a given isotopic space-time $\hat{E}$ and isosymmetry $\hat{G}$ characterized by the isounit $\hat{1}$, our isotopic techniques permit the identification of corresponding isodual quantities characterized by

$$\hat{1}^d = -\hat{1}. \tag{8.1}$$

The first point we would like to indicate is that these novel notions are not identifyable in contemporary theoretical physics, because they necessarily require the use of generalized units, and this is the reason why they have escaped identification until now.

We shall now illustrate the property pointed out earlier that *invariance under isoduality $\hat{1} \Rightarrow \hat{1}^d = -\hat{1}$ appears to be a new universal law of Nature* , We would like also to indicate that, contrary to first impressions, this latter invariance is independent from inversions $r \Rightarrow r' = \hat{P}^* r = Pr = - r$.

For this purpose, let us consider a given *isospace-time*

$$\hat{\Re}_t \times T * \hat{E}(r,\hat{G},\hat{\Re}), \qquad \hat{\Re}_t = \Re_t \hat{1}_t, \qquad \hat{\Re} = \Re\, \hat{1}, \tag{8.2a}$$

$$\hat{1}_t = B_4^{-2} > 0, \quad \hat{1} = \text{diag.}\,(\hat{G}^{-1}, \hat{G}^{-1}), \quad \hat{G} = \text{diag.}\,(B_1^2, B_2^2, B_3^2), \tag{8.2b}$$

with *isogalilean symmetry*

$$\hat{G}(3.1) = \hat{G}^{\hat{1}}(3.1) = [\hat{O}_\theta(3) \otimes \hat{T}_{r^\circ}(3)] \times [\hat{T}_{v^\circ}(3) \times \hat{T}_{t^\circ}(1)]\,, \tag{8.3}$$

and related *isosymmetry transformations*

$$\begin{cases} t \Rightarrow t' = t + t^\circ \tilde{B}_4^{-2}, & \text{isotime translations} \quad (8.4a) \\ r_i \Rightarrow r'_i = r_i + r^\circ_i\, \tilde{B}_i^{-2}, & \text{isospace translations} \quad (8.4b) \\ r_i \Rightarrow r'_i = r_i + t^\circ\, v^\circ\, \tilde{B}_i^{-2}, & \text{isoboosts} \quad (8.4c) \\ r_i \Rightarrow r'_i = \hat{R}(\theta) * r, & \text{isorotations} \quad (8.4d) \\ r \Rightarrow r' = \hat{P} * r = - r, & \text{isoinversions} \quad (8.4e) \end{cases}$$

Then, the *isodual isospace-time* is given by *(Sect. II.3)*

$$\Re^d{}_t \times T * \hat{E}^d(r,\hat{G}^d,\Re^d), \qquad \Re^d{}_t = \Re_t \hat{1}^d{}_t, \qquad \Re^d = \Re\, \hat{1}^d, \tag{8.5a}$$

$$\hat{1}^d{}_t = - B_4^{-2} < 0, \quad \hat{1}^d = \text{diag.}\,(\,\hat{G}^{d-1}, - \hat{G}^{d-1}) = - \hat{1} < 0, \tag{8.5b}$$

$$\hat{G}^d = \text{diag.}\,(- B_1^2, - B_2^2, - B_3^2), \tag{8.5c}$$

while the *isodual isogalilean symmetry* has the structure

$$\hat{G}^d(3.1) = \hat{G}^{\hat{1}^d}(3.1)\ [\hat{O}^d{}_\theta(3) \otimes \hat{T}^d{}_{r^\circ}(3)] \times [\hat{T}^d{}_{v^\circ}(3) \times \hat{T}^d{}_{t^\circ}(1)]\,, \tag{8.6}$$

and admits related *isodual isosymmetry transformations*

$$\begin{cases} t \Rightarrow t' = t - t^\circ \hat{B}_4^{-2}, & \text{isodual isotime transl.} \quad (8.7a) \\ r_i \Rightarrow r'_i = r_i - r^\circ_i \hat{B}_i^{-2}, & \text{isodual isospace transl.} \quad (8.7b) \\ r_i \Rightarrow r'_i = r_i - t^\circ v^\circ \hat{B}_i^{-2}, & \text{isodual isoboosts} \quad (8.7c) \\ r_i \Rightarrow r'_i = \hat{R}^d(\theta) * r, & \text{isodual isorotations} \quad (8.7d) \\ r \Rightarrow r' = \hat{P}^d *^d r = -r, & \text{isodual isoinversions} \quad (8.7e) \end{cases}$$

Proposition III.5.1 essentially establishes that *any system which is invariant under isotransformations (8.4), is also invariant under their isoduals (8.7)* . To illustrate this important property, consider the quadratically damped particle (7.10), i.e.,

$$m \ddot{r} + \gamma \dot{r}^2 = 0 \tag{8.8}$$

with Hamilton-isotopic representation (7.11), i.e.,

$$R^\circ = (p, 0), \quad T = \text{diag.}\,(\hat{\delta}, \hat{\delta}), \quad H = p\hat{\delta}p/2m, \tag{8.9a}$$

$$\hat{\delta} = \exp\{2\gamma\, r \,/\, m\} \tag{8.9b}$$

It is then easy to see that the *isodual representation*

$$R^{\circ d} \equiv R^\circ = (p, 0), \quad T^d = \text{diag.}\,(\hat{\delta}^d, \hat{\delta}^d), \quad H^d = p\hat{\delta}^d p/2m, \tag{8.10a}$$

$$\hat{\delta}^d = -\exp\{\, 2\,\gamma\, r \,/\, m \tag{8.10b}$$

yields exactly the same equations of motion (8.8). The same situation occurs for all other cases, as the interested reader can verify.

Note the crucial role of the isogalilean invariance of systrems (8.8) under isounit (8.9b) for the isodual symmetry to hold. In fact, it is easy to see that, in case the original isogalilean invariance is not verified, the same holds for its isodual.

The independence of the isodual symmetry from the isoinversions is also easily see, e.g., by comparing Eq.s (8.4e) and (8.7e).

As a complementary comment, the system with *antidamping force*

$$m \ddot{r} - \gamma \dot{r}^2 = 0 \tag{8.11}$$

is physically different then system (8.8), because the energy decreases in time for the latter (dissipation), while increasses in time for the former (nonconservation).[6]

As a result of this physical difference, system (8.11) admits the isorepresention

$$R^{\circ} = (p, 0), \quad T' = \text{diag.}\,(\hat{\delta}', \hat{\delta}'), \quad H' = p\hat{\delta}'p/2m, \tag{8.12a}$$

$$\hat{\delta}' = \exp\{-2\gamma r / m\} \tag{8.12b}$$

from which one can see that

$$\hat{\delta}' = \exp\{-2\gamma r / m\} \neq \hat{\delta}^{d} = -\exp\{2\gamma r / m\} \tag{8.13}$$

Thus, the transition from dissipative system (8.8) to its antidissipative image (8.11) is not representable via isoduality. Note that system (8.11) is also invariant under its own isoduality transformation,

$$\hat{\delta}' \Rightarrow \hat{\delta}'^{d} = -\exp\{-2\gamma r / m\}. \tag{8.14}$$

III.9: CONCLUDING REMARKS

The most salient property of the isotopic techniques in general, and of the isogalilean relativities in particular, is their capability to unify at the abstract level conventional, linear, local and Hamiltonian systems with thir broadest possible nonlinear, nonlocal and nonhamiltonian generalizations.

In the concluding remrks of this chapter we would like to illustrate this property in more details because it is fundamental in understanding that the axiomatic structure of conventional relativities is preserved in their entirety under our isotopies, not only at the Galilean level of this chapter, but also at the relativistic and gravitational levels of the subsequent chapters.

Consider the simplest possible Galilean system, the free particle in conventional space-time $\Re_t \times E(r,\delta,\Re)$, with *canonical variational principle*

$$\delta A = \delta \int_{t_1}^{t_2} (R^{\circ}\, da - H\, dt) = 0, \tag{9.1a}$$

[6] These systems occur when extended particles of sufficient small masses move in sufficiently turbulent gases, in which case their speed increases in time, as established by systems of our physical reality, such as baloons in our atmosphere, etc.

$$R^\circ = (R^\circ_\mu) = (p, 0), \qquad a = (a_\mu) = (r, p), \quad \mu = 1, 2, ..., 6 \tag{9.1b}$$

$$H = p^2/2m = p\,\delta\,p\,/\,2m = p_i\,\delta_{ij}\,p_j\,/\,2m, \quad i, j = 1, 2, 3, \tag{9.1c}$$

which is geometrically based on the familiar *canonical one-form*

$$\Phi_1 = R^\circ\,da - H\,dt = R^\circ_\mu\,da^\mu - H\,dt = p_k\,dr_k - H\,dt \tag{9.2}$$

and is manifestly invariant under the Galilei's symmetry G(3.1). Consider now the isotopic liftings of the above system, the *isocanonical representations*

$$\delta\,\hat{A} = \delta \int_{t_1}^{t_2} (R^\circ * da - H \odot dt) = 0, \tag{9.3a}$$

$$R^\circ = (R^\circ_\mu) = (p, 0), \qquad a = (a_\mu) = (r, p), \tag{9.3b}$$

$$H = p^{\hat{2}}/2m = p\,\hat{\delta}\,p\,/\,2m \tag{9.3c}$$

generated by the *iso-one forms* (Sect. II.9) [7]

$$\hat{\Phi}_1 = \Re^\circ * da + H \odot dt =$$

$$= R^\circ_\mu\,T_r{}^\mu{}_\nu(t, a, \dot{a}, \ddot{a}, ...)\,da^\nu - H\,T_t(t, a, \dot{a}, \ddot{a}, ...)\,dt =$$

$$p_k\,T_{r\,ki}(t, r, p, \dot{p}, ...)\,dr_i - H\,T_t(t, r, p, \dot{p}, ...)\,dt. \tag{9.4}$$

with underlying isocotangent bundle[8]

$$\hat{\Re}_t \times T^*\hat{E}(r,\hat{\delta},\hat{\Re}): \quad \hat{R}_t = R\,\hat{1}_t, \quad \hat{1}_t = T_t^{-1} = b_4^{-2} \tag{9.5a}$$

$$\hat{\delta} = \text{diag.}\,(b_1^2, b_2^2, b_3^2), \tag{8.5b}$$

$$\hat{\Re} = \Re\,\hat{1}_r\,, \qquad \hat{1}_r = T_r^{-1} = \text{diag.}\,(\hat{\delta}^{-1}, \hat{\delta}^{-1}) \tag{9.5c}$$

---

[7] The attenfive reader has noted that we have performed here the transition from the even-dimensional isosymplectic geometry of one-isoforms $R^\circ * da$ of Sect. II.9 to a broader off-dimensional geometry of one-isoforms $R^\circ * da - H \odot dt$ inclusive of the time component, which can be called *contact-isotopic geometry* or *isocontact geometry* for short.

[8] The reader should keep in mind from Sect.s II.7 and II.9 that isospace (9.5) is that of the representation of the system via a variational principle (via one-isoforms) which, as such, *is not* the space of its isosymmetries (requiring two-isoforms), and this explains the reason for the use of the b's in isometrics (9.5b), rather than the B's of the isogalilean symmetry of Sect. III.6.

The invariance of the systems under the isogalilean symmetry $\hat{G}(3.1)$ then follows, with the consequential applicability of our isogalilean relativities.

The most effective way to illustrate the ultimate unity of physical and mathematical thought between the conventional and isogalilean relativities is by nothing that, by conctruction, the canonical one-form (9.2) coincides with its isotopes (9.4). In fact, all distinction are evidently lost at the abstract level between the algorithms $R°da - Hdt$ and $R°*da - H\odot dt$.

The most effective way of illustrating the physical differences between the conventional and isogalilean relativities is by nothing that, while Eq.s (9.3) represent the free particle in vacuum, their isotopes (8.) represent instead extended particles moving within a physical medium, resulting in nonconservative conditions.

In fact, isorepresentation (9.3) includes as particular case examples (7.10), 7.13), (7.15) and similar ones, with examples

$$\hat{\delta} = \pm \exp.\{2\gamma r / m\}: \qquad m\ddot{r} + \gamma\dot{r}^2 = 0, \tag{9.6a}$$

$$\hat{\delta} = \pm \exp\{-2\gamma r / m\}: \qquad m\ddot{r} - \gamma\dot{r}^2 = 0, \tag{9.6b}$$

$$\hat{\delta} = \pm \exp \gamma r \int_\sigma d\sigma\, \mathcal{F}(r)\}: \qquad \ddot{r} + \gamma\dot{r}^2 \int_\sigma d\sigma\, \mathcal{F}(r) = 0, \tag{9.6c}$$

Note that the only similarity between the free Galilean particle and its isotopic images is the absence of potential forces. This illustrate the central methodological aspect of these volumes, the representation of nonpotential-nonhamiltonian forces and effects[9] via the generalization of the trivial unit $I = \text{diag.}\,(1,1,1)$ of contemporary use into the isounits $\hat{I} = \hat{I}(t, r, p, \dot{p}, ...)$.

Note also that, while the Galilean system is unique, Eq.s (9.5) represent an infinite number of geometrically equivalent, but physically different systems, as shown in Eq.s (9.6). This illustrates the reason for our continued use of the plural in "isogalilean relativities".

The free Galilean particle is evidently the simplest conceivable system. A more general class is given by systems of N particles invariant under the Galilei's group G(3.1), which can be represented via the canonical variational principle

---

[9] We here recall the teaching of Hamilton (1834) according to which there exist effects in Nature which are not representable via his function $H = T + V$, as it is the case, e.g., for the deformation of a given shape, which led to the submission of his celebrated equations with external terms (Chapter I).

$$\delta A = \delta \int_{t_1}^{t_2} (R^\circ da - H dt) = 0, \qquad (9.7a)$$

$$R^\circ = (R^\circ_\mu) = (p, 0), \qquad a = (a_\mu) = (r, p), \qquad (9.7b)$$

$$H = p^2/2m + V = p_{ia}\, \delta_{ij}\, p_{ja} / 2m + V(r_{ab}) \qquad (9.7c)$$

$$r_{ab} = [(r_{ia} - r_{ib})\, \delta_{ij}\, (r_{ja} - r_{jb})]^{\frac{1}{2}} \qquad (9.7d)$$

$\mu = 1, 2, \ldots 6N, \quad a, b = 1, 2, \ldots, N, i, j = 1, 2, 3.$

Their isotopic generalizations are given by the isocanonical representations

$$\delta \hat{A} = \delta \int_{t_1}^{t_2} (R^\circ * da - H \odot dt) = 0, \qquad (9.8a)$$

$$R^\circ = (R^\circ_\mu) = (p, 0), \qquad a = (a_\mu) = (r, p), \qquad (9.8b)$$

$$H = \hat{p}^2/2m + V = p_{ia}\, \hat{\delta}_{ij}\, p_{ja} / 2m + V(\hat{r}_{ab}) \qquad (9.8c)$$

$$\hat{r}_{ab} = [(r_{ia} - r_{ib})\, \hat{\delta}_{ij}\, (r_{ja} - r_{jb})]^{\frac{1}{2}} \qquad (9.8d)$$

whose invariance under the isogalilean symmetries $\hat{G}(3.1)$ is now familiar, with consequential applicability of the isogalilean relativities.

The most important physical difference between systems (9.5) and (9.8) is that the former represent one particle in an external medium, thus resulting in a nonconservative system. On the contrary, Eq.s (9.8) characterize a new class of compositise systems, called closed nonhamiltonian, veryfying conventional total conservation laws, while the internal forces are nonhamiltonian.

In different terms, systems (9.8) represents the "closure" of systems (9.5) with their environment, including the presence of conventional potential interactions. The preservation of the ten, conventional, Galilean, total quantities is ensured by the now familiar preservation of the generators in the isotopies $G(3.1) \Rightarrow \hat{G}(3.1)$.

Again, closed selfadjoint systems (9.7) and their closed nonhamiltonian generalization (9.8) coincide at the abstract, realization free level by construction.

The reader should keep in mind that isorepresentations (9.8) are still a particular case of the isotopic techniques. In fact, the R-functions are the canonical one, $R = R^\circ = (p, 0)$. The most general possible isotopic generalizations of canonical systems (9.5) is given by the Birkhoff-isotopic representation (Sect. II.7)

$$\delta \hat{A} = \delta \int_{t_1}^{t_2} (R * da - B \odot dt) = 0, \qquad (9.9a)$$

$$R \neq R^\circ = (R^\circ_\mu) = (p, 0), \qquad (9.9b)$$

where the Birkhoiffian B is restricted to be an isogalilean invariant. The latter systems are not studied in these volumes because excessively general, as well as of unknown operator image at this time.[10] At any rate, the simpler Hamilton-isotopic systems (9.5) or (9.8) are amply sufficient for our needs.

Despite that, the most general possible isotopic systems (9.9) continue to coincide with the canonical ones (9.5) at the abstract level under the conditions assumed.

This abstract unity is at the foundations of all subsequent steps of our analysis, and it is essential to understand later on in Chapter V the preservation of the isoparallel transport and isogeodesic character in the transition from Galilean systems (9.1) or (9.7) to their respective isogalilean generalizations (9.3) or (9.8), of course, when formulated in a suitable, structural generalization of the Riemannian geometry.

But, by far, the most intriguing aspect of the analysis of this chapter is that ***the isogalilean relativities do not need experimental verification in our classical environment, for the evident reason that they are constructed from given equations of motion,*** as illustrated in Sect. II.7 and III.7. As a result, they do provide the form-invariant description of the systems considered by construction.

The proposal submitted in Chpter VII deal, specifically, with the test of the isogalilean relativity in particles physics.

[10] This is due to the fact that the isotopic Hamilton-Jacobi equations for systems (9.8) are given by the forms (III.2.10), i.e.,

$$\frac{\partial \hat{A}}{\partial t} + H = 0, \qquad \frac{\partial \hat{A}}{\partial r_{ia}} = p_{ja} \, T_{s\ ja\ ia}, \qquad \frac{\partial \hat{A}}{\partial p_{ia}} = 0, \qquad (a)$$

which, via the use of the isotopic mapping $\hat{A} \Rightarrow i\, \hat{I} \log \psi(t, r)$, $\hat{I} = \hbar\, T^{-1}$, $\hbar = 1$, yield the *isoschrödinger's equations* of *hadronic mechanics* (II.6.24), i.e.,

$$i \frac{\partial}{\partial t} \psi(t, r) = H * \psi(t, r) = H\, T \psi\, (t, r) = \hat{E} * \psi(t, r) = E\, \psi(t, r), \qquad \hat{E} \in \hat{\mathfrak{R}} \quad (b)$$

In turn, this sets the foundations of the operator image of the isogalilean symmetries and relativities studied in Santilli (1989). In the transition to the more general, Birkhoff-isotopic systems (9.9), $\partial \hat{A} / \partial p_{ia} \neq 0$, and no consistent operator image is known at this writing (because it would require the generalization $\psi(t, r) \Rightarrow \psi(t, r, p)$).

APPEN.DIX III.A: TWO-BODY AND THREE-BODY CLOSED NONHAMILTONIAN SYSTEMS

As indicated in the main text of this chapter, a main implications of the isogalilean relativities is the characterization of a new composite sysrtem, with intriguing possibilities of novel advances in particle physics. In this appendix we shall study these novel systems in more details

In tSect. III.7 we have: reviewed the notion of losed-isolated systems with potential (selfadjoint) and nonhamiltonian (nonselfadjoint) internal forces, called *closed nonselfadjoint systems*; outlined the generalized analytic, algebraic and geometrical methods for their treatment; and identified their invariance under the Galilei-isotopic symmetries $\hat{G}(3.1)$.

In Sect. III.8 we have presented a generalized notion of particle, called *isoparticle*, characterized by the isogalilean relativities.

In this appendix we shall study the generalization of Kepler's two-body and three-body systems suggested by these advances.

The reader should be aware that the classical notion of closed nonselfadjoint systems is centered in the existence of an interior medium which is responsible for the contact nonhamiltonian interactions. In turn, such a medium is classically created by a large number of constituents, as in Jupiter's structure.

In this appendix we shall study closed nonselfadjoint systems in their smallest possible number of constituents N = 2 and 3, in which case the interior medium is evidently absent. The internal forces are then merely expressed by the condition of *contact interaction* among the constituents, that is, the (extended) constituents must be in physical contact among each other as a necessary condition to have two- and three-body closed nonselfadjoint systems.

Whenever such a physical contact is removed, and the constituents move freely in space, the systems considered reacquire their Keplerian selfadjoint structure.

It is intriguing to anticipate since now that, in the transition to an operator treatment, the interior medium exists also for the case of two- and three-bodies closed nonselfadjoint systems, which therefore acquire a deeper meaning. In fact, in these cases each constituent is an extended wavepacket moving *within* the wavepackets of the remaining constituents resulting in nontrivial isounits of type (II.6.26).

In different terms, at the classical level of this appendix, the constituents can indeed be in mutual physical contact, but evidently without mutual penetration. This results in the *lack* of underlying medium and, therefore, in special forms of the nonselfadjoint forces. In the particle case, instead, we do have indeed total mutual penetration of the wavepackets of the constituents, in which case each constituent is the medium of the others.

Finally, an equivalent way of defining closed nonselfadjoint systems is by noting that they are *closed systems of isoparticles*. In fact a closed system of conventional particles can only be selfadjoint and Galilean, while a closed system of isoparticles implies the existence, by definition, of internal nonselfadjoint forces.

In considering the content of this section, the reader should therefore keep in mind that the constituents of the generalized Kepler's systems are not conventional particles, but generalized isoparticles.

In conclusion, the analysis of this appendix should be essentially considered a rudimentary classical and nonrelativistic basis for a number of possible, future developments, such as the study of closed nonhamiltonian systems with a large number of constituents; the study of their operator counterpart; or the study of the systems as bound states of isoparticles.

TWO-BODY CLOSED NONSELFADJOINT SYSTEMS. Two-body closed nonselfadjoint systems were first studied in the original proposal (Santilli (1978b), pp. 622-633), which also identified their stable configuration. The systems were then studied in details in Santilli (1982a). The first Birkhoffian representation of the systems was reached by Jannussis, Mijatovic and Veljanoski (1991).

Consider systems (III.2.3). For the case of two particles, motion is necessarily in a plane, say $k = 1, 2$ (= x, y), and the systems become

$$\begin{cases} M\ddot{R} = 0, & \text{(A.1a)} \\ m\ddot{r} = F^{SA} + F^{NSA}(r, \dot{r}, \ddot{r}), & \text{(A.1b)} \end{cases}$$

where

$$M = m_1 + m_2, \tag{A.2a}$$

$$m = \frac{m_1 m_2}{m_1 + m_2}, \tag{A.2b}$$

$$r = r_1 - r_2, \qquad R = \frac{m_1 r_1 + m_2 r_2}{m_1 + m_2}, \tag{A.2c}$$

Conditions (III.2.4) on the NSA forces then become

$$F_1^{NSA} = - F_2^{NSA} \stackrel{def}{=} - F^{NSA}, \tag{A.3}$$

with general solution=

$$F^{NSA} = g\,[r_1^{(2n)} - r_2^{(2n)}], \tag{A.4}$$

where g = cost, and $r^{(2n)}$ represents the 2n-th derivative. For $n = 1$, the only admissible stable orbit is then the circle, in which case the nonselfadjoint force assumes form (3.4.11) of Santilli (1978b), i.e.,

$$F^{NSA} = g\,\ddot{r}, \tag{A.5}$$

with equations of motion (3.4.12) (*loc. cit.*), which we can write

$$\hat{m}_1\,\ddot{r} = -k\,(r_1 - r_2) / |r_1 - r_2|^3, \tag{A.6a}$$

$$\hat{m}_2\,\ddot{r} = +k(r_1 - r_2) / |r_1 - r_2|^3, \tag{A.6b}$$

$$\hat{m}_1 = b^2\, m_1, \quad \hat{m}_2 = b^2\, m_2, \quad b^2 = (m - g) / m > 0, \quad g < m, \tag{A.6c}$$

namely, *the closed nonselfadjoint generalization of the two-body Kepler's system essentially provides a renormalization of the masses of the constituents within a purely classical context.*

What is remarkable is that this occurs already at the classical nonrelativistic level of this appendix and prior to any operator counterpart.

The reader is warned against the appraisals of the above results via old notions that are inapplicable to the physical conditions considered. Specifically, if one keeps thinking at dimensionless points, the above results are evidently inconsistent because elliptic orbits are admissible too, as well know.

However, the systems under consideration represent, by central assumption, *extended particles under mutual contact interactions*, such as two spheres in mutual contact (or, operationally, two extended wavepackets moving one inside the other). It is then evident that two spheres (or, for that matter, two extended objects of arbitrary shape) can rotate one with respect to the other under mutual contact only in a circle (and exactly the same situation is expected for wavepackets, as shown in Santilli (1978b), (1990)).

Stated differently, one can indeed think at point-like particles for systems (A.1), with the understanding that such a conception directly implies the loss of the nonselfadjoint forces, with consequential recovering of elliptic trajectories (see Fig. III.A.1 for more details).

Finally, the emergence of an acceleration-dependent force should not be dismissed lightly, because it appears to have rather intriguing connections with Mack's Principle, as well as with the Ampère-Neumann electrodynamics (see in this respect Assis (1990), Graneau (1990) and quoted works).

We consider now *Jannussis' representation* of system (A.1), which can be written in terms of the Pfaffian

$$\hat{A} = \int_{t_1}^{t_2} dt \, \{p_{ia} b^2 \dot{r}_{ia} - [p_{ia} b^2 p_{ia} / 2m_a + k / |r_1 - r_2|]\}, \quad a = 1,2 \qquad (A.7)$$

which is naturally written in our isoeuclidean spaces $\hat{E}(r,\hat{\delta},\hat{\mathfrak{R}})$.

In conclusion, the two-body nonselfadjoint generalization of the conventional Kepler's system is characterized by the simplest possible isotopic lifting of a Lie algebra, that provided by the *scalar isotopy* of the canonical Lie tensor

$$\omega^{\mu\nu} \Rightarrow \Omega^{\mu\nu} = b^{-2} \, \omega^{\mu\nu}, \qquad (A.8)$$

Despite the simplicity of the isotopy, the physical implications in particle physics are rather intriguing on a number of counts (see Figure III.A.1).

THREE-BODY CLOSED NONSELFADJOINT SYSTEMS. Their existence and consistency was also identified in the original proposal by Santilli (1978b), and then reviewed in Santilli (1982a). Their first detailed study was provided by Jannussis and his collaborators (1991) (see below).

The motion of the three-body closed nonselfadjoint systems in their general configuration occurs in three dimension and the subsidiary constraints become essential.

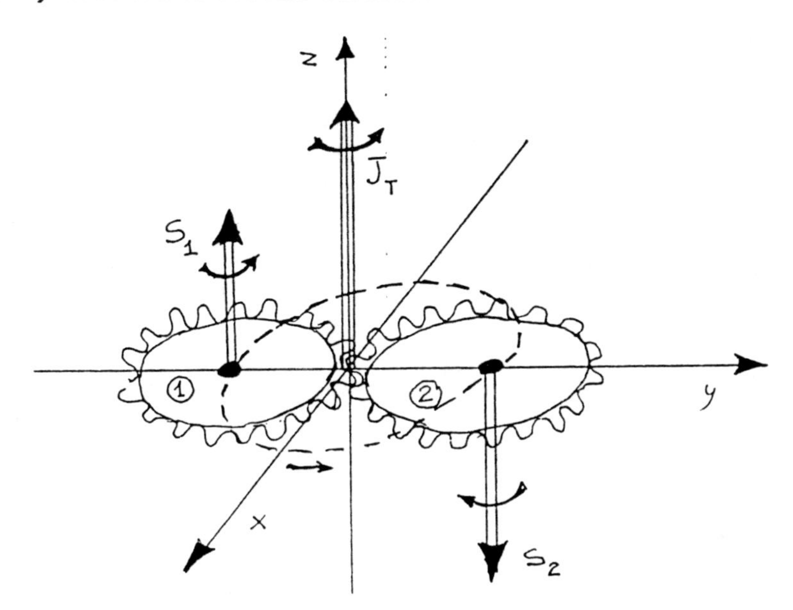

FIGURE III.A1: A symbolic view of the *"gear model"* for the representation of closed nonselfadjoint two-body systems introduced in Santilli (1978b), Sect. 5. The model was suggested because effective in identifying the stable configurations under contact mutual interactions, classically and operationally. In fact, gears can rotate one around the other only in a circle, as represented by Eq.s (A.4) Moreover, when an intrinsic angular momentum is added, the gear model provides a visualization of the property that *only singlet states are stable*. In fact, gears rotates "in phase" one inside the other in a singlet coupling, as shown in the figure. For triplet states, instead, we have an unstable system because the gears should rotate one against the other, thus creating mutually resistive forces which would push the gears one away from the other, by therefore rendering null the nonselfadjoint forces. In the transition to operator settings, the gear model is equally effective because in this case we have extended wavepackets rotating one around the other in conditions of large mutual penetration. The mutually resistive forces for triplet couplings then persist, resulting again in the singlet state with circular orbit as the only stable state. The model was used in (*loc. cit.*) to study the structure of the π^0 particle as a *"compressed positronium"*, i.e., as a closed nonselfadjoint system of one electron compressed inside the positron and rotating one with respect to the other at mutual distances of the order of 1F (= 10^{-13} cm) in a singlet state. It essentially emerged that such a model is capable of providing a quantitative representation of *all* the total characteristics of the π^0 (total energy, meanlife, spin, charge radius, space and charge parity, etc.), thanks to the "renormalization" of the masses which is implicit in Eq.s (A.6). On the contrary, and this should be stressed here, the same model is inconsistent within the context of the conventional quantum mechanics for numerous reasons, such as: the inability to represent the total energy of the π^0 with very light constituents; the impossibility of achieving the relatively high meanlife of the particle; etc. (see (*loc. cit.*) for details). The studies presented in this volume are intended as a Newtonian basis for a reconsideration of the above operator models we hope to present at some future time. Their primary objective, as one can see, is to attempt the identification of the hadronic constituents with suitably generalized forms of ordinary particles which can be freely produced in the spontaneous decays, along lines historically established for the preceding nuclear and atomic structures.

The study of these systems requires a step-by-step generalization of the (rather vast) structure of the conventional three-body Kepler systems (for an excellent analytic treatment of the latter, see Hagihara (1970)). Evidently, this task cannot be performed in this appendix. We shall therefore content ourselves with the identification of the most stable configurations without subsidiary constants.

The equations of motion in their second-order form in $T^*E(r,\delta,\Re)$ can be written

$$\begin{cases} m_1\ddot{r}_1 = -\dfrac{m_1 m_2}{r_{12}^3}(r_1 - r_2) - \dfrac{m_1 m_3}{r_{13}^3}(r_1 - r_3) + F_1^{NSA}, & \text{(A.9a)} \\ m_2\ddot{r}_2 = -\dfrac{m_2 m_1}{r_{21}^3}(r_2 - r_3) - \dfrac{m_2 m_3}{r_{23}^3}(r_2 - r_1) + F_2^{NSA}, & \text{(A.9b)} \\ m_3\ddot{r}_3 = -\dfrac{m_3 m_1}{r_{31}^3}(r_3 - r_1) - \dfrac{m_3 m_2}{r_{32}^3}(r_3 - r_2) + F_3^{NSA} & \text{(A.9c)} \end{cases}$$

where

$$r_{ij} = |r_i - r_j|, \quad i, j = 1, 2, 3. \qquad (A.10)$$

and the NSA forces are restricted by conditions (III.2.4) to verify the identities

$$\sum_a F_a^{NSA} = 0, \quad \sum_a p_a \times F_a^{NSA} = 0, \quad \sum_a r_a \wedge F_a^{NSA} = 0. \qquad (A.11)$$

$$a = 1, 2, 3.$$

A straightforward generalization of the two-body solution of Eq.s (A.4) with $n = 1$ to the three-body case, leads to the following realization of the NSA forces (apparently introduced for the first time in Santilli (1988a))

$$F_1^{NSA} = c(\ddot{r}_1 - \ddot{r}_2) + d(\ddot{r}_1 - \ddot{r}_3), \qquad (A.12a)$$

$$F_2^{NSA} = c(\ddot{r}_2 - \ddot{r}_3) + d(\ddot{r}_2 - \ddot{r}_1), \qquad (A.12b)$$

$$F_3^{NSA} = c(\ddot{r}_3 - \ddot{r}_1) + d(\ddot{r}_3 - \ddot{r}_2), \qquad (A.12c)$$

where, for simplicity, we have ignored higher (even-order) derivatives; the quantities c and d are assumed to be constants; and we shall put hereon $c = d$.

The first stable configuration is that with one particle at rest at the center of the system with solution

$$r_1 = \text{const.}, \quad r_2 = \text{constants}, \quad r_3 = \text{const.} \quad \text{for} \quad m_1 \neq m_3 \qquad (A.13a)$$

$$r_1 = \text{cnost.}, \quad r_2 = 0, \quad r_3 = \text{cnost.} \quad \text{for} \quad m_2 = m_3. \qquad (A.13b)$$

The system then rotates rigidly around its center-of-mass.

Moreover the centers-of-mass of the three bodies must be on a straight line. This is the *nonselfadjoint extension of the restricted three-body problem.* For further properties, see Fig.III.A.2.

The next stable configuration is that of the celebrated *Lagrange triangle* (Hagihara (*loc. cit*)), which is also described by forces (A.12) with solution

$$r_1 = \text{const.} \equiv 0, \qquad r_2 = \text{const.} \equiv 0, \qquad r_3 = \text{const.} \equiv 0. \tag{A.14}$$

Again, we essentially have contact interactions forcing the three bodies one against the other. The system then rotates rigidly around its center-of-mass. On historical grounds, we should recall the

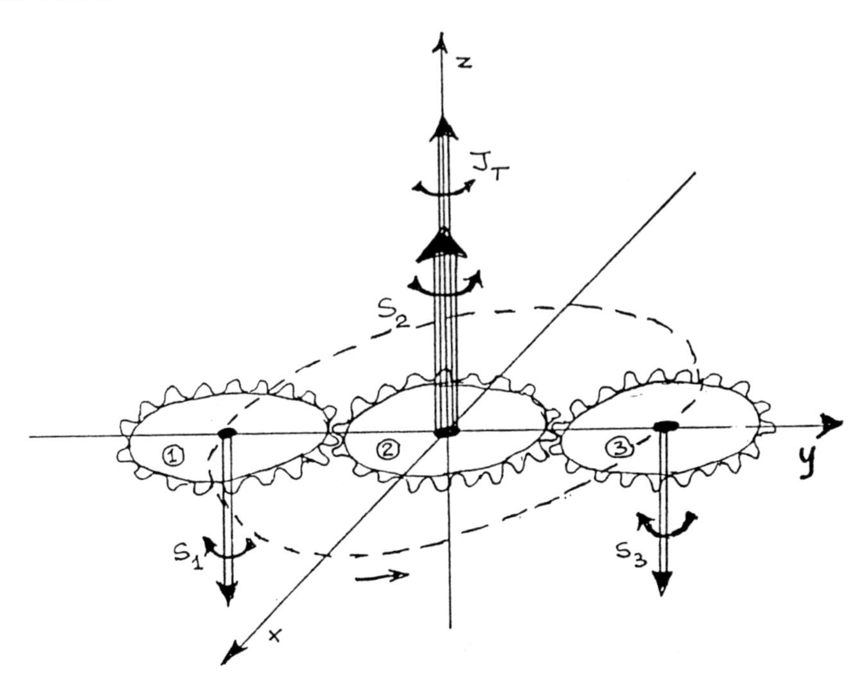

FIGURE III.A.2: The *"gear-model"* for a stable configuration of the three-body, closed, nonselfadjoint systems submitted in Santilli (1988a). The configuration here depicted is the simplest possible one consisting of the three bodies under mutual physical contact, disposed along a straight line, and rotating around the center-of-mass of the system, with the peripheral bodies having equal masses (for simplicity but without loss of generality). The configuration allows the introduction of a new notion of nucleus, called the *isonucleus* , which can essentially have an unrestricted (non-null) mass under contact nonselfadjoint interactions. By comparison, the conventional nucleus can occur in Kepler's systems only when its mass is much larger than the mass of the peripheral constituents. In fact, contact interactions merely force the peripheral bodies against the central body which, as such, can have an arbitrare mass, including a mass much smaller than those of the peripheral bodies. For stability, the motion must necessarily occur along a straight line for much of the same reasons of the conventional Kepler's case (Hagihara (1970)). In fact, whenever the centers-of-mass of the three particles are no longer along a straight line, the configuration changes into that of *Lagrange's historical triangle* (see Figure III.A.3). The conceptual value of the gear model becomes visible when adding the intrinsic angular momenta. In fact, the pairing of particles cannot be any longer a singlet or a triplet, as in the conventional Kepler's case, but can only be of singlet type in order to prevent mutual resistive forces and their consequential instabilities. Moreover, the gear model becomes particularly valuable in special cases, e.g., when the central body has a null intrinsic angular momentum. In this case, the peripheral bodies must necessarily have non-null intrinsic angular momenta in order to be able to have a stable rotation around the stationary central body. Note that the notion of isonucleus persists for closed nonselfadjoint systems of more than three particles. The above results are essentially unchanged when passing to the operator bound state of three extended wavepackets in conditions of total mutual immersion. In this case, the configuration of this figure appears to be relevant for the study of Rutherford's historical conception of the neutron as a *"compressed hydrogen atom"* (Rutherford (1920)) according to which the peripheral electron is compressed (say, in the core of a star) inside the proton, all the way down to the center of the hyperdense medium in the proton structure (for preliminary operator studies ofthe model see Santilli (1990)). Note that *Rutherford's configuration of the electron at the center of the proton is permitted by our notion of isonucleus for a closed nonselfadjoint three-bodysystem, but strictly prohibited by the conventional three-body system.* A central objective of these studies is to permit a quantitative treatment of Rutherford's historical hypothesis, evidently following the operator formulation of the theory. The hope is the possibility of identifying the d-quark with Rutherford's electron e (although both in a mutated form $\hat{d}$ and $\hat{e}$, see Sect. III.7) which, as such, can be freely produced in the spontaneous decays $\hat{d} \Rightarrow e + \bar{\nu}_e$. It is hoped that, in this way, the reader begins to see the reasons for the necessary prior study of the Newtonian setting.

analytic difficulties Lagrange's faced in achieving the stable triangular configuration (A.14). It is then significative to point out that these difficulties are readily resolved by contact interactions. For further comments see Fig. III.A.3.

The above two cases exhaust all possible configurations of three-body isogalilean systems with stable individual orbits[11]. The remaining

[11] These systems therefore exhaust all possible cases when the constituents are "isoparticles" as per Definition III.7.1, that is, generalized particles which can be effectively characterized by the Lie-isotopic formulations because with conserved

cases are predictably more involved than the corresponding conventional cases, and cannot possibly be treated in this note. They

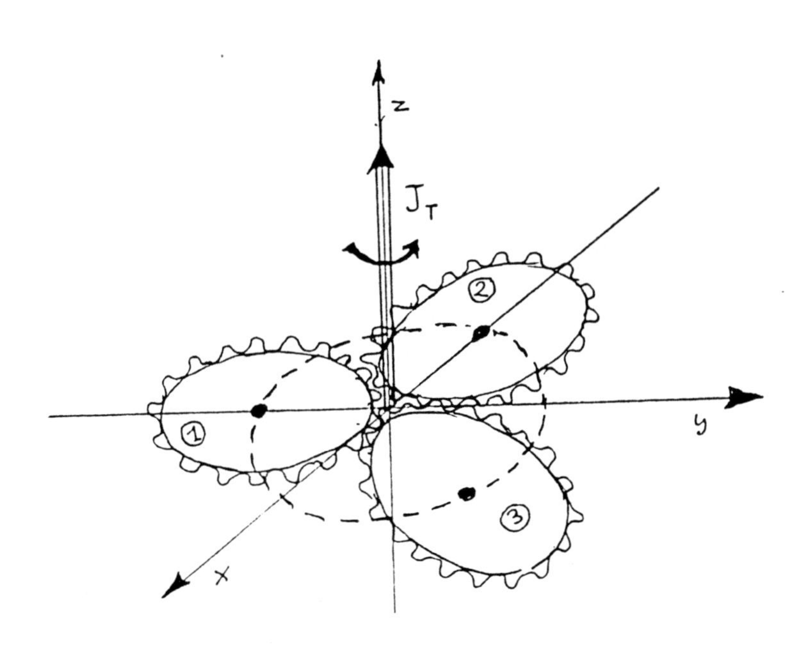

FIGURE III.A.3: The *"gear model"* of the three-body, closed, nonselfadjoint systems in *Lagrange's historical triangle configuration* (Santilli (1988a)). In this case, the three bodies are forced into mutual contact and rotate rigidly around the center-of-mass of the system, each orbit being a circle. The conceptual value of the gear model becomes evident in this case, because it illustrates the fact that *a necessary condition for stability is that all three constituents have a null intrinsic angular momentum.* Intriguingly, the individual angular momenta are null, but the total angular momentum of the system is evidently not null. The configuration of this figure is significant for the study of heavier hadrons as closed three-body isogalilean bound states, as we hope to illustrate in subsequent studies.

---

energy. When the constituents are in general unstable orbits, they are no longer isoparticles, but "isobiparticles" as per Definition II.E.1, namely, they require the more general Lie-admissible formulations for the adequate treatment of their nonconservative conditions.

can be best illustrated via *Jannussis' (1991) Birkhoffian representation*

$$B = \frac{p_1^2}{2m_1} + \frac{p_2^2}{2m_2} + \frac{p_3^3}{2m_3} - \frac{m_1 m_2}{r_{12}} - \frac{m_1 m_3}{r_{13}} - \frac{m_2 m_3}{r_{23}} = E_{tot} \qquad (A.15a)$$

which represents the nonselfadjoint forces of clear resistive character

$$F_1^{NSA} = \gamma(\dot{r}_2 - \dot{r}_3), \quad F_2^{NSA} = \gamma(\dot{r}_3 - \dot{r}_1), \quad F_3^{NSA} = \gamma(\dot{r}_1 - \dot{r}_2). \qquad (A.16)$$

Note the bona-fide isotopic structure (II.2.5) of Jannussis action (A.15). This implies that the Lie-isotopic tensor of the theory is no longer a trivial scalar multiple of the canonical one, but a more complex quantity of the type

$$(\hat{\Omega}^{\circ\mu\nu} = ((\omega^{\mu\alpha})\,(\hat{1}_{2\alpha}{}^{\nu}). \qquad (A.17)$$

Moreover, closedness requires in this case the essential, irreducible constraint

$$r_1 \wedge r_2 + r_2 \wedge r_3 + r_3 \wedge r_1 = 0, \qquad (A.18)$$

For additional details in this intriguing case, we refer the interested reader to Jannussis, Mijatovic and Veljanoski (*loc., cit*).

CHAPTER IV: ISOTOPIC GENERALIZATIONS OF EINSTEIN'S SPECIAL RELATIVITY

IV:1: STATEMENT OF THE PROBLEM

As limpidly expressed in the historical contributions by Lorentz (1904), Poincaré (1905) Einstein's (1905), and others (see, e.g., Pauli (1921) and quoted historical papers), the body of formulations today known as *Einstein's special relativity* was conceived for the description of

1. particles which can be well approximated as being massive points;

2. while moving in the homogeneous and isotropicvacuum (empty

space) under action-at-a-distance, potential forces;

3. under the conditions that quantum mechanical and gravitational effects on curved spaces can be ignored.

Subsequent, rather vast, epistemological, theoretical and experimental studies proved the special relativity to be exact, in the above specified conditions, beyond any reasonable scientific doubt.

To avoid major misrepresentations of this work, the reader should keep in mind that:

I) The exact validity of the special relativity under the historical conditions of its original conception shall be tacitly implied throughout the entire course of this volume;

II) This volume deals with physical conditions fundamentally different than those of the original conception; and, last but not least

III) even when inapplicable,[12] the special relativity shall be tacitly assumed hereon to remain approximately validity.

The above original conception of the special relativity was kept in the well written papers and treateses in the topic in the first half of this century. For instance, in Bergmann (1942) one can read the title of Chapter VI: "Relativistic mechanics of massive points".

With the passing of time and the growing, well known, successes of the special relativity, the physical arena of original conception was progressively abandoned, up to the presentation of contemporary papers and books in the field according to which the relativity is assumed as being universal, that is applicable everywhere and at any time, for whatever physical conditions exist in the Universe.

Yet, authoritative doubts on the exact validity of the special relativity under physical conditions *different* then those of the original conception have been expressed since the early part of this century, classically and quantum mechanically.

As an example, in regard to the interior of strongly interacting particles, Fermi (1949) expressed

[12] It is appropriate here to recall that Einstein did not claim Gailei's relativity to be "violated" at relativistic speeds, but merely "inapplicable" because referred to physical conditions substantially beyond those of its original conception. This sound teaching is preserved in these volumes. We shall therefore say that Einstein's special (or general) relativity is "inapplicable" for, and not "violated" by nonlinear, nonlocal and nonhamiltonian systems of the interior problem because conceived for fundamentally different physical conditions.

"doubts as to whether the usual concepts of geometry hold for such small region of space."

The understanding is that doubts in the geometry imply necessary, consequential doubts in the applicable relativity.

The open legacy of Fermi and other Founding Fathers of contemporary physics was based on the ultimate *nonlocal* nature of the strong interactions, as expected under conditions of deep mutual penetrations of the wavepackets of particles. In fact, the nonlocality of the theory implies a breakdown of the geometrical structure of the special relativity, let alone its physical laws (e.g., local commutation rules).

The successes of the quark theory on the hadronic structure (see, e.g., the reprints of the original contributions edited by Lichtenberg and Rosen (1980)) then contributed to the general assumption of the exact validity of the special relativity also in the hadronic structure. In fact, a central assumption of the theory is that *quarks are point-like*, thus restoring the conditions of the original conceptions of the special relativity.

Despite that, hadron physics has remained afflicted by a number of still unresolved conceptual and technical problems, which are so fundamental to warrant a reinspection of the underlying physical laws. On technical grounds, the quark theory has been unable to achieve full consistency because of the lack of *true quark confinement*, i.e., a confinement not only with an infinite potential barrier, but also with an explicitly computed, and identically null probability of tunnel effects of free quarks.

In fact, explicit calculations done by Chatterjee and Gautam (1986) as well as others, show that the nonrelativistic probability of the quark constituting, say, a proton or a neutron to tunnel into free conditions is non-null, contrary to experimental evidence.

By repeating these calculations, one can see that, whether relativistically or nonrelativistically, the nonnull probability of tunnel effects of free quarks is a necessary consequence of the assumed exact validity of Heisenberg's uncertainty principle for the interior problem. In fact, when the quark is on the infinite potential barrier, the probability that it also exists in the outside of the barrier simply cannot be rendered null.

It is evident that, by assuming the exact validity of the *same* laws in the *interior and and exterior* of hadrons, the probability of tunnel effects for free quarks is not expected to be identically null at all levels of the theory, including the quantum field theoretical level.

On conceptual grounds, the difficulties are perhaps deeper. In fact, the hadronic constituents can indeed have a *point-like charge* of the type possessed by the electrons, as experimentally supported by tests conducted by Bloom *et al.* (1969). However, *"point-like wavepackets"* are only figments of academic imagination and they do not exist in Nature.

In the physical reality, the hadronic constituents are "freelike"[13] particles possessing a wavelength of the order of the size of all hadrons (about 1F = 10^{-13} cm). This results in necessary conditions of total mutual overlapping of their wavelengths with consequential internal nonlocal effects, exactly along the historical legacy by Fermi (1949), Bogoliubov (1960) and other Founders of particle physics.

But above all, the greatest uneasiness created by quark theories and their underlying special relativity lies in their reduction of the entire universe to a strictly *linear and local* theory. In fact, the complexity of the universe in general, and of strong interactions in particular, is certainly such to require, sooner or later, nonlinear and nonlocal formulations. Even assuming that the historical nonlocal legacy is somewhat by-passed via local approximations, a local theory of strong interactions should be at least nonlinear[14] to avoid an excessive simplification of Nature.

We can therefore conclude our introductory comments on the particle profile by saying that the historical open legacy on the nonlocality of the strong interactions, rather than having been

[13] This is evidently due to the asymptotic freedom of the hadronic constituents. Intriguingly, the operator formulation of the Lorentz-isotopic relativity presented in this chapter provides a direct interpretation of this setting, trivially, because nonlocal nonhamiltonian internal effects have no energy by conception, that is, no binding energy (Santilli (1988a).

[14] Classically, the nonlinearity is usually referred to the coordinates r and/or velocities $\dot{r}$. Quantum mechanically, the nonlinearity is usually referred in the contemporary literature to that in the wavefunction ψ and its conjugate $\psi^\dagger$. We here refer to the nonlinearity permitted by *Animalu's isounit*

$$\hat{1} = \hat{1}(t, r, \dot{r}, \ddot{r}, ..., \psi, \psi^\dagger, \partial\psi, \partial\psi^\dagger, ...),$$

that is, ***nonlinearity in all local coordinates, wavefunctions and their conjugates, as well as their derivatives of arbitrary order***. The reader should be aware that the experiments on the possible "nonlinearity of quantum mechanics" conducted until now (see the recent tests by Bollinger *et al.* (1989), Chupp *et al.* (1990), Walsworth *et al.* (1990) and quoted preceding tests), which resulted to be all negative, are *inapplicable* to the context here considered. In fact, these experiments are all referred to the *atomic* structure, while we treat, specifically, the *hadronic* structure. Also, the "nonlinearity" studied in the experiments is solely referred to the wavefunction, while we are more interested in the nonlinearity typical of interior dynamical problems, that on the derivatives of the wavefunction.

resolved, has been left fundamentally open by the current quark theory, despite its clear successes. In fact, as soon as the extended character of the wavepackets of quarks is admitted, their mutual penetration with the inherent nonlinear and nonlocal effects studied in this volume becomes consequential.

In Definition III.7.2 we have introduced the notion of *nonrelativistic isoquark* (and weshall extend it to *relativistic isoquarks* in Sect. IV.7) to permit the inclusion of short-range nonlocal and nonhamiltonian interactions, without losing the SU(3) symmetry.

In this chapter we shall therefore study the classical generalization of Einstein's special relativity which is applicable under the most general possible nonlinear, nonlocal and nonhamiltonian interactions, as preparatory grounds for a subsequent operator study of the historical open legacy on strong interactions.

Classically, the limitations of the special relativity were also well known in the early part of this century, but then progressively ignored. In fact, the *contact interactions* of our Newtonian reality (e.g., those experienced by a satellite during re-entry in Earth's atmosphere) were knwon to be structurally outside the representational capabilities of the special relativity, e.g., because of their *zero range*, or, equivalently, of their evident *instantaneous character*.[15]

Moreover, the reader should keep in mind that

---

[15] The reader with a young mind of any age should be warned against dismissing instantaneous forces in classical mechanics just because not predicted by Einstein's special relativity. The reality is in fact the opposite, i.e., the inability of the special relativity to predict instantaneous forces is precisely a central reason for its inapplicability to interior dynamical problems. In particular, *a classically instantaneous force/zero-range force is a classical approximation of the short range interactions originating in the overlapping of wavelengths at the particle level.* On technical grounds, the special relativity can *only* represent variationally selfadjoint forces (Helmholtz (1887), Santilli (1978e)). These are necessarily action-at-a-distance forces for which the "instantaneous" character is inapplicable, and the conventional formulation of the special relativity follows. However, there exist in Nature also contact nonselfadjoint forces for which the "action-at-a-distance" character has no physical meaning and which can occur only at the instant of contact, as established by experimental evidence of our everyday life. These latter forces are fundamentally outside the representational capabilities of Einstein's special (and general) relativities and are the subject of study of these volumes. A pragmatic rule to identify the applicable relativity is the following. If a force admits a potential energy (selfadjointness), then it is strictly Einsteinian. On the contrary, if a force does not admit a potential energy (nonselfadjointness), then it is outside the arena of true applicability of Einstein's special relativity.

the notion of "zero-range" interactions is precisely a classical approximation of the "nonlocal" quantum mechanical effects expected under conditions of overlapping of the wavelengths of particles.

Stated differently, the overlapping of wavelengths is an event over a finite region of space at the particle level, but it becomes an event over a null region of space at a first classical approximation. In such a transition, however, the nonhamiltonian character of the event, that is, its lack of a potential energy, persists.

There is no need to conduct elaborate experiments to identify the inapplicability of the special relativity in our classical environment, but only observe (and admit) physical reality, such as spinning tops with monotonically decaying angular momentum, charged spheres experiencing deformations of their shapes, satellites during re-entry on decaying trajectories, etc. According to the special relativity, none of these systems are permitted, trivially, because of the necessary exact rotational invariance under which the angular momentum must be conserved. We can therefore say that

The insistence on the exact validity of the special relativity in the interior dynamical problem implies the necessary acceptance of the perpetual motion in a physical environment.

We have to insist that there is no known, *technically* established way to avoid this conclusion, because the inconsistency Theorems I.3.1 and I.3.2. prevent the simplistic reduction of the decaying classical orbits to idealistic, stable, elementary orbits, as well as for the other reasons indicated in Chapter I.

Also, it was well known in the first part of this century that the *deformability* of extended bodies, and the theory of elasticity at large, are fundamentally outside the technical capabilities of the special relativity, trivially, because one of its pillars, the rotational symmetry, is known to be applicable only for *rigid bodies.*[16] Regrettably, this sound scientific attitude was progressively suppressed, up to current literature in the field where the incompatibility of the theory of elasticity with Einstein's special

[16] This author still remembers his college teacher of special relativity stressing this point as one of the limitations of the special relativity and, thus, performing a true scientific function in stimulating the minds of young students. This true teaching function should be compared with the current "teaching" of the special relativity in contemporary colleges, as the final and terminal theory applicable under the most unimaginable conditions of the Universe, which causes evident scientific *damages* because it prevents or otherwise discourages creativity in young students.

relativity is generally ignored.

We have therefore to conclude that

The insistence on the exact validity of the special relativity in our classical environment implies the necessary acceptance that the objects of our physical reality are perfectly rigid.

Again, we must insist that there is no known, *technically* established way to avoid this conclusion, that is, we are aware of no compatibility between the theory of elasticity and the special relativity which has been published in a refereed Journal.

Finally, the analytic structure of the interior dynamical problem of our physical environment was known to be also *nonlocal,* i.e., of a type analytically equivalent to those of the historical legacy on the strong interactions. In fact, the contact interactions experienced by a satellite during re-entry are precisely of nonlocal-integral type, i..e., of the same analytic character of the forces expected by hadronic constituents in conditions of total mutual overlapping of their wavepackets.

At a deeper analysis, the interactions of our physical reality were known since the origin of analytic mechanics to be nonlinear, nonlocal and nonhamiltonian, as studied earlier in this volume. In fact, Lagrange (1788) and Hamilton (1834) formulated their analytic equations with *external terms,* precisely to represent the contact interactions of our reality.

In turn these contact interactions imply a breakdown, not only of the geometry and topology of the special relativity, but also of its Lie structure. In fact, the entire conventional Lie's theory is inapplicable under the contact interactions of our everyday experience, let alone that of the Lorentz group.

We can therefore conclude our introductory comments on the classical profile by saying that

The physical reality of our environment establishes the existence on clear experimental grounds of nonlinear, nonlocal and nonlagrangian-nonhamiltonian forces which are beyond the representational capabilities of Einstein's special relativity and, more precisely, beyond the physical conditions of its original conception.

Moreover, the classical conditions of inapplicability of the special relativity are analytically equivalent to the historical open legacy on the ultimate nonlocality of the strong interactions.

As indicated in the Preface, this author dedicated his research life to the study of the limitations and possible generalizations of the special relativity with respect to: a) the classical profile; b) the operator formulation; c) the inter-relations between the above two profiles; d) the nonrelativistic limit; and e) the gravitational extention.

The nonrelativistic studies have been reviewed in the preceding chapters, including the study of:

A) the inequivalence of the exterior and interior dynamical problems (Sect. I.2);

B) the impossibility of "eliminating" the nonlinear, nonlocal and nonhamiltonian forces of our physical reality via the reduction of macroscopic bodies to ideal point-like constituents in stable orbits (Sect. I.3);

C) the achievement of compatibility between the local-potential, exterior problem, and the nonlocal-nonhamiltonian interior problem via the notion of *closed nonselfadjoint systems* (Sect. III.2), with the explicit study of the most stable configurations for the generalized two-body and three-body cases (Appendix III.A);

D) an outline of the *Lie-isotopic theory* (Sect. II.6);

E) the geometrization of the interior three-dimensional physical media via the *isoeuclidean spaces* (Sect. II.3), and the identification of a particular geometry, called *isosymplectic geometry* (Sect. II.9) which permits the treatment of nonlocal interactions;

F) the construction of classical, nonlinear, nonlocal and nonhamiltonian realizations of an infinite family of *isorotational, isoeuclidean* and *isogalilei symmetries* (Sect.s III.3, III.4 and III.5);

G) the construction of the infinite family of isotopic liftings of Galilei's relativity under the name of *Galilei-isotopic relativities* (Sect. III.6);

H) The direct representation by the isogalilean relativities of all infinitely possible (volume preserving) *deformations* of the original shape of the particle considered (Sect.s III.7 and III.8); and,

I) the emergence of a consequential, generalized notion of extended-deformable particle, called *isoparticle* (Sect. III.7) or, more generally, *isobiparticle* (Appendix II.E).

These and all other foundational aspects of the nonrelativistic profile of the studies, such as the *Birkhoffian mechanics* (II.7) and the *Lie-admissible theory* (Sect.s I.4, II.5 and Appendix II.A) shall be hereon tacitly assumed as known.

In this chapter we shall review *the classical relativistic generalization of the above nonrelativistic setting.* In particular, our primary objectives are:

α) Construct the infinite family of classical, Lie-isotopic generalizations of the Lorentz and Poincaré symmetries, under the tentative names of *Lorentz-isotopic* (or *isolorentz*) and *Poincaré-isotopic* (or *isopoincaré*) *symmetries* , respectively, for the most general possible nonlinear, nonlocal and nonhamiltonian systems;

β) Construct the elements of the corresponding, expected generalizations of Einstein's special relativity for the *classical* interior problem (only), under the tentative name of *Lorentz-isotopic relativities* , or *isospecial relativities* , applicable within inhomogeneous and anisotropic physical media; identify the primary deviations from the conventional relativity; and propose suitable *classical experiments* for the verification of the special relativity in interior dynamical conditions, or its disproof in favor of covering relativities;

γ) Under the conditions that the isospecial relativities:

γ-a) coincide with the conventional relativity at the abstract, realization-free level, thus achieving an ultimate unity of conceptual, physical and mathematical thought;

γ-b) admit the Galilei-isotopic relativities at the nonrelativistic level; and

γ-c) are locally admitted by suitable isotopies of Einstein's gravitation, to be studied in the next chapter.

The number of authors who have conducted relativistic studies preceding my efforts is so large to prohibit a comprehensive list. An outline of the primary references known to this author has been provided in Sect. I.5. We here limit ourselves to recall that the first relativistic considerations of *Lie-isotopic character* were submitted in Santilli (1982b) on the maximal causal speed under contact zero-range interactions; the first construction of the *Lie-isotopic generalization of the Lorentz symmetry* with the structural foundations of the *Lorentz-isotopic relativities* were submitted by Santilli (1983a) for the nonlinear, nonlocal and nonhamiltonian, matrix formulation, and therefore of operator character; the classical formulation of the generalized relativities for nonliner, nonlocal and nonhamiltonian interactions was reached in Santilli (1988c) which constitute the basis of this review. Additional studies can be found in Santilli (1991c, d).

The mathematical foundations of the classical generalized relativities, the Lie-isotopic theory, first appeared in Santilli (1978a) and then expanded in Santilli (1982a) for the linear and nonhamiltonian, but local interactions. Their generalization to classical nonlocal settings first appeared in Santilli (1989a, b) and was subsequently expanded in Santilli (1989g, h).

Additional studies important for this chapter are those by: De Sabbata and Gasperini (1982), who studied the maximal causal speed in the interior of hadrons; Aringazin (1989), who proved the "direct universality" of the Lorentz-isotopic relativities; Mignani (1990), who studied the application of the isotopic redshift law to the problem of the quasars' redshift; and Cardone et al. (1991), who provided a phenomenological study on the application of the Lorentz-isotopic relativities to the behaviour of the meanlife of unstable hadrons with speeds.

Additional studies on a true *generalization* of the special relativity were conducted by by Bogoslovski ((1977), (1984)) for the case of anisotropic space-time, here called *Bogoslovski's special relativity*. The study was conducted via the *conventional Lie's theory*, and the anisotropy was referred to space-time itself . [By comparison, we use the covering Lie-isotopic formulations, while we study anisotropy *and* inhomogenuity referred, specifically, to physical media].

A generalization of the Lorentz transformations was achieved by Edwards (1963) and Strel'tsov (1990) via the assumption of a time anisotropy with consequential differentiation between forward and backward speeds of lights.

Recami and Mignani (1972) proposed a generalized transformation, called *superluminal*, which maps time-like four-vector into space-like ones, and which has particularly intriguing algebraic implications for our studies (Sect. 5). Additional references will be quoted dusring the course of our analysis.

III.2: CLOSED, RELATIVISTIC, NONHAMILTONIAN SYSTEMS

In Sect. III.2 we studied *nonrelativistic, closed, nonselfadjoint systems.* These are systems such as Jupiter which, when seen from the outside, verify all conventional total conservation laws and space-time symmetries (exterior dynamical problem). Nevertheless, their interior trajectories are intrinsically nonlinear, nonlocal and nonhamiltonian (interior dynamical problem).

In this section we shall introduce the notion of *relativistic, closed, nonselfadjoint systems.* Let us begin by introducing the conventional *Minkowski space*

$$M(x,\eta,\Re): \quad x = (x^{\mu}) = (r, x^4), \quad x^4 = c_o t, \;\; r \in E(r,\delta,\Re), \tag{2.1a}$$

$$\eta = (\eta_{\mu\nu}) = \text{Diag.}\,(1,1,1,-1)\,, \;\; \mu, \nu = 1, 2, 3, 4, \tag{2.1b}$$

$$x^2 = x^{\mu}\,\eta_{\mu\nu}\,x^{\nu} = x^1x^1+x^2x^2+x^3x^3 - tc_o{}^2t, \tag{2.1c}$$

$$ds^2 = -\,dx^{\mu}\,\eta_{\mu\nu}\,dx^{\nu} = \text{inv} \tag{2.1d}$$

where: the x's represent the local coordinates; the *velocity of light in vacuum* is indicated hereon with the symbol c_o; ds^2 is the invariant separation; and all contractions are in the reals $\Re$.

We now introduce in M a system of N point-like particles denoted with the index a = 1, 2, ..., N, and suppose that they constitute a composite system with conventional (local-differential) relativistic forces $K^{a\mu}{}_{SA}(x, p)$ verifying the conditions of *variational selfadjointness* (SA) (Santilli (1978e)) for the existence of a potential, as well as the conventional Poincaré symmetry P(3.1).

Then, the composite systems characterized by the special relativity are given by the class of relativistic, closed selfadjoint systems on $M(x,\eta,\Re)$

$$m_{oa}\frac{du^{a\mu}}{ds} = K^{a\mu}{}_{SA}(x,p), \qquad u^{a\mu} = dx^{a\mu}/ds \tag{2.2a}$$

$$\frac{d}{ds}\,P^{\mu} = \frac{d}{ds}\,\sum_a p^{a\mu} = 0, \tag{2.2b}$$

$$\frac{d}{ds} J^{\mu\nu} \quad \frac{d}{ds} \sum_a J^{a\mu\nu} = 0, \tag{2.2c}$$

$$\mu, \nu = 1, 2, 3, 4, \qquad a = 1, 2, \ldots, N$$

where P^{μ} are the generators of translations of P(3.1) and $J^{\mu\nu}$ are the generators of the Lorentz subgroup O(3.1).

Note that global stability of systems (2.2) is characterized by the stability of each individual constituent. In particular, each individual constituent is rotationally invariant and possesses a conserved angular momentum.

As a result, systems (2.2) are ideally suited for a relativistic treatment of exterior dynamical systems such as the planetary or the atomic systems.

Suppose now that systems (2.2) are generalized to represent N *extended particles* moving within a physical medium, as it is the case, say, for Jupiter's structure assumed to be constituted by extended molecules in a relativistic gaseous state. This implies the additional presence of contact interactions of each particle with the medium consisting of all remaining particles, which we can write $F^{a\mu}{}_{NSA}$, where NSA represents *variational nonselfadjojntness*, i.e., the violation of the integrability conditions for the existence of a potential (Santilli (*loc. cit.*)), and have an arbitrary, generally nonlinear and nonlocal (e,.g., integral) dependence on the local variables x, and p, their derivatives $\dot{x}$ and $\dot{p}$ with respect to the independent parameter s, as well as any needed additional quantity, $F^{a\mu}{}_{NSA} = F^{a\mu}{}_{NSA}(x, p, \dot{p}, \ldots)$.

Suppose that the latter forces violate, as a central condition, the Poincaré symmetry via one of the various mechanisms we encountered in the preceding chapter (e.g., the *isotopic, selfadjoint, canonical, semicanonical and essentially nonselfadjoint breakings* of Sect. A-12, Santilli (1982a)).

Our *relativistic, closed, nonselfadjoint systems* can then be written

$$m_{oa} \frac{d\, u^{a\mu}}{ds} = K^{a\mu}{}_{SA}(x, p) +$$

$$+ K^{a\mu}{}_{NSA}(x, p, \dot{x}, \dot{p}, \ldots) + \int_{\sigma} d\sigma\, \mathcal{F}^{a\mu}{}_{NSA}\, x, \dot{x}, p, \dot{p}, \ldots) \tag{2.3a}$$

$$\frac{d}{ds} P^{\mu} = \frac{d}{ds} \sum_a p^{a\mu} = 0, \tag{2.3b}$$

$$\frac{d}{ds} J^{\mu\nu} = \frac{d}{ds} \sum_a J^{a\mu\nu} = 0, \qquad (2.3c)$$

$$\mu, \nu = 1, 2, 3, 4, \quad a = 1, 2, \ldots, N$$

where the nonlocal term is a relativistic extension of the corresponding newtonian term in Eq.s (III.1.1).

Note that the total conservation laws (2.3b) and (2.3c) are the same as in the conventional systems [2.2), but they are no longer necessarily verified along the individual trajectories and, for this reason, they are now *subsidiary constraints* to the equations of motion (2.2a).

The primary similarities between systems (2.2) and (2.3) is that they show no visible difference when inspected from an outside observer. In fact, both systems verify the total Lorentzian conservation laws, and the trajectories of the center-of-mass of both systems verify the Poincaré symmetry.

Despite that, the physical difference between systems (2.2) and (2.3) are rather deep. In fact, the trajectories of each individual constituent is unstable and nonconservative by construction, as ensured by the forces $K^{a\mu}{}_{NSA}$, the equations of motion are no longer invariant under the Poincaré symmetry, and the internal dynamics is generally nonlinear, nonlocal and nonhamiltoniam. Therefore, systems (2.3) imply the violation of the Poincaré symmetry , and therefore of Einstein's special relativity, at all its structural levels:

1) *topologically* , the nonlocal character of the internal forces imply the breakdown of the fundamental Zeeman's topology of the special relativity;

2) *analytically* , we have the inapplicability of the canonical theory underlying conventional relativistic formulations; and

3) *algebraically* , we have the inapplicability of the entire, conventional Lie's theory for the very construction of the Poincaré symmetry.

Systems (III.1.1) represent our Newtonian model of Jupiter's structure, while systems (2.3) represent our relativistic extension of the same structure. For a gravitational treatment, see the subsequent chapter.

Systems (III.1.1) also constitute our primitive Newtonian form of a conceivable structure model of hadrons as closed nonselfadjoint systems. Systems (2.3) above then constitute their classical relativistic version.

A classical conceptual guidance for our study is therefore provided by Jupiter's structure, while a particle counterpart can be given by the hadronic structure conceived as generalized bound states of particles wavelengths in conditions of total mutual immersion, with consequential internal nonlinear, nonlocal and nonhamiltonian forces precisely of type (2.3).

It is easy to see that, exactly as it was in the nonrelativistic level, systems (2.3) admit unconstrained solutions. In fact, given system (2.2a) verifying conditions (2.2b) and (2.2c), its implementation to system (2.3) holds iff

$$\sum_a F^{a\mu}_{NSA} = 0, \quad \mu = 1,2,3,4, \tag{2.4a}$$

$$\sum_a p_{a\mu} F^{a\mu}_{NSA} = 0, \tag{2.4b}$$

$$\sum_a (x_a \wedge F_{aNSA})^\mu = 0, \tag{2.4c}$$

where we have absorbed the nonlocal forces in the nonselfadjoint ones for notational convenience. Eq.s (2.4) constitute *nine independent conditions.* Therefore, for given Poincaré-invariant forces $F^{a\mu}_{SA}$, a solution in the Poincaré-noninvariant forces $F^{a\mu}_{NSA}$ always exist for N > 1 the case N = 2 being a particular case because its space motion occurs in a plane (see Appendix III.A). The case N = 1 is impossible because a free particle can experience no force, whether selfadjoint or not.

Of paramount inportance is the need that the exterior behavior of composite systems (2.3) verifies Einstein's special relativity to such an extent to provide no indication whatever of its generalized internal structure. Specifically,

We here impose as fundamental condition for consistency, that a composite particle, such as a proton in a particle accelerator, when assumed to possess a closed nonselfadjoint internal structure, verifies the totality of the Einsteinian laws.

The above condition can be realized even in a stricter form with the more general class of relativistic, closed nonselfadjoint systems

$du^{a\mu}$

$$m_{oa} \frac{}{ds} = K^{a\mu}{}_{SA}(x, p) +$$

$$+ F^{a\mu}{}_{NSA}(x, p, \dot{x}, \dot{p}, ...) + \int_\sigma d\sigma \, \mathcal{F}^{a\mu}{}_{NSA}(x, \dot{x}, p, \dot{p}, ...) \quad (2.5a)$$

$$\frac{d}{ds} P^\mu = \frac{d}{ds} \sum_a p^{a\mu} = 0, \quad \frac{d}{ds} J^{a\mu\nu} = \frac{d}{ds} \sum_a J^{a\mu\nu} = 0, \quad (2.5b)$$

$$\frac{d}{ds} (P^\mu \eta_{\mu\nu} P^\nu) = 0, \quad (2.5c)$$

$$\frac{d}{ds} (W^\mu \eta_{\mu\nu} W^\nu) = 0, \quad W^\mu = \tfrac{1}{2} \epsilon^{\mu\alpha\beta\gamma} J^{\alpha\beta} P^\gamma, \quad (2.5d)$$

$$ds = \dot{X}^\mu \eta_{\mu\nu} \dot{X}^\nu = -1, \quad X^\mu = \sum_a x^{a\mu}, \quad (2.5e)$$

where the reader will recognize as subsidiary constraints: the *conventional, Einsteinian, total conservation laws* (2.5a); the *Casimir invariants* of the Poincaré symmetry, Eq.s [2.5.c) and (2.5.d); and *the validity of all the preceding conditions on a conventional Minkowski space,* Eq.s (2.5e).

Despite the rather considerable number of subsidiary constraints, systems (2.5) still admit unconstrained solutions in the Poincaré noninvariant forces. In fact, the total number of subsidiary constraints is now thirteen, which can be reduced to ten via Eq.s (2.4). A solution therefore always exist for $N > 1$.

In this way we have proven the following

LEMMA IV.2.1: The center-of-mass trajectory of relativistic, closed nonselfadjoint systems (2.5) verifies the Poincaré symmetry P(3.1) and Einstein's special relativity in their totality.

To state it differently, no detection whatever of the generalized interior structure can be detected from the outside of systems (2.5). As a result,

COROLLARY IV.2.1.1: The current exterior experimental evidence according to which hadrons verify the special relativity exactly, say, when in a particle accelerator, does not constitute evidence that the hadronic constituents must necessarily verify the same relativity.

A primary objective of this section is to identify the generalized symmetries and relativity laws that are *directly applicable* to systems (2.3) or (2.5), that is, applicable in the frame of the outside observer without any transformation of the local coordinates.

In this respect it should be mentioned that (see also the corresponding Newtionian case III.2) the relativistic formulation of the Lie-Koening theorem (Hill (1967)) does indeed allow the reduction of Eq.s (2.3a) to Poincaré-invariant equations (2.2a). However, this occurs only for local (as well as regular and analytic) systems and, as such, the reduction is not established for nonlocal systems.

Moreover, even when the reduction of Eq.s (2.3a) to (2.2a) holds, the transformations are necessarily nonlinear and noncanonical (*loc. cit.*). As such, they violate Einstein's special relativity because of the evident lack of preservation of inertial frames.

Finally, the transformed frames have a purely mathematical meaning (e.g., they imply frames in hyperbolic or transcendental trajectories).

Because of these reasons, the transformation of Poincaré-noninvariant systems into Poincaré-invariant forms shall be strictly prohibited in this volume. Only *after* the generalized symmetries and physical laws have been identified for systems (2.3) or (2.5) *in the physical frame of the observer*, then the use of the transformation theory may have a physical relevance.

IV.3: MINKOWSKI-ISOTOPIC SPACES

A fundamental methodological tool in the construction of the isotopies of Einstein's special relativity is the *geometrization of the inhomogeneuity and anisotropy of interior physical media*, which is done via the *Minkowski-isotopic spaces*, or *isominkowski spaces* of Sect. II.3. These isospaces also constitutes the first step in the quantitative study of relativistic, closed, nonhamiltonian system, inasmuch as they permit the identification of their internal carrier space.

Isominkowski spaces were originally introduced in Santilli (1983a) and then studied in more details in Santilli (1985a), (1988c), (1991a, c, d) under the conditions that the generalized interior spaces:

a) represent the generally inhomogeneous, anisotropic and nonlocal character of interior physical media;

b) admit the conventional Minkowski space as a particular case, and

c) preserve linearity locality of the transformation theory of the conventional Minkowski space in their isolinear and isolocal version, respectively, while the underlying transformation theory is intrinsically nonlinear and nonlocal.

It is important to review these results, particularly for the nonlocal profile. Recall that at the Newtonian level the transition from motion in vacuum (exterior dynamical problem) to motion within a physical medium (interior dynamical problem) is represented by an isotopy of the Euclidean metric, $\delta \Rightarrow \hat{\delta}$.

Our first assumption is therefore that the transition from relativistic motion in empty space to motion within a physical medium, is represented by an isotopy of the Minkowski metric

$$\eta = (\eta_{\mu\nu}) \quad \Rightarrow \quad \hat{\eta} = (\hat{\eta}_{\mu\nu}) = T\eta = (T_{\mu}{}^{\alpha})\,(\eta_{\alpha\nu}), \tag{3.1}$$

which is also called a *mutation* in the language of Sect. II.3. The new metric $\hat{\eta}$ is also called the *isometric* and T is called the *isotopic element*.

The sole conditions requested by the Lie-isotopic theory on the isometric $\hat{\eta}$ (or, equivalently, on the isotopic element T) are *nonsingularity* (invertibility) and *Hermiticity* (symmetric and real valuedness). The important point is that the functional dependence of the isometric is left completely unrestricted by the isotopies themselves, with the understanding that specific isosymmetries may imply specific restrictions.

We shall therefore assume that the interior metric $\hat{\eta}$ depends, in a generally nonlinear and nonlocal way, in the coordinates x, the velocities $u = dx/ds$, the index of refraction n, the density μ, the temperature τ, and any needed additional quantity (such as the accelerations $a = du/ds$),

$$\hat{\eta}_{\mu\nu} = \hat{\eta}_{\mu\nu}(x, u, a, \mu, \tau, n, ...) \tag{3.2a}$$

$$T_{\mu}{}^{\nu} = T_{\mu}{}^{\nu}(x, u, a, \mu, \tau, n, ...). \tag{3.2b}$$

The condition of nonsingularity implies the existence everywhere in the interior space of the inverse

$$\hat{1} = T^{-1} \tag{3.3}$$

which is here assumed as the the *isounit* of the theory (Sect. II.6).

We remain with the central conditions of achieving isolinearity and isolocality. This can be done from the results of Sect. II.3 via the following

DEFINITION IV.3.1 (Santilli (1983a)): The Minkowski-isotopic (or isominkowski) spaces $\hat{M}(x,\hat{\eta},\hat{\Re})$ are given by the infinite family of possible isotopes of the conventional Minkowski space $M(x,\eta,\Re)$, Eq.s (IV.2.1), with isotopic separation

$$\hat{M}(x,\hat{\eta},\hat{\Re}): \quad x = (x^{\mu}) = (r, x^4), \quad r \in \hat{E}(r,\hat{\delta},\hat{\Re}), \quad x^4 = c_o t, \tag{3.4a}$$

$$x^{\hat{2}} = (x^{\mu}\ \hat{\eta}_{\mu\nu}\ x^{\nu})\ \hat{1} \in \hat{\Re}, \quad \hat{\eta} = T\eta, \quad \hat{1} = T^{-1}, \tag{3.4b}$$

$$ds^{\hat{2}} = (-dx^{\mu}\ \hat{\eta}_{\mu\nu}\ dx^{\nu})\ \hat{1} = inv. \tag{3.4c}$$

where: the local coordinates x are unchanged as a central condition of isotopy, c_o represents the speed of light in vacuum; $ds^{\hat{2}}$ is the isotopic invariant; $\hat{E}(r,\hat{\delta},\hat{\Re})$ represents the isoeuclidean spaces of Sect. II.3; and the quantity $\hat{\Re}$ is the isofield

$$\hat{\Re} = \{\hat{N}\ |\ \hat{N} = N\ \hat{1},\ N \in \Re,\ \hat{1} = T^{-1}\}, \tag{3.5}$$

with elements $\hat{N}$ called isonumbers, verifying the isomultiplication law

$$\hat{N}_1 * \hat{N}_2 = (N_1 N_2)\ \hat{1} \tag{3.6}$$

and the ordinary sum $\hat{N}_1 + \hat{N}_2 = (N_1 + N_2)\ \hat{1}$.

As now familiar, the achievement of isolinearity is centrally dependent on the lifting of the field, $\Re \Rightarrow \hat{\Re}$. In fact, for a given, conventionally linear and local transformation $x' = Ax$ in $M(x,\eta,\Re)$, the corresponding transformations in $\hat{M}(x,\hat{\eta},\hat{\Re})$ are given by the *modular-isotopic action*

$$x' = A*x = ATx, \qquad T = \text{fixed}, \tag{3.7}$$

The above *isotransformations* are called *isolinear* (*isolocal*) in $\hat{M}(x,\hat{\eta},\hat{\Re})$ when A is conventionally linear (local) in $M(x,\eta,\Re)$, in which case they are linear (local) at the abstract, realization-free level. Nevertheless, the transformations are generally nonlinear (nonlocal) when projected in $M(x,\eta,\Re)$,

$$x' = AT(x, u,. a,..)x \tag{3.8}$$

By using Propositions II.4.1, II.4.2 and II.4.3, we therefore have the following

LEMMA IV.3.1: Under sufficient topological conditions, for any given nonlinear and/or nonlocal transformations

$$x \Rightarrow x'(x,..) \,, \tag{3.9}$$

in Minkowski space, there always exists an isotopic lifting

$$Mx,\eta,\Re) \Rightarrow \hat{M}(x,\hat{\eta},\hat{\Re}) \,, \tag{3.10a}$$

$$\hat{\eta} = T\eta, \quad \text{det. } T \neq 0, \quad T = T^{\dagger} \,, \quad \hat{\Re} = \Re \hat{1}, \quad \hat{1} = T^{-1}, \tag{3.10b}$$

under which the transformations can be identically written in an isolinear and/or isolocal form

$$x \Rightarrow x'(x,...) \equiv A{*}x = ATx = A\,T(x,...)\,x. \tag{3.11}$$

The reader should keep in mind the above property because the characterization of hadrons via our Lie-isotopic symmetries implies a *de facto* nonlinear and nonlocal treatment of strong interactions, although such to recover the linearity and locality of the electromagnetic interactions at the abstract coordinate-free level.

The reader should also keep in mind that the above isolinearity and isolocality are permitted by our isotopies of the Minkowski spaces, and, more particularly, by our use of the isofield (3.5). Also, the actual numbers of the theory are the ordinary numbers N because, from composition law (3.6), the isomultiplication of any quantity Q by an isonumber $\hat{N}$ coincides with the conventional one, i.e.,

$$\hat{N}{*}Q \equiv NQ. \tag{3.12}$$

In the following analysis we shall encounter the following quantities:

a) *isofourvectors* , which are ordinary four-vector although defined on $\hat{M}(x,\hat{\eta},\hat{\Re})$ and, therefore, with isocomposition (3.4b);

b) *isothreevectors* , which are also ordinary three-vectors, although defined on isospace $\hat{E}(r,\hat{\delta},\hat{\Re})$, where $\hat{\eta}$ is evidently restricted to the space component of the isometric $\hat{\eta}$; and

c) *isoscalas* , which are elements of $\hat{\Re}$. However, in all practical calculations we shall ignore isofields and use ordinary fields, owing to identities of the type

$$\tilde{\eta}_{\mu\nu} * \tilde{x}^{\nu} \equiv \hat{\eta}_{\mu\nu} x^{\nu}, \qquad \tilde{\eta} = \hat{\eta}\hat{I}, \qquad \tilde{x} = x\hat{I} \tag{3.13}$$

Evidently, there exist infinitely many isotopes of the Minkowski space because of the infinite possibilities of different interior physical media. In order to focus our attention on the most important ones, we introduce the following

DEFINITION IV.3.2 (Santilli (1988c)): The Minkowski isotopic spaces $\hat{M}(x,\hat{\eta},\hat{\Re})$ are classified into:
Spaces of Class I, denoted $\hat{M}^{I}$, *when the isometrics $\hat{\eta}$ preserve the topological properties of the Minkowsi metric, i.e., when the isotopic element T is positive-definite,*

$$T > 0, \tag{3.14}$$

and the spaces have null curvature, i.e., the Christoffel symbols of the second kind are identically null,

$$\Gamma^{\rho}_{\mu\nu} = \tfrac{1}{2}\, \hat{\eta}^{\rho\sigma} (\hat{\eta}_{\mu\sigma,\nu} + \hat{\eta}_{\sigma\nu,\mu} - \hat{\eta}_{\mu\nu,\sigma}) \equiv 0, \tag{3.15a}$$

$$\hat{\eta}_{\mu\nu,\sigma} = \partial\, \hat{\eta}_{\mu\nu} / \partial\, x^{\sigma}, \quad \mu, \nu, \rho, \sigma = 1, 2, 3, 4, \tag{3.15b}$$

Spaces of Class II, denoted $\hat{M}^{II}$, *when the spaces are still flat, i.e., verify conditions (3.15) above, but the isometrics $\hat{\eta}$ do not necessarily possess the topological structure of the original Minkowski metric η , i.e., the isotopic element has an arbitrary signature*

$$\text{Sig. } T = (\pm 1, \pm 1, \pm 1, \pm 1); \qquad (3.16)$$

Spaces of Class III, denoted $\hat{M}^{III}$, *when they are curved, i.e., they violate some or all conditions (3.15).*

$$\hat{\eta} = \hat{\eta}(x, ...), \qquad \partial\hat{\eta}/\partial x \neq 0, \qquad \Gamma^{\rho}_{\mu\nu} \neq 0. \qquad (3.17)$$

The Lie-isotopic analysis of this chapter will be conducted for isospaces $\hat{M}^{III}$ because they are the most general possible and therefore inclusive of all others, i.e.,

$$\hat{M}^{I} \subset \hat{M}^{II} \subset \hat{M}^{III}, \qquad (3.18)$$

Notice that isospaces M^{III} are considerably broader than conventional Riemannian spaces, e.g., because they depend on the velocities and are structurally nonlocal. Similarly, isospaces $\hat{M}^{III}$ are broader than the Finslerian spaces, evidently because of their general inhomogenuity; etc. Isospaces $\hat{M}^{III}$ are the fundamental spaces of our gravitational analysis, and they will be studied in detail in the next chapter.

Isospaces $\hat{M}^{II}$ will have a primary mathematical function, e.g., for the classification of all possible isotopies of the Lorentz group.

It is evident that all physically achievable mutations of the Minkowski metric are of Class I. In fact, suppose that an extended test particle originates its motion in the exterior space M with the familiar topology Sig η = (+, +, +, −). Suppose that the particle at a given point in its trajectory penetrates within a physical medium. Then, no physical event generated by the interior dynamics can possibly change the original topology.

Therefore, isotopies of the type Sig. η = (+, +, +, −) $\Rightarrow$ Sig. $\hat{\eta}$= (+, +, −, −) belonging to isospaces $\hat{M}^{II}$, are not physically realizable, resulting in fundamental condition (3.9) for isospaces $\hat{M}^{I}$. For these and other reasons, isospaces $\hat{M}^{I}$ are the fundamental isospaces for all our relativistic treatments of interior dynamics.

We now pass to the physical interpretation of the isospaces. For this purpose, the reader should first keep in mind the relationship between the original and the isotopic spaces.

PROPOSITION III.3.1 (Santilli 1983a): All infinitely possible isotopes $\hat{M}(x,\hat{\eta},\hat{\Re})$, $\hat{\eta} = T\eta$, $\hat{\Re} = \Re\hat{I}$, *of the conventional Minkowsky spaces* $M(x,\eta,\Re)$ *are locally isomorphic to the latter,* $\hat{M}(x,\hat{\eta},\hat{\Re}) \approx M(x,\eta,\Re)$, *when the isotopic element is the inverse of the isounit,* $\hat{I} = T^{-1}$, *and* $\hat{I} > 0$.

Stated differently, *our isotopies of the conventional Minnkowski space preserve the conventional axiomatic structure of space-time*. As we shall see in the next chapter, this mathematically simple property has rather important physical implications, e.g., for the geometric unification of the special and general relativities.

In fact, the isometrics $\hat{\eta}$ of the above proposition can indeed represent a *conventional* Riemannian metric g(x), thus permitting the representation of Einstein's gravitation as an isotope of the special relativity.

As a matter of fact, this occurrence is of considerable conceptual guidance in the identification of the physical meaning of our isominkowski spaces because they confirm their *geometrization of physical media, as a step inclusive of, but structurally more general than gravitation*.

To state this important point more explicitly, we know today that *the conventional Minkowski space-time $M(x,\eta,\Re)$ is modified by the presence of gravitation, resulting in the Riemannian spaces $R(x,g,\Re)$. Along similar conceptual lines, our isotopic theories indicate that the conventional space $M(x,\eta,\Re)$ is also modified by the presence of a physical medium, resulting in a broader space $\hat{M}(x,\hat{\eta},\hat{\Re})$ which may or may not inclusde gravitation, depending on the case at hand*

In fact, isominkowski spaces can readily achieve objectives a) at the beginning of this section and, in particular, they can provide a direct representation of:

a-1) the *inhomogenuity* of physical media, e.g., via a dependence of the isometric on the locally varying density;

a-2) the *anisotropy* of physical media, e.g., via a factorization in the isometric of a preferred direction in the medium itself such as that caused by intrinsic rotations; and

a-3) the *nonlocality* of the interior problem via integral realizations of the isounit.

As we shall see in more details in Sect. IV.10 and Chapter VII, the above properties generally result in a *space-time anisotropy*, e.g., $b_4 \neq b_3$, which is testable with current technology in a number of way.

To achieve quantitative predictions suitable for experiments, we shall proceed in stages, by beginning with the study in this section of isospaces of Class I.

Under the nonsingularity, Hermiticity and positive-definitness of the isounits, the isometrics $\hat{\eta}$ can always be diagonalized, resulting in the particular realization

$$\hat{M}^I(x,\hat{\eta},\hat{\mathfrak{R}}): \quad x = (x^\mu) = (r, x^4), \quad r \in \hat{E}(r,\hat{\delta},\hat{\mathfrak{R}}), \quad x_4 = c_o t, \tag{3.19a}$$

$$\hat{\eta} = \text{Diag.}\ (b_1{}^2, b_2{}^2, b_3{}^2, -b_4{}^2) = T\eta, \tag{3.19b}$$

$$T = \text{Diag.}\ (b_1{}^2, b_2{}^2, b_3{}^2, b_4{}^2) > 0, \tag{3.19c}$$

$$b_\mu = b_\mu(x, u, a, \mu, \tau, n, ...) > 0 \tag{3.19d}$$

$$x^{\hat{2}} = x^1 b_1{}^2 x^1 + x^2 b_2{}^2 x^2 + x^3 b_3{}^2 x^3 - t b_4{}^2 c_o{}^2 t, \tag{3.19e}$$

$$ds^{\hat{2}} = -dx^1 b_1{}^2 dx^1 - dx^2 b_2{}^2 dx^2 + dx^3 b_3{}^2 dx^3 - dx^4 b_4{}^2 dx^4 = \text{inv}. \tag{3.19f}$$

where, for simplicity, we shall hereon ignored the multiplicative term $\hat{I}$ in the latter two equations.

Our problem is now that of identifying the possible physical meaning of the quantities b_μ called *characteristic b-quantities of the medium considered*. First, we have to consider the following two primary alternatives as in the isoeuclidean case of Sect. III.3:

A) *local relativistic description* of a trajectory within a physicical medium, in which case the b's have the nonlinear and nonlocal dependence on the needed local variables indicated earlier, and are called *characteristics b-functions*; or

B) *global relativistic description* of the effect of a physical medium, such as the propagation of light through the entire Earth's atmosphere, in which case the b-quantities can be effectively averaged into constants as in Eq.s (III.3.56) and are called *characteristic b-constants* of the medium considered.

The physical interpretation in each of the above cases can be best done by considering the b-quantities as a natural generalization of the original, corresponding quantities in $M(x,\eta,\mathfrak{R})$.

For this purpose, consider the following alternative formulation

$$b_\mu = \frac{1}{n_\mu}, \quad \mu = 1, 2, 3, 4 \tag{3.20a}$$

$$c = c_o\, b_4 = c_o / n_4, \tag{3.20b}$$

$$x^{\hat{2}} = x^1 \frac{1}{n_1^2} x^1 + x^2 \frac{1}{n_2^2} x^2 + x^3 \frac{1}{n_3^2} x^3 - t \frac{c_o^2}{n_4^2} t\,. \tag{3.20c}$$

where rthe n's can be assumed to be functions (costants) for local (global) descriptions.

Then, by recally that c_o represents the speed of light in vaccum, *the most natural meaning of the generalized quantity* $c = c_o b_4 = c_o/n_4$ *is the representation of the speed of light (or, more generally, of any electromagnetic wave) in the physical medium considered,* under the evident assumption that it is transparent to electromagnetic waves.

In this case we have the simple geometrical interpretation $b_4 = 1/n_4$, where n_4 is the local function representing index of refraction, when interested in the speed of light at one given point in the medium considered, or it is averaged to a constant, when interested in global aspects (e.g., the average speed of light when passing through our entire atmosphere).

However, at the classical level, physical media are generally opaque to light as well as to all electromagnetic waves although still permitting relativistic motions (this is the case, say, for metals which are opaque to light but not to electrons). In this latter case the term $\hat{\eta}_{44} = - b_4^2$ has a purely geometrical meaning, conceptually similar to the geometrical meaning of the element g_{44} of a Riemannian geometry (see Sect. IV.10 for more details).

The space elements $\hat{\delta}_{kk} = b_k^2$ have the same interpretation of the corresponding elements in the isoeuclidean space $\hat{E}(r,\hat{\delta},\hat{\mathfrak{R}})$ (see the remarks at the end of Sect. III.3), such as the representation of:

1) the actual shape of the object considered;

2) the infinitely possible deformations of the original shape; or

3) the nonlinear, nonlocal and nonhamiltonian interactions of the interior dynamics.

For these and all other aspects of the space component of the isometrics, we refer the reader to Chapter III.

Similarly, in Sect. III.7 we have introduced the notion of *nonrelativistic isoparticle,* as an extended particle under both potential (selfadjoint) and nonhamiltonian (nonselfadjoint) interactions which can acquire an infinite number of intrinsic characteristics depending on the local physical conditions, and which are represented by the infinite number of *isogalilean symmetries* $\hat{G}(3.1)$. Another objective of the isominkowski space is to permit the identification of the relativistic notion of isoparticle (Sect. IV.7).

A number of additional meanings of isospaces $\hat{M}^I(x,\hat{\eta},\hat{\Re})$, and of isometrics $\hat{\eta}$ will emerge in specific applications of the theory, e.g., in particle physics, superconductivity, etc. At this point, we would like to indicate the *application of isominkowski spaces for the geometrization of hadrons, as suggested by phenomenological studies on the behavior of the meanlife of unstable hadrons with speed.*

As well known, it appeared for decades that such a behavior is at variance with the Einsteinian behavior precisely in the expectation that the internal structure is nonlocal (see, e.g., Kim (1978) and quoted papers).

The first phenomenological predictions of departures from the special relativity are those by Blockhintsev (1964), Redei (1966), and others.

Kim (*loc. cit.*) formulated specific predictions of violation at given energies via the use of quantum field theory and the loss of the canonical commutation rules caused by nonlocal internal interactions. He also pointed out the expectation that the (still unresolved) origin of violation of discrete symmetries may also be the same as that of the anomalous meanlife, and both due to the internal nonlocality of hadrons.

More recently, H.B.Nielsen and I. Picek (1983) studied the Higgs sector of the spontaneous symmetry breaking for the interior of pions and kaons within the context of unified gauge theories, and suggested the following modification of the Minkowski metric

$$\hat{\eta} = \{(1 - \alpha/3), (1 - \alpha/3), (1 - \alpha/3), -(1 + \alpha)\}. \tag{3.21}$$

where the parameter α, called *"Lorentz-asymmetry parameter"*, has the following value for pions

$$\alpha = (-3.79 \pm 1.37) \times 10^{-3}, \tag{3.22}$$

and the *different* value for kaons

$$\alpha = (+0.61 \pm 0.17) \times 10^{-3}. \tag{3.23}$$

The above structure constitutes a clear illustration of our

isominkowski spaces (Santilli (1983a)), for the case of a global description where the characteristic b-quantities are averaged into costants, yielding $b_1 = b_2 = b_3 = 1 - \alpha/3$, $b_4 = 1 + \alpha$, with numerical values (3.22) and (3.23).

It should be stressed that values (3.22) and (3.23) are a mere *first approximation for low energies* . In fact, the subsequent measures by Aronson *et al.* (1983) for kaons at energies 30-100 GeV confirmed the expected *nonlinear* dependence of the meanlife with the speed (as well as for other quantities). Thus, when averaged to constants via techniques of type (III.3.56), the results by Aronson *et al.* (*loc. cit.*) are different than those by Nielsen and Picek (*loc. cit.*).

We learn in this way that the globalization of the characteristic b-functions into constants for hadrons should be referred to a specific energy range, because is not expected to be necessarily valid for arbitrary energies.

However, the more recent experiments by Grossman *et al.* (1987) in the range 100-350 GeV have shown no apparent deviation of the behavior of the meanlife with speed from the Einsteinian predictions.

The issue is predictably far from being settled, as elaborated by Cardone *et al.* (1992a, b), who have shown that *the experiments by Grossman et al. (loc. cit.) and those by Aronson et al (loc. cit.) are endered compatible by our isominkowskian representations of the kaons*.

These experimental aspects will be considered in Chapter VII. In this section we would like to point out that our isotopic liftings of the special relativity offer a number of new possibilities for the study of the problem, in addition to that of compatibilities of the indicated diverging data.

First, let us recall that *our nonselfadjoint models (IV.2.5) verify all Einsteinian laws in the exterior center-of-mass behavior, including the behavior of the meanlife with energy* (Corollary IV.2.1.1). Thus, *hadronic structures represented via models (IV.2.5) verify all the data of the experiment by Grossman et al. (1987).*

But the interior structure of model (IV.2.5) is generalized. Thus, our studies indicate the possibility that

> *All phenomenological and experimental investigations on the noneinsteinian behavior of the meanlife of unstable hadrons with energy reviewed above, may characterize generalized* ***interior*** *metrics $\hat{\eta}$ in such a way to be compatible with the exact,* ***exterior,*** *Einsteinian behavior along the notion of*

closed, relativistic, nonhamiltonian systems considered here.

In different terms, the phenomenological and experimental studies considered above do indeeed achieve a generalized interior structure of nonselfadjoint type, but they should be complemented with

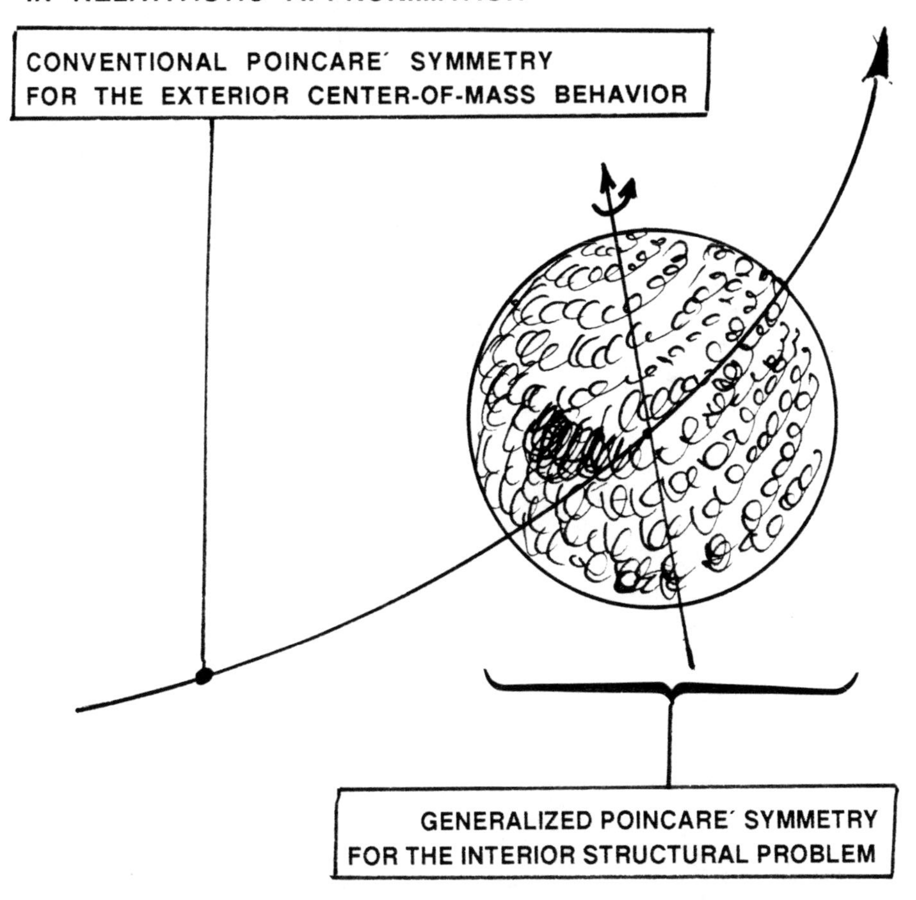

FIGURE IV.1: As recalled in Fig. I.1.1, the origin of all contemporary relativities can be identified with Galilei's first visual inspection of the Jovian system back in 1609. Exactly the same situation occurs for the isogalilean relativities of Chapter III, as well as for the isospecial relativities studied in this chapter, as majestically permitted today by close-up pictures of Nasa's Planetary Missions. This time, however, the attention is focused on Jupiter's structure conceived as a relativistic gas of extended molecules, each one in motion within the medium composed by all others. This results in the closed, relativistic, nonhamiltonian systems of Sect. IV.2 which permit compatibility between conventional relativities of the center-of-mass behavior of Jupiter in the Solar system, and generalized relativities for Jupiter's structure. A primary objective of this chapter is to provide technical means for a quantitative treatment of the relativistic effects caused by such an interior medium, which is mathematically rigorous as well as effective for practical calculations. This is achieved via the isominkowski spaces (Definition IV.3.1) for the characterization of inhomogeneous and anisotropic interior media and, in particular, by the isotopies of the Minkowski metric $\eta \Rightarrow \hat{\eta} = T\eta$, $T = \text{diag.}(b_1^2, b_2^2, b_3^2, b_4^2) > 0$, where the isotopic element T or, equivalently, the b-quantities, have an explicit dependence, in general, on all possible local variables and quantities. A hierarchy of isometrics then emerges as conceivable:

1) isometrics $\hat{\eta}$ with constant b-quantities, evidently intended to represent the interior medium in first approximation;

2) isometric $\hat{\eta}$ with b-functions on the coordinates, e.g., to provide a representation of the change of the density with the distance from the center;

3) isometrics $\hat{\eta}$ with b-functions with an explicit dependence in the velocities, as a first dynamical representation of the drag forces typical of interior motions;

4) isometric $\hat{\eta}$ with b-functions with a nonlinear and nonlocal dependence on all possible variables and quantities, as a representation of the interior dynamics; and, last but not least,

5) isometrics $\hat{\eta}$ with a general nonlinear and nonlocal b-functions, although averaged over to b-constants.

We are here referring to the fact thatthe description of specific, local events, such as the trajectory of a space-ship during penetration in Jupiter's atmosphere, must have an explicit nonlinear and/or nonlocal dependence on all local quantities, as well known. Isometrics of type 3) and 4) above are then appropriate. However, there may exist other events, such as the propagation of light within inhomogeneous and anisotropic (transparent) media, which do not necessarily need a nonlinear and nonlocal dependence, but which can be well treated via a suitable global average of the isometrics 3) and 4) into b-constants. This is exactly the case for the Doppler redshift predicted by our isospecial relativities for motion of light within physical media (Sect. IV.9), as apparently verifiable with classical experiments (Chapter VII)

subsidiary constraints to achieve compatibility with the Einsteinian center-of-mass behaviour.

The first way to illustrate this possibility is with the studies by Nielsen and Picek (1983) which result in generalized isometric (3.21). The violation of Einstein's special relativity under such a metric is evident and need no comments.

Prior to our studies on isotopies, generalized metrics of type (3.21) have only one possible interpretation, that they imply a necessary noneinsteinian behavior of the meanlife with speed and other consequences studies later on in this chapter (loss of the Lorentz symmetry).

By keeping in mind the open character of the final experimental resolution whether or not the meanlife of unstable hadrons is Einsteinian, our isotopies establish that the studies by Blockhintsev ()1964), Redei (1966), Kim (1978), Nielsen and Picek (1983), Aronson *et al.* (1983), and other can characterize interior, generalized isometrics irrespective of whether the exterior behaviour of the meanlife is Einsteinian or not.

Moreover, the papers considered illustrate the fact that, *while the Minkowski space is unique, it admits infinitely many different isotopes*. In fact, *the isometrics change in the transition from from pions, Eq.s (3.21), to kaons, Eq.s (3.22)*. This result can be also reached independently from the behavior of the meanfile. In fact, all hadrons have approximately the same charge radius R = 1F which does not appreciably increase with mass. This implies that different hadrons have different densities, resulting in different interior physical media and, therefoire, differenmt isominkowskian represen-tations.

We can therefore state that

> *The isometrics or, equivalently,the isounits of hadrons generally vary from particle to particle* (Santilli (1983a)).

In turn, this illustrates the infinite number of different isogalileanalilean generalizations of the Galilei's symmetry of the preceding chapter, and indicates the expectation of the corresponding existence of an infinite number of isopoincaré generalizations of the Poincaré symmetry, as treated in Sect. IV.6.

Thus, the possible representation of hadrons as closed nonselfadjoint systems implies (from phenomenological data such as the increase of density with mass, the different local values of the spontaneous symmetry breakings in unified gauge theories, and other data) the apparent existence of different models (IV.2.5) for different hadrons, with different isounits and, consequently, different isosymmetries.

As a final comment, let us recall that in conventional special special relativities physical systems are characterized by only one function, the Hamiltonian, while in our isospecial relativities physical systems are characterized by the Hamiltonian and the underlying isounit.

We should then recall that *the special relativity cannot possibly predict the numerical value of a Hamiltonian, but provides only general laws for its characterization* . Exacly along the same lines, *the isospecial relativities cannot possibly predict the numerical values of the Hamiltonian and of the isounit, but provide only general laws for their characterization* . This can be seen from a more general viewpoints according to which *no geometry can possibly predict the numerical value of its own unit*, which must therefore be identified via experiments.

In summary, *our isotopic liftings $\eta \Rightarrow \hat{\eta} = T\eta$ of the conventional Minkowski metric η can be first interpreted as a form of geometrization of the physical media of the interior problem (Sect. IV.10), whose explicit meaning depends on the characteristics of the medium at hand, and generally varies from case to case. Moreover, except for nonsingularity and Hermiticity, the isometrics $\hat{\eta}$ have the most general conceivable or otherwise known, nonlinear, nonlocal and nonhamiltonian dependence on all possible variables and quantities, thus achieving the unification of all existing phenomenological generalizations in one geometrical structure. Finally, owing to their arbitrariness, the isometrics $\hat{\eta}$ include as particular cases the conventional Riemannian, Finslerian, Minkowskian or other metrics.*

IV.4: RELATIVISTIC LIE-ISOTOPIC FORMULATIONS

After having identified the internal carrier space of systems (IV.2.3) or (IV.2.5), our second step is the identification of the mathematical methods needed for their effective treatment.

These methods are evidently given by the *relativistic Lie-isotopic formulations* , hereinafter called *isorelativistic formulations* , in their algebraic, geometric and analytic branches, first introduced in Santilli (1988c), which is the basis of this review. For additional studies, one may consult Santilli (1991c,d).

In the following we shall first outline the relativistic Lie-isotopic formulations for local systems, and then point out their extension for nonlocal interactions. We shall also first present the geometric profile and then the algebraic and analytic aspects, because geometry provides the most direct formulation of the integrability conditions for the existence of the desired algebraic and analytic structures. A knowledge of Sect.s II.6, II.7 and II.9 is a necessary pre-requisite for the content of this section.

The *geometrical foundations* of isorelartivistic mechanics in *local-differential approximation* are provided by the relativistic generalization of the symplectic geometry on the *conventional* cotangent bundle $T^*M(x,\eta,\Re)$. By using the local coordinates of the unified form

$$a = (a^i) = (x,p) = (x^\mu, p^\mu), \quad i = 1, 2, ..., 8, \quad \mu = 1, 2, 3, 4 \tag{4.1}$$

the main structural notions are the *relativistic Pfaffian one-form*

$$\Theta_1 = R_i \, da^i , \quad R_i = R_i(a), \tag{4.2}$$

and the *relativistic, exact, symplectic two-form*

$$\Omega_2 = d\Theta_1 = d(R_i \, da^i) = \Omega_{ij} \, da^i \wedge da^j, \tag{4.3}$$

where

$$\Omega_{ij} = \partial_i R_j - \partial_j R_i, \quad \partial_i = \partial / \partial a^i, \tag{4.4}$$

is the nowhere degenerate *covariant Birkhoff's tensor*, with corresponding *contact extensions* on $\Re_s \times T^*M(x,\eta,\Re)$ not considered here for brevity (see footnote[7] p. 91).).

A relativistic vector-field Γ on $T^*M(x,\eta,\Re)$ is then called a (global) *Birkhoffian vector field* when it verifies the rule

$$\Omega_2 \rfloor \Gamma = - dB. \tag{4.5}$$

where B is called the *relativistic Birkhoffian*, in order to differentiate it from the *relativistic Hamiltonian* H.

The *analytic foundations* are characterized by the *general relativistic Pfaffian principle*, here expressed with respect to a generic independent variable s in the *semiautonomous form* (see below),

$$\delta \hat{A} = \delta \int_1^2 ds [R_i(a) \, \dot{a}^i - B(s, a, ..)] = 0, \quad \dot{a} = da/ds, \tag{4.6}$$

where one can recognize in the integrand the Pfaffian one-form (4.2) (while its contact extension would represent the entire integrand).

Principle (4.6) readily yields the following *relativistic covariant Birkhoff's equations*

$$\Omega_{ij}(a)\,\frac{da^j}{ds} = \frac{\partial B(s, a, ...)}{\partial a^i} \qquad (4.7)$$

with consequential *relativistic, Birkhoffian, Hamilton-Jacobi equations*

$$\frac{\partial \hat{A}}{\partial s} + B = 0, \qquad \frac{\partial \hat{A}}{\partial a^i} = R_i. \qquad (4.8)$$

The remaining aspects of the relativistic Birkhoffian mechanics then follows and must be here necessarily assumed for brevity.

The *algebraic foundations* are characterized by the following *relativistic covariant Birkhoff's equations*

$$\frac{d a^i}{d s} = \Omega^{ij}(a)\,\frac{\partial B(s, a, ...)}{\partial a^j}, \qquad (4.9)$$

where

$$\Omega^{ij} = (\,|\,\Omega_{rs}\,|^{-1}\,)^{ij}, \qquad (4.10)$$

is called the *relativistic contravariant Birkhoff's tensor.*

The algebraic structure of the theory is then characterized by the brackets

$$[A \,\hat{,}\, B] = \frac{\partial A}{\partial a^i}\,\Omega^{ij}(a)\,\frac{\partial B}{\partial a^j}, \qquad (4.11)$$

which are *Lie-isotopic brackets* in the sense of verifying the axioms of the Lie-isotopic algebras (Sect. II.5), i.e.

$$[A \,\hat{,}\, B) + [B \,\hat{,}\, A] = 0, \qquad (4.12a)$$

$$[[A \,\hat{,}\, B] \,\hat{,}\, C] + [[B \,\hat{,}\, C] \,\hat{,}\, A] + [[C \,\hat{,}\, A] \,\hat{,}\, B] = 0, \qquad (4.12b)$$

as guaranteed by the symplectic character of two-form (4.3) (see Sect. II.9).

The exponentiated form of Eq.s (4.9) is then readily given by

$$a' = \{e^{s_o \Omega^{ij}(a)(\partial_j B)(\partial_i)}\} a, \tag{4.13}$$

and results to be a *Lie-isotopic group* (Sect. II.6). The relativistic formulation of the Lie-isotopic theory then follows.

By ignoring for a moment the subsidiary constraints (see also Appendix IV.B), the conventional relativistic formulations are recovered identically for $R = R^\circ = (p, 0)$, for which Birkhoffian one-forms (4.2) and two-forms (4.3) recover the conventional, relativistic, canonical, one- and two-forms, respectively,

$$\Theta_{1|R=R^\circ} = \theta_1 = p_i \, da^i, \quad R^\circ = (p, 0), \tag{4.14a}$$

$$(\Omega_{ij})_{|R=R^\circ} = (\omega_{ij}) = \begin{pmatrix} 0 & -I \\ I & 0 \end{pmatrix}, \quad (\Omega^{ij})_{|R=R^\circ} = (\omega^{ij}) = \begin{pmatrix} 0 & I \\ -I & 0 \end{pmatrix} \tag{4.14b}$$

Birkhoff's equations (4.7) and (4.9) then assume the conventional relativistic Hamiltonian form in our unified notation (again without the multipliers)

$$\omega^{ij} \frac{\partial a^j}{ds} = \frac{\partial H(s, a, ...)}{\partial a^i}, \tag{4.15a}$$

$$\frac{da^i}{ds} = \omega^{ij} \frac{\partial H(s, a, ...)}{\partial a^j}, \tag{4.15b}$$

while the Lie-isotopic brackets (4.11) assume the conventional canonical form

$$[A \hat{,} B]_{|R=R^\circ} = \frac{\partial A}{\partial a^i} \omega^{ij} \frac{\partial B}{\partial a^j}$$

$$= \frac{\partial A}{\partial x^\mu} \eta_{\mu\nu} \frac{\partial B}{\partial p^\nu} - \frac{\partial B}{\partial x^\mu} \eta_{\mu\nu} \frac{\partial A}{\partial p^\nu}, \tag{4.16}$$

The above relativistic formulations are, however, *inadequate* for our needs on a number of counts, and they do not constitute the desired "isorelativistic" formulations. First, they are based on the conventional symplectic geometry. As such, *the formulations can indeed represent nonlinear and nonhamiltonian systems, but only in local approximation.* This is evidently due to the strictly local topology of the assumed geometry discussed earlier.

Second, we need formulations which can be readily "hadronized" into an operator form with wavefunctions independent from the linear momenta, while action (4.6) is excessively general and would lead to "wavefunctions" dependent on the linear momenta (see the comments in footnote[10], p. 94).

Third, we need formulations defined on our isominkowski space $\hat{M}_1(x,\hat{\eta},\hat{\mathfrak{R}})$, while the preceding formulations are on a conventional space.

Fourth, we need a classical relativistic geometry which permits the treatment of nonlocal interactions and, as such, it is the true geometrical counterpart of the Lie-isotopic algebras.

Finally, we need isorelativistic formulations which contain explicitly identified isounits directly in the geometric two-forms, the algebraic brackets and the analytic equations, so as to readily identify the particular equations of motion considered, by assigning the Hamiltonian and the isounit.

All these objectives can be achieved via the *relativistic isosymplectic geometry* first introduced in Santilli (1988c).

Consider the iso*relativistic phase space (cotangent bundle)* $T^*\hat{M}_1(x,\hat{\eta},\hat{\mathfrak{R}})$ characterized by the *isorelativistic one-isoforms*

$$\hat{\Theta}^\circ_1 = \theta_1 \times \hat{T}_1 = (R^\circ \times \hat{T}_1)_i \, \hat{d}a^i = \hat{R}^\circ_i(a)\, \hat{d}a^i = p_\mu \, T_1{}^\mu{}_\nu(s, x, p,..)\, \hat{d}x^\nu, \quad (4.17a)$$

$$\hat{R}^\circ = (p_\mu, 0) = (pT_1, 0), \quad (4.17b)$$

where all nonlocal, as well as nonlinear and nonhamiltonian interactions are factorized in the isotopic element $\hat{T}_1$.

The *isorelativistic two-isoforms* then hold iff all nonlinear, nonlocal and nonhamiltonian terms are factorizable, this time, in an isotopic element $\hat{T}_2$ multiplied by the the relativistic canonical two-form along structure (II.9.86), i.e.,

$$\hat{\Phi}_2 = \hat{d}\hat{\Phi}_1 = \frac{\partial (R^\circ_{i_1} T^{i_1}{}_{j_1})}{\partial a^{i_2}} T^{i_2}{}_{j_2} \, \hat{d}a^{j_1} \hat{\wedge} \hat{d}a^{j_2} =$$

$$= \left(\frac{\partial R^\circ_{i_1}}{\partial a^{i_2}} T^{i_1}{}_{j_1} T^{i_2}{}_{j_2} + R^\circ_{i_1} \frac{\partial T^{i_1}{}_{j_1}}{\partial a^{i_2}} T^{i_2}{}_{j_2} \right) \hat{d}a^{j_1} \hat{\wedge} \hat{d}a^{j_2} =$$

$$= \tfrac{1}{2}\delta^{j_1 j_2}_{\ \ k_1 k_2} \left(\frac{\partial R^\circ_{i_1}}{\partial a^{i_2}} T^{i_1}_{\ j_1} T^{i_2}_{\ j_2} + R^\circ_{i_1} \frac{\partial T^{i_1}_{\ j_1}}{\partial a^{i_2}} T^{i_2}_{\ j_2} \right) \hat{d}a^{k_1} \hat{\wedge} \hat{d}a^{k_2} =$$

$$= \tfrac{1}{2} [\, \omega_{i_1 i_2} T^{i_1}_{\ j_1} T^{i_2}_{\ j_2} + (R^\circ_{i_1} \frac{\partial T^{i_1}_{\ j_1}}{\partial a^{i_2}} T^{i_2}_{\ j_2} - R^\circ_{i_2} \frac{\partial T^{i_2}_{\ j_2}}{\partial a^{i_1}} T^{i_1}_{\ j_1}) \hat{d}a^{j_1} \wedge \hat{d}a^{j_2}$$

$$\stackrel{\text{def}}{=} \tfrac{1}{2} \omega_{j_1 k} T_2{}^{k}{}_{j_2} \hat{d}a^{j_1} \wedge \hat{d}a^{j_2},$$

while being defined on isospace

$$T^*\hat{M}_2(x,\hat{\eta},\hat{\mathfrak{R}}) \equiv T^*\hat{M}(x,\hat{g},\hat{\mathfrak{R}}): \qquad \hat{g} = T_2\eta, \qquad \hat{1}_2 = \hat{T}_2^{-1}, \qquad (4.19a)$$

$$T_2 = (T_2{}^{k}{}_{j_2}) = \qquad (4.19b)$$

$$= \{ \omega^{j_1 k} [\omega_{i_1 i_2} T^{j_1}_{\ j_1} T^{i_2}_{\ j_2} + (R^\circ_{i_1} \frac{\partial T^{i_1}_{\ j_1}}{\partial a^{i_2}} T^{i_2}_{\ j_2} - R^\circ_{i_2} \frac{\partial T^{i_2}_{\ j_2}}{\partial a^{i_1}} T^{i_1}_{\ j_1})] \}$$

where the new metric $\hat{g}$ must be computed from the old metric $\hat{\eta}$ via rules (4.18) (see Sect. II.9 for details), and T_2 is assumed to be symmetric $T_2{}^{i}{}_{j} = T_{2j}{}^{i} = T_{2i}{}^{j}$.

As one can see, structure (4.18) allows the preservation of the local-differential topology of the symplectic geometry characterized by its canonical two-form ω, which can represent all conventional relativistic, local, differential and potential models of contemporary physics. The factorized isotopic terms $\hat{T}_2$ can then represent an infinite class of nonlinear, nonlocal and nonhamiltonian interactions, in essentially the same way as it occurred at the Newtonian level of the preceding chapter.

Analytically, the symplectic-isotopic geometry implies a *restriction* of Birkhoffian systems, from the most general possible Pfaffian principle (4.6) on $T^*M(x,\eta,\mathfrak{R})$, to the particular form on $T^*\hat{M}_1(x,\hat{\eta},\hat{\mathfrak{R}})$

$$\delta\hat{A}^\circ = \delta\int_1^2 [\hat{R}^\circ(a) * da - B(s, a,...) \odot ds]$$

$$= \delta\int_1^2 [p_\mu T_1{}^{\mu}{}_{\nu}(x, p,...) dx^\nu - B(s, x, p, ...) T_t ds] = 0. \qquad (4.20)$$

The *isorelativistic, covariant, Hamilton-isotopic equations* are then given by

$$\omega_{ir} \, T_2{}^r{}_j(a) \frac{d a^j}{d s} = \frac{\partial B(s, a, ...)}{\partial a^i} \tag{4.21}$$

where the new tensor Ω°_{ij} is given by Eq.s (4.18).

The *isorelativistic, contravariant Hamilton-isotopic equations*, under the assumption of diagonal isounits $\hat{1}_2 = (\hat{1}_2{}^i{}_j) = \mathrm{diag.}\{ ({}_2{}^{\mu}{}_{\nu}), (\hat{1}_2{}^{\mu}{}_{\nu}) \}$, is evidently given by

$$\frac{da^i}{ds} = \omega^{ik} \, \hat{1}_{2k}{}^j \frac{\partial B}{\partial a^j} = \begin{cases} \dfrac{dx^{\mu}}{ds} = \hat{1}_2{}^{\mu}{}_{\nu}(s,x,p,...) \dfrac{\partial B(s, x, p)}{\partial p_{\nu}}, \\ \dfrac{dp_{\mu}}{ds} = - \hat{1}_{2\mu}{}^{\nu}(s,x,p,..) \dfrac{\partial B(s, x, p)}{\partial x^{\nu}}, \end{cases} \tag{4.22}$$

where one can recognize *the identification of the relativistic isounit $\hat{1}_2$ of the Lie-isotopic theory directly in the structure of the analytic equations.*

The algebraic structure is then characterized by the brackets

$$[A \hat{,} B] = \frac{\partial A}{\partial a^i} \, \omega^{ir} \, \hat{1}_{2r}{}^j(a) \, \frac{\partial B}{\partial a^j}$$

$$= \frac{\partial A}{\partial x^{\mu}} \hat{1}_2{}^{\mu}{}_{\nu} \frac{\partial B}{\partial p_{\nu}} - \frac{\partial B}{\partial x^{\mu}} \hat{1}_2{}^{\mu}{}_{\nu} \frac{\partial A}{\partial p_{\nu}}, \tag{4.23}$$

whose Lie-isotopic structure is ensured by the symplectic-isotopic character of two-forms (4.18). The *fundamental relativistic isotopic brackets* are the given by

$$([a^i \hat{,} a^j]) = (\hat{\Omega}^{\circ ij}) = \begin{pmatrix} [x^{\mu} \hat{,} x^{\nu}] & [x^{\mu} \hat{,} p_{\nu}] \\ [p_{\mu} \hat{,} x^{\nu}] & [p_{\mu} \hat{,} p_{\nu}] \end{pmatrix} = \begin{pmatrix} (\delta^{\mu\nu}) & \hat{1}_2{}^{\mu}{}_{\nu}) \\ - (\hat{1}_{2\mu}{}^{\nu}) & (\delta_{\mu\nu}) \end{pmatrix}, \tag{4.24}$$

and they predictably play a fundamental role in the operator formulation of isorelativistic theories (Santilli (1989), (1991d)).

The exponentiated structure of time evolution (4.22) is given by

$$a' = \{ e^{s_o \omega^{ik} \hat{1}_{2k}{}^{j}(a)\,(\partial_j B)\,(\partial_i)} \hat{1}_2 \} * a. \tag{4.25}$$

which exhibits more clearly the Lie-isotopic structure of the underlying group as well as of the transformation theory (Sect. IV.3).

Finally, Eq.s (4.8) become the *relativistic isotopic Hamilton-Jacobi equations*

$$\frac{\partial \hat{A}^\circ}{ds} + B = 0, \quad \frac{\partial \hat{A}^\circ}{\partial x^\mu} = p_\alpha T_1{}^\alpha{}_\mu, \qquad \frac{\partial \hat{A}^\circ}{\partial p^\mu} = 0, \tag{4.26}$$

namely, *the isotopic action $\hat{A}^\circ$ is independent from the velocities,* as needed for the operator formulation of the theory.

The latter occurrence can be better understood on geometric grounds. In fact, the one-isoforms in the integrand of principle (4.20) can be written

$$p_\mu T_1{}^\mu{}_\nu da^\nu \equiv p_\mu\, dx^\nu = p^\mu\, dx_\nu, \tag{4.27}$$

because of the general properties of isominkowski spaces $\hat{M}(x,\hat{\eta},\mathfrak{R})$

$$p_\mu x^\nu \equiv p_\mu \hat{\eta}^{\mu\nu} x_\nu \equiv p_\mu\, \hat{1}_2{}^\mu{}_\alpha\, \eta^{\alpha\nu} x_\nu \equiv$$

$$\equiv p^\mu\, \hat{\eta}_{\mu\nu}\, x^\nu \equiv p^\mu\, \eta_{\mu\alpha}\, T_1{}^\alpha{}_\nu\, x^\nu, \tag{4.28a}$$

$$(\, \hat{\eta}^{\mu\nu}\,) = (\hat{\eta}_{\alpha\beta}\,)^{-1}, \quad \hat{\eta}_{\mu\alpha}\, \hat{\eta}^{\alpha\nu} = \delta_\mu{}^\nu. \tag{4.28b}$$

Equations $\partial \hat{A}^\circ / \partial p^\mu \equiv 0$ then follow from property (4.27), as in the conventional case, because all the nonlocal as well as velocity dependent terms of the integrand are embedded in the isometric of the theory.

Note that properties (4.28) arise from, and illustrate rather clearly the local isomorphisms $M(x,\eta,\mathfrak{R}) \approx \hat{M}(x,\hat{\eta},\hat{\mathfrak{R}})$.

Again, as it was the case for the nonrelativistic formulations, *the relativistic isosymplectic geometry is the true geometrical structure underlying the relativistic Lie-isotopic algebras* . In fact, in both geometry and algebra we can incorporate all nonlocal terms in the explicitly identified isounit.

Next, we *restrict* relativistic, closed Birkhoffian system to be well defined in $T^*\hat{M}_2(x,\hat{g},\hat{\mathfrak{R}})$, i.e., to admit the representation in terms of reduced action (4.20) or, equivalently, in terms of the reduced Lie-isotopic equations (4.22).

This can be achieved via one of the various methods of Sect.s II.7 and II.9 for the construction of the Birkhoffian representation from given equations of motion.

The reader should be aware in this respect that these methods generally result in the so-called *nonautonomous relativistic representations* on $T^*M(x,\eta,\Re)$

$$\delta A = \delta \int_1^2 ds\, [R_i(s, a, ...)\, \dot{a}^i - B(s, a, ...)] = 0, \qquad (4.29)$$

that is, representations with R = R(s, a,..) and B = B(s, a,..), even when the original equations are not expl;icitly dependent on the independent parameter s.

But, *the relativistic nonautonomous Birkhoff's equations characterized by principle (4.27) do not admit any consistent algebraic structure,* let alone that of the Lie-isotopic algebras exactly as it happen for the nonrelativistic case of Appendix II.A.

This implies the *first necessary restriction* of systems (IV.2.3a) or (IV.2.5a) of admitting a reduction to the *semiautonomous representations* (4.6), that is, those with R = R(a) and B = B(s, a, ...). The reduction can be readily accomplished via one of the several degrees of fredom of Birkhoffian mechanics, such as via the *relativistic Birkhoffian gauge transformations*

$$R'_i(a) = R_i(s, a,..) + \frac{\partial G(s, a, ...)}{\partial a^i}, \qquad (4.30a)$$

$$B'(s, a, ...) = B(s, a, ...) - \frac{\partial G(s, a, ...)}{\partial s}, \qquad (4.30b)$$

Once representation (4.6) is reached, the *second necessary restriction* is that of admitting only systems (IV.2.3a) with reduced representations (4.20). This can alse be done via gauges (4.28) and other means. Note that this step identifies the isometric $\hat{\eta}$ of the reduced Pfaffian one-form.

The *third restriction* is that of admitting only reduced representations (4.20) with a symplectic-isotopic structure (4.18), by identifying in this way the isometric $\hat{g}$ of the reduced Lie-isotopic brackets (4.23). This third step can also be done via the use of the degrees of freedom of the theory.

Once the representation of systems (IV.2.3a) is reached via the Lie-admissible equations (4.22) on $T^*\hat{M}(x,\hat{g},\hat{\Re})$, we are finally equipped for the study of their space-time symmetries.

The following comments appear recommendable. First, we should note that, *while Birkhoffian mechanics is directly universal for all regular, local, analytic and nonhamiltonian systems (IV.2.3a) (Santilli (1982a)), its isorelativistic form characterized by Eq.s (4.22) is not expected to be directly universal.*

However, the attentif reader has noted that in the preceding presentation we have introduced the simplest possible realization of the Birkhoffian-isotopic and symplectic-isotopic formulations, those of Hamiltonian-isotopic character with the factorization of the canonical form ω, while the most general possible formulations is that with the factorization of the Birkhoffian forms Ω (see Sect.s II.7 and II.9 for details).

The latter, more general mechanics is expected to possess a direct universality for all possible systems (IV.2.5) in their full nonlocal form. This property is only conjectured here for separate technical treatment at some future time.

The relativistic Hamilton-isotopic formulations presented in this section are amply sufficient for our needs. In fact, they are characterized by a conventional Hamiltonian H which represents all conventional relativistic models of current use, plus our isounits $\hat{1}_2$ which can represent all needed nonlinear, nonlocal and nonhamiltonian interactions.

We should also note that the analytic representations of conventional relativistic theories are *degenerate* (because the Lagrangian is homogeneous of second degree in the velocities, see Dirac (1964)) and *constrained* (because of invariant constraint (2.1d)).

By comparison, isorelativistic representations (4.22) are *regular* (because the Lagrangian is not necessarily homogeneous of second degree in the velocities) but *constrained* (because of the subsidiary constraints (3.4d) *as well as* (2.3b) and (2.3c)).[17]

As a result of this situation, our relativistic, closed, nonhamiltonian systems generally require only the methods of *Lagrange's multipliers* (Appendix A) and not necessarily *Dirac's methods for subsidiary constraints* (Appendix B).

The consequences are rather important for symmetries and conservation laws. In fact, unlike Dirac's constraints, Lagrange's multipliers can be ignored in the study of the symmetries, evidently because they essentially add regular (nondegenerate) degrees of

[17] Explicitly, conventional Lagrangians are of the type $L = (-\dot{x}^{\mu}\eta_{\mu\nu}\dot{x}^{\nu})^{\frac{1}{2}}$, while the isotopic Lagrangians are of the generalized type $\hat{L} = (-\dot{x}^{\mu}\hat{\eta}_{\mu\nu}x^{\nu})^{\frac{1}{2}} = [-\dot{x}^{\mu}\hat{\eta}_{\mu\nu}(x, \dot{x}, \ddot{x},...)\dot{x}]^{\frac{1}{2}}$, as we shall see in more details later on.

freedom in the brackets.

Finally, the reader should be aware that *the isosymplectic geometry has been formulated in this section for isominkowski spaces of the most general possible Class III,* that is, for generally *curved isospaces* $\hat{M}(x,\hat{g},\mathfrak{R})$ whose isometrics $\hat{g}$ are not necessarily positive-definite. Such a general formulation will be primarily useful in the next chapter. In this chapter we shall use the formulation on isospaces of Class I, in which case we shall tacilty imply the additional restriction $\hat{g} > 0$.

IV.5. CONSTRUCTION OF THE LORENTZ-ISOTOPIC SYMMETRIES

The infinite family of *Lorentz-isotopic symmetries*, also called *isolorentz symmetries* $\hat{O}(3.1)$ on isominkowski spaces, was constructed for the first time in Santilli (1983a) in their most general possible abstract form, in particular, as abstract isotopes of the compact orthogonal symmetry O(4) on a 4-dimensional Euclidean space with metric δ = diag. (1,1,1,1). Their classical realization was presented for the first time in Santilli (1988c), which is followed in this review. Additional studies can be found in Santilli (1991c).

The construction of $\hat{O}(3.1)$ as isotopes of O(4) is mathematically most effective, e.g., for the unification of all possible simple six-dimensional Lie groups of Cartan's classification (see later on in this section). However, the approach implies isometrics $\hat{\eta} = T\delta$ characterized by the (nonsingular and Hermitean) isotopic elements T which are not positive-definite. In turn, this context creates un-necessary problems in the operator formulation of the theory (whereby the conventional positive-definite unit of quantum mechanics has to be replaced by an isounit of unnecessarily undefined topology with evident problematic aspects, e.g., in the measurement theory).

In this section we shall instead construct the isolorentz symmetries $\hat{O}(3.1)$ in such a way that their isounit can be positive-definite. This requires their necessary construction as isotopes of the conventional Lorentz symmetry O(3.1).

In particular, we shall first construct the isolorentz symmetries in the most general possible isominkowski spaces of Class III, and then restrict the analysis to isospaces of Class I. This sets the basis for a more adequate treatment of further advances, such as gravitation or operator formulations, as we shall see.

DEFINITION IV.5.1 (Santilli (1983a), (1988c)): The abstract Lorentz-isotopic (or isolorentz) symmetries are defined on the Minkowski-isotopic spaces of unspecified physical interpretation

$$\hat{M}^{III}{}_2(x,\hat{\eta},\hat{\Re}) = \hat{M}(x,\hat{g},\hat{\Re}): \ x = (x^\mu) = (r, x^4), \ r \in \hat{E}_2(r,\hat{G},\hat{\Re}), \ x^4 = c_o t \tag{5.1a}$$

$$x^{\hat{2}} = x^\mu \hat{g}_{\mu\nu} x^\nu = x^1\hat{g}_{11}x^1 + x^2\hat{g}_{22}x^2 + x^3\hat{g}_{33}x^3 + x^4\hat{g}_{44}x^4, \tag{5.1b}$$

$$\hat{g} = T_2\, \eta = \text{Diag.}\,(\hat{g}_{11}, \hat{g}_{22}, \hat{g}_{33}, \hat{g}_{44}), \tag{5.1c}$$

$$\eta = \text{diag.}\,(1, 1, 1, -1), \quad T_2 = \text{diag.}\,(\hat{g}_{11}, \hat{g}_{22}, \hat{g}_{33}, -\hat{g}_{44}), \tag{5.1d}$$

$$\hat{g}_{\mu\mu} = \hat{g}_{\mu\mu}(x, \dot{x}, \mu, \tau, n, ...) \neq 0 \ \text{ and real-valued}, \tag{5.1e}$$

$$\hat{\Re} = \Re\, \hat{1}_2, \ \hat{1}_2 = T_2^{-1}, \tag{5.1.f}$$

and are given by the isotransformations

$$x' = \hat{\Lambda} * x = \hat{\Lambda}\, T_2(x,\dot{x},...)\, x, \ T_2 = \text{fixed}, \tag{5.2}$$

under the conditions that they form a simple six-dimensional Lie-isotopic group $\hat{O}(3.1)$ with isotopic laws

$$\hat{\Lambda}(w) * \hat{\Lambda}(w') = \hat{\Lambda}(w') * \hat{\Lambda}(w) = \hat{\Lambda}(w+w'), \ w \in \Re; \tag{5.3a}$$

$$\hat{\Lambda}(0) = \hat{\Lambda}(w) * \hat{\Lambda}(-w) = \hat{1}_2 = T_2^{-1}, \tag{5.3b}$$

and leave invariant isoseparation (5.1b).

The above transformations are called "abstract" because the individual elements $\hat{g}_{\mu\mu}$ can be either positive or negative, thus characterizing either compact or noncompact groups. The definition has been conceived to be effective for the classification of all possible isotopes $\hat{O}(3.1)$, as well as because directly applicable to gravitational models (evidently for a well defined topology, e.g., $\hat{g}_{kk} > 0$, $\hat{g}_{44} < 0$).

The (necessary and sufficient) conditions for isotransformations (5.2) to leave invariant isoseparation (5.1b) are given by

$$\hat{\Lambda}^t \hat{g}\, \hat{\Lambda} = \hat{\Lambda}\, \hat{g}\, \hat{\Lambda}^t = \hat{g}^{-1}, \tag{5.4}$$

or, equivalently,

$$\hat{\Lambda}^t T_2 \eta \hat{\Lambda} = \hat{\Lambda} T_2 \eta \hat{\Lambda}^t = \hat{1}_2 \eta, \qquad (5.5)$$

To obtain the conditions in a more explicit form, suppose that the original Lorentz transformations $x' = \Lambda x$ are realized with the familiar expressions $x'^\mu = \Lambda^\mu{}_\alpha x^\alpha$, e.g., as in Schweber (1962). Then, the isotopic element and isounit can be written

$$T_2 = (T_2{}^\alpha{}_\beta) = (T_{2\beta}{}^\alpha), \qquad \hat{1}_2 = (\hat{1}_{2\alpha}{}^\beta) = (\hat{1}_2{}^\beta{}_\alpha), \qquad (5.6a)$$

$$\hat{1}_{2\alpha}{}^\beta T_{2\beta}{}^\gamma = \delta_\alpha{}^\gamma. \qquad (5.6b)$$

Lifting (5.2) can be written

$$x'^\mu = \hat{\Lambda}^\mu{}_\alpha T^\alpha{}_\beta x^\beta, \qquad (5.7)$$

and conditions (5.4) can be written explicitly

$$\hat{\Lambda}_\alpha{}^\beta T_\beta{}^\rho \eta_{\rho\sigma} \hat{\Lambda}^\sigma{}_\tau = \hat{1}_\alpha{}^\delta \eta_{\delta\sigma}. \qquad (5.8)$$

THEOREM IV.5.1 (Santilli (loc. cit.)): The abstract isolorentz symmetries on isospaces $\hat{M}^{III}(x,\hat{g},\hat{\Re})$ leave invariant the isoseparations

$$x'^{\hat{2}} = x'^\mu \hat{g}_{\mu\nu} x'^\nu = x^\mu \hat{g}_{\mu\nu} x^\nu = x^{\hat{2}}, \qquad (5.9)$$

or, more explicitly,

$$x'^\mu \hat{g}_{\mu\nu}[x(x',p',..), p(x', p',..)] x'^\nu \equiv x^\alpha \hat{g}_{\alpha\beta}(x,p,...) x^\beta. \qquad (5.10)$$

with nonsingular, Hermitean, sufficiently smooth and diagonal isometrics

$$\hat{g} = T_2 \eta, \quad \eta \in M(x,\eta,\Re), \qquad (5.11)$$

under the sole condition that the isounits $\hat{1}_2$ are the inverse of the mutation elements T_2. All the infinitely possible isosymmetries $\hat{O}(3.1)$ admit the connected semisimple subgroups

$$S\hat{O}(3.1): \quad \det.(\hat{\Lambda}\,\hat{g}) = +1, \qquad (5.12)$$

as well as the discrete invariant subgroups

$$\hat{\phi}(3.1): \quad \text{Det.}\ (\hat{\Lambda}\hat{g}) = -1, \tag{5.13}$$

and possess the following classical realization in the isocotangent bundle $T{*}\hat{M}^{III}(x,\hat{g},\hat{\Re})$ *with local coordinates* $a = (a^i) = (x^\mu, p^\mu)$, $\mu = 1, 2, ..., 4$, $i = 1, 2, ..., 8$:

1) the same (ordered set of) parameters of the conventional symmetry O(3.1), i.e., the Euler's angles θ *and Lorentz boosts* w

$$u = (w_k) = (\theta, w), \quad k = 1, 2, ..., 6, \tag{5.14}$$

2) the same (ordered set of) generators of O(3.1)

$$J = (J_k) = (J_{\mu\nu}) = (J_\chi, L_\chi), \tag{5.15a}$$

$$J_{\mu\nu} = x_\mu p_\nu - x_\nu p_\mu, \quad J_i = \epsilon_{ijl} J_{ji}, \quad L_\chi = J_{\chi 4}, \tag{5.15b}$$

$$k = 1, 2, ..., 6, \quad \mu, \nu = 1, 2, 3, 4, \quad \chi = 1, 2, 3,$$

3) the isocommutation rules of the Lie-isotopic algebra $\hat{\mathbf{O}}(3.1)$ *of* $\hat{O}(3.1)$ *in terms of brackets (IV.4.23)*

$$\hat{\mathbf{O}}(3.1): \ [J_{\mu\nu} \hat{,} J_{\alpha\beta}] = \hat{g}_{\nu\alpha} J_{\beta\mu} - \hat{g}_{\mu\alpha} J_{\beta\nu} - \hat{g}_{\nu\beta} J_{\alpha\nu} + \hat{g}_{\mu\beta} J_{a\nu}, \tag{5.16}$$

4) with local isocasimir invariants

$$\hat{C}^{(o)} = \hat{I}_2, \quad \hat{C}^{(1)} = (J_{\mu\nu} J^{\mu\nu})\, \hat{I}_2, \quad \hat{C}^{(2)} = (\epsilon_{\mu\nu\alpha\beta} J^{\mu\nu} J^{\alpha\beta})\, \hat{I}_2 \tag{5.17}$$

5) the Lie-isotopic group for the connected component

$$S\hat{O}(3.1): \quad a' = \hat{\Lambda}(u) * x$$

$$\stackrel{\text{def}}{=} \{[\textstyle\prod_k e_{|\xi}^{u_k \omega^{iq} \hat{I}_{2q}{}^{j}) (\partial_j J_k) (\partial_i)}] \hat{I}_2\} * a = \hat{S}_{\hat{g}}(u)\, x, \tag{5.18}$$

6) the invariant discrete subgroup $\hat{\phi}(3.1)$ *characterized by the isoinversions*

$$\hat{\phi}(3.1): \quad \hat{P}*x = P\,x = (-r, x^4), \tag{5.19a}$$

$$\hat{T}*x = T x = (r, -x^4), \quad (PT)x = (-r, -x^4), \tag{5.19b}$$

where P and T are the ordinary inversions, and

7) isosymmetries $\hat{O}(3.1)$ admit as maximal compact forms the orthogonal group in four dimension O(4) as well as all its infinitely possible isotopies.

PROOF. Conditions (5.4) are verified iff

$$[\mathrm{Det}\,(\hat{\Lambda})]^2 = [\mathrm{Det}\,\hat{1}_2]^2 = [\mathrm{Det}\,(\hat{g}^{-1})]^2, \tag{5.20}$$

Properties (5.12) and (5.13) then follow. The preservation of the conventional parameters (5.14) and generators (5.15) is a central feature of the Lie-isotopic theory (Sect. II.6). Isocommutation rules (5.16) trivially follow from the use of the Lie-isotopic product (IV.4.23) for generators (5.15). Isocasimir invariants (5.17) then hold, *locally*, in the sense elaborated for the $\hat{O}(3)$ symmetry in Sect. III.3. Lie-isotopic group structure (5.18) holds as an expansion in the conventional associative algebra ξ of vector-fields. Isoinversions are the same as those of the original abstract derivation (Santilli (1983a)). Finally, the maximal compact subgroup O(4) trivially holds for T_2 = diag. (1,1,1,−1), $\hat{g} = T_2\eta$ = diag. (1,1,1,1). Q.E.D.

In the above presentation, we have assumed the reader is familiar with a number of aspects of the Lie-isotopic theory we cannot possibly review here, such as: the fact that at the level of the abstract (matrix) representation the isounits $\hat{1}$ are an intrinsic part of the isocasimirs, while in classical realizations only their coefficients isocommute with all generators; the isotopic generalization of the *Poincaré-Birkhoff-Witt Theorem* (which ensures the existence and consistency of group structure (5.18)); the isotopic formulation of the *Baker-Campbell-Hausdorff Theorem* (which ensures that products (5.18) still belong to $S\hat{O}(3.1)$); the general theorem on isosymmetries of Sect. II.8 (which ensures the construction of $\hat{O}(3.1)$ from the sole knowledge of the original symmetry and of the new metric); etc.

The following comments are now in order:

1) While the Minkowski space $M(x,\eta,\mathfrak{R})$ with trivial unit I is unique, there exist infinitely many possible isospaces $\hat{M}^{III}(x,\hat{g},\hat{\mathfrak{R}})$ with isounits $\hat{1}_2 = \hat{T}_2^{-1}$ because they represent the infinitely many possible, interior, inhomogeneous and anisotropic physical media (Sect. IV.3);

2) While the isolorentz O(3.1) is unique, there exist infinitely many

possible isolorentz symmetries $\hat{O}(3,1)$ characterized by the infinitely many possible isounits $\hat{I}_2$, which all possess the same dimension and simplicity of O(3.1);

3) In the same way as the Lorentz invariance cannot identify the explicit value of a Lagrangian, the invariance under the isolorentz symmetries cannot identify the isometrics, which must be computed from the given local physical conditions of the interior medium at hand;

4) While the Lorentz transformations are unique, there exist an infinite number of different isolorentz transformations (see below for examples), characterized by the Lie-isotopic structure (5.18);

5) Each one of the infinitely possible isolorentz transformations can be computed in an explicit finite form via expansion (5.18), whose convergence is assured by the assumed topological conditions (and essentially reduces to that of the conventional expansions), with the understanding that the explicit computation of the infinite series is not expected to be necessarily simple;

6) Each of the infinitely many isolorentz transformations can be computed via the sole knowledge of the old parameters and generators and of the new metric (or, equivalently, of the new unit);

7) The isolorentz transformations are formally isolinear and isolocal on $\hat{M}^{III}(x,\hat{g},\hat{\Re})$, but generally nonlinear and nonlocal in $M(x,\eta,\Re)$;

8) The lifting of the conventional symmetry O(3.1) into the isotopes $\hat{O}(3.1)$ implies the generalization of the *structure constants* of the conventional formulation of Lie's theory into the *structure functions* of the Lie-isotopic theory (Sect. II.6);

9) Except for the needed nonsingularity and Hermiticity (invertibility, reality and symmetric characters), the isolorentz symmetries $\hat{O}(3.1)$ leave completely unaffected the explicit functional dependence of the isometrics;

10) The classical realization of the isolorentz symmetries can indeed admit nonlocal (integral) forms, provided that they are all embedded in the isounit $\hat{I}_2$, as permitted by the underlying isosymplectic geometry (Sect. IV.4);

12) The isometrics $\hat{g}$ of isosymmetries $\hat{O}(3.1)$ can be, as particular cases, *conventional* Riemannian metrics. Therefore, Theorem IV.5.1 provides methods for the explicit construction of the (generally nonlinear) symmetries of conventional gravitational metrics such as the Schwarzschild's metric (see next chapter for details).

We now study the conditions for the local isomorphism $\hat{O}(3.1) \approx O(3.1)$. Note that, even though isocommutation rules (5.18) appear to coincide with the conventional commutation rules of O(3.1) (see, e.g., Eq. (30), p. 41 of Schweber (1962)), they are generally *different,* e.g., because the topology of the isometric $\hat{g}_{\mu\nu}$ is different than that of the Minkowski metric η.

The following property was proved for the abstract case, and the same proof trivially holds for the classical realization of this section.

THEOREM IV.5.2 (loc. cit.): All abstract isolorentz symmetries $\hat{O}(3.1)$ on isospaces $M^{IIi}(x,\hat{g},\hat{\Re})$ with invariant separation (5.1b) are locally isomorphic to the conventional Lorentz symmetry O(3.1) under the sole condition that the isometrics $\hat{g} = T_2\, \eta$ possess the same topological properties of the Minkowski metric η, e.g., whenever the isotopic elements T_2 or the isounits $\hat{I}_2 = T_2^{-1}$ are positive-definite; otherwise, depending on the topology of the isounits, the Lorentz-isotopic symmetries $\hat{O}(3.1)$ are locally isomorphic to any other simple six-dimensional group of Cartan's classification, such as O(4) or O(2.2).

Note that the positive definiteness of the isotopic element T_2 holds in a number of conventional gravitational models. The Lie-isotopic theory has therefore the remarkable capacity of pointing out the following property which is here presented in preparation of our gravitational studies of the next chapter.

COROLLARY IV.5.2.1 (loc. cit.): Einstein's Gravitation or any other gravitational theory (not necessarily Riemannian) with metric $\hat{g} = T\eta$, $T > 0$, admits the conventional Lorentz symmetry as a general[18] isotopic symmetry.

We now pass to the construction of the explicit form of the *abstract isolorentz transformations.* The general form of the

[18] The term "general" is here referred to the symmetry of a full gravitational line element, rather than its local/tangent symmetry, and it is distinguished from the term "global" in its conventional topological sense.

transformations for the case of the $\hat{O}(3)$ subgroups has been computed in Sect. III.3, and it will be tacitly implied hereon.

We shall therefore restrict our attention only to the *abstract isolorentz boosts*, also called *isoboosts*, in the (x^3, x^4)-plane with parameter w and generator J_{34}. Their most general possible form computed from the Lie-isotopic group (5.18) for isometrics (5.1c) is given by (Santilli (*loc. cit.*))

$$x' = \hat{\Lambda}(w) * x = \hat{S}_{\hat{g}}(w)\, x$$

$$= \begin{pmatrix} 1 & 0 & 0 & 0 \\ 0 & 1 & 0 & 0 \\ 0 & 0 & \cos(w\, \hat{g}_{33}^{\frac{1}{2}}\, \hat{g}_{44}^{\frac{1}{2}}) & -(\hat{g}_{44}/\hat{g}_{33})^{\frac{1}{2}} \sin(w\, \hat{g}_{33}^{\frac{1}{2}}\, \hat{g}_{44}^{\frac{1}{2}}) \\ 0 & 0 & (\hat{g}_{33}/\hat{g}_{44})^{\frac{1}{2}} \sin(w, \hat{g}_{33}^{\frac{1}{2}}\, \hat{g}_{44}^{\frac{1}{2}}) & \cos(w\, \hat{g}_{33}^{\frac{1}{2}}\, \hat{g}_{44}^{\frac{1}{2}}) \end{pmatrix} \begin{pmatrix} x^1 \\ x^2 \\ x^3 \\ x^4 \end{pmatrix} \tag{5.21}$$

Note that the elements of the isometric are completely unrestricted in their functional dependence in the above derivation. We have therefore proved the following

LEMMA IV.5.1 (loc. cit.): The abstract isolorentz transformations in the (3,4)-plane on isominkowski spaces of Class III, Eq.s (5.1), are given, in their broadest possible form, by

$$\begin{cases} x'^1 = x^1, \\ x'^2 = x^2, \\ x'^3 = x^3 \cos(w\, \hat{g}_{33}^{\frac{1}{2}}\, \hat{g}_{44}^{\frac{1}{2}}) - x^4 (\hat{g}_{44}/\hat{g}_{33})^{\frac{1}{2}} \sin(w\, \hat{g}_{33}^{\frac{1}{2}}\, \hat{g}_{44}^{\frac{1}{2}}), \\ x'^4 = x^3 (\hat{g}_{33}/\hat{g}_{44})^{\frac{1}{2}} \cos(w\, \hat{g}_{33}^{\frac{1}{2}}\, \hat{g}_{44}^{\frac{1}{2}}) + x^4 \cos(w\, \hat{g}_{33}^{\frac{1}{2}}\, \hat{g}_{44}^{\frac{1}{2}}) \end{cases} \tag{5.22}$$

where the isometric elements $\hat{g}_{33}$ and $\hat{g}_{44}$ have the most general possible nonlinear and nonlocal dependence on all needed variables and quantities

$$\hat{g}_{33} = \hat{g}_{33}(x, \dot{x}, \mu, \tau, n, \ldots), \quad \hat{g}_{44} = \hat{g}_{44}(x, \dot{x}, \ddot{x}, \mu, \tau, n, \ldots), \tag{5.23}$$

subject only to the conditions of being sufficiently smooth, nowhere null and real valued.

The following particular cases are important.

COROLLARY IV.5.1.1 (loc. cit.): The abstract isolorentz transformations admit, as particular cases, the conventional rotations in four dimensions, trivially, for

$$\hat{g}_{11} = \hat{g}_{22} = \hat{g}_{33} = \hat{g}_{44} = 1, \tag{5.24}$$

and the conventional Lorentz transformations (see also below) for

$$\hat{g}_{11} = \hat{g}_{22} = \hat{g}_{33} = -\hat{g}_{44} = 1, \tag{5.25}$$

It is evident that the abstract Lorentz-isotopic transformations also admit an infinite family of isorotations and Lorentz-isotopic transformations for nontrivial values of the isometric elements.

Note that, according to the above results, *the conventional rotations in four-dimensions are Lorentz-isotopic transformations in an isospace of Class III with isometric (5.24).*

We now pass to the study of the particular subclass of isolorentz transformations that are physically relevant for the isotopies of the special relativity.

DEFINITION IV.5.2 (loc. cit.): The abstract $\hat{O}(3.1)$ symmetries (or transformations) are called "general Lorentz-isotopic (or isolorentz) symmetries" when they are defined in the most general possible isospaces of Class I of the diagonal form

$$\hat{M}^I{}_2(x,\hat{\eta},\hat{\mathfrak{R}}) \equiv \hat{M}^I(x,\hat{g},\hat{\mathfrak{R}}) : \; x^{\hat{2}} = x^\mu \hat{g}_{\mu\nu} x^\nu$$

$$= x^1 \hat{b}_1{}^2 x^1 + x^2 \hat{b}_2{}^2 x^2 + x^3 \hat{b}_3{}^2 x^3 - x^4 \hat{b}_4{}^2 x^4, \tag{5.26a}$$

$$\hat{g} = T_2\, \eta, \quad \eta = \text{diag.}\,(1, 1, 1, -1), \tag{5.26b}$$

$$T_2 = \text{diag.}\,(\hat{b}_1{}^2, \hat{b}_2{}^2, \hat{b}_3{}^2, \hat{b}_4{}^2) > 0, \tag{5.26c}$$

$$\hat{b}_\mu = \hat{b}_\mu(x, \dot{x}, \mu, \tau, n, \ldots) > 0, \tag{5.26d}$$

$$\hat{1}_2 = T_2{}^{-1} > 0; \tag{5.26e}$$

and they are called "restricted" when defined on isospaces of Class I of the constant diagonal form

$$\hat{M}^I_2(x,\hat{\eta},\Re) \equiv \hat{M}^I(x,\hat{\eta},\hat{\Re}) : \quad x^{\hat{2}} = x^\mu \hat{\eta}_{\mu\nu} x^\nu, \tag{5.27a}$$

$$\hat{\eta} = T_2 \eta = \text{diag.} (b_1^2, b_2^2, b_3^2, -b_4^2), \tag{5.27b}$$

$$T_2 = \text{diag.} (b_1^2, b_2^2, b_3^3, b_4^2) > 0, \tag{5.27c}$$

$$b_\mu = \text{constants} > 0, \tag{5.27d}$$

Similar definitions hold for the general and special isoransformations on isospaces of Class II.

The reader should keep in mind the notation here adopted: whenever using generic elements $\hat{g}_{\mu\mu}$ we are referring to the "abstract" Lorentz-isotopic transformations; whenecver using the elements $\hat{b}_\mu^{\ 2}$ we are referring to the "general" isolorentz transformations; and, finally, whenever using the elements $b_\mu^{\ 2}$ we are referring to the "special" isotransformations.

Since we now deal with a physically identified space, we can assume for parameter w its conventional physical meaning given by a speed v along the third axis, w = v. By recomputing again infinite series (5.18) for isometric (5.26b), the general Lorentz-isotopic transformations on $\hat{M}^I(x,\hat{g},\Re)$ can be written

$$x' = \hat{\Lambda}(v) * x = \hat{S}_{\hat{g}}(v)\, x =$$

$$= \begin{pmatrix} 1 & 0 & 0 & 0 \\ 0 & 1 & 0 & 0 \\ 0 & 0 & \cosh(v\,\hat{b}_3\,\hat{b}_4) & -(\hat{b}_4/\hat{b}_3)\sinh(v\,\hat{b}_3\,\hat{b}_4) \\ 0 & 0 & -(\hat{b}_3/\hat{b}_4)\sinh(v\,\hat{b}_3\,\hat{b}_4) & \cosh(v\,\hat{b}_3\,\hat{b}_4) \end{pmatrix} \begin{pmatrix} x^1 \\ x^2 \\ x^3 \\ x^4 \end{pmatrix} \tag{5.27}$$

But the functional dependence of the $\hat{b}$-quantities is completely unrestricted in the above derivation. We have therefore proved the following

LEMMA IV.5.2.2 (loc. cit.) : The general isolorentz transformations in isospaces $\hat{M}^I_2(x,\hat{g},\hat{\Re})$, in their broadest possible form characterized by a (local) speed v along the third axis, are given by

$$\begin{cases} x'^1 = x^1, & (5.28a) \\ x'^2 = x^2, & (5.28b) \\ x'^3 = x^3 \cosh(v\,\hat{b}_3\,\hat{b}_4) - x^4 \dfrac{\hat{b}_4}{\hat{b}_3} \sinh(v\,\hat{b}_3\,\hat{b}_4), & (5.28c) \\ x'^4 = -x^4 \dfrac{\hat{b}_3}{\hat{b}_4} \sinh(v\,\hat{b}_3\,\hat{b}_4) + x^4 \cosh(v\,\hat{b}_3\,\hat{b}_4). & (5.28d) \end{cases}$$

Note that we have eliminated in the above derivation the isotopic character of the transformations for simplicity of expressions. Nevertheless, the mathematically correct form remains that of isotransformations (5.2).

The proof that isotransformations (5.28) are indeed a particular case of broader isotransformations (5.22), is an instructive exercise for the interested reader, because it implies delicate topological aspects in the transition from compact to noncompact settings.

The local isomorphism between isotransformations (5.28) and the conventional Lorentz transformations is evident, owing to the positive-definitness of the $\hat{b}$-quantities. In fact, we have the following property.

THEOREM IV.5.3 (loc. cit.): All infinitely possible, general, isolorentz symmetries $\hat{O}(3.1)$ on isominkowski spaces of Class I, Eq.s (5.26), are locally isomorphic to the conventional Lorentz symmetry $O(3.1)$.

It is then evident that *the general isolorentz symmetries constitute isotopic coverings of the Lorentz symmetry,* in the sense that: a) the formers are constructed with methods broader than those of the latter; b) the formers represent physical conditions broader than those of the latter; and c) the formers all contain the latter as a particular case.

The following property has rather intriguing and novel physical implications for the isotopies of both the conventional and general relativities.

LEMMA IV.5.3 (loc. cit.): All infinitely possible general or special isolorentz transformations in the (3-4)-plane of isominkowski spaces of Class I can be written in the form

$$\begin{cases} x'^1 = x^1, & (5.29a) \\ x'^2 = x^2, & (5.29b) \\ x'^3 = \hat{\gamma}\,(x^3 - \beta x^4), & (5.29c) \\ x'^4 = \hat{\gamma}\,(x^4 - \hat{\beta} x^3), & (5.29d) \end{cases}$$

where[19]

$$\beta = v/c_o, \quad \hat{\beta} = \frac{v b_3}{c_o b_4}, \quad \hat{\beta}^2 = \hat{\beta^2} = \frac{v^k b_k{}^2 v^k}{c_o b_4{}^2 c_o}, \tag{5.30a}$$

$$\cosh\,(v\, b_3\, b_4) = \hat{\gamma} = |1 - \hat{\beta^2}|^{-\frac{1}{2}}, \tag{5.30b}$$

$$\sinh\,(v\, b_3\, b_4) = \hat{\beta}\,\hat{\gamma}. \tag{5.30c}$$

The above formulation of the isolorentz tranformations is important inasmuch as it exhibits rather clearly their abstract identity with the conventional Lorentz transformations (owing to the identity $\hat{\beta}^2 = \hat{\beta^2}$). Thus, at the level of realization-free formulations, all distinctions between isotopic and conventional Lorentz transformations cease to exist, exactly as desired.

The most important difference between the general and special isotransformations is pointed out by the fopllowing

COROLLARY IV.5.3.1 (loc. cit.): While the general isolorentz transformations are nonlinear and nonlocal in all variables, the special isotransforations are linear and local in all variables.

This property will have evidently relevant physical implications in the liftings of the special relativity, e.g., to preserve the inertial character of the observers whenever needed.

The following additional property should be recalled here.

---

[19] The author would like to thank E. Ferrari of the Phys. Dept. "G. Marconi" of the Univ. "La Sapienza" in Rome, Italy, for bringing to his attention the original erroneous form $\hat{\gamma} = (1 - \hat{\beta}^2)^{-\frac{1}{2}}$ during a seminar delivered at the Math. Dept. "G. Castelnuovo" of the same University, thanks to a kind invitation by Prof. G. Caricato, which is here acknowledged with sincere gratitude. In fact, certain physical media imply $\hat{\beta}^2 > 1$ even for speeds $v \ll c_o$ (see Sect. IV.10), thus implying imaginary values of $\hat{\gamma}$. This aspect was resolved during discussions with R. Mignani of the same Phys. Dept. following the seminar, and resulted in the expression $\hat{\gamma} = |1 - \hat{\beta}^2|^{-\frac{1}{2}}$ based on the paper Recami-Mignani (1972). Prior to these discussions, the general form of the isolorentz transformations was thought to be (5.28). These discussions essentially implied that all possible isotransformations (5.28) can be cast in form (5.29), including those with $\hat{\beta}^2 > 1$. In turn, this has important physical implications discussed later owing to the abstract identity of isotransformations (5.29) with the convenbtional ones.

COROLLARY IV.5.3.2 (loc. cit.): The general (or special) isoorentz symmetries $\hat{O}(3.1)$ on isospaces $\hat{M}^I(x,\hat{g},\hat{\Re})$ can reconstruct as exact at the isotopic level all conventional breakings of the Lorentz symmetry, under the sole condition that the underlying generalized metrics $\hat{g} = T_2\,\eta$ preserve the topology of the conventional Minkowski metric η, i.e., $T_2 > 0$.

An example is provided by the deformation of the Minkowski metric worked out by H. B. Nielsen and I. Picek (1983) in the interior of pions and kaons, Eq.s (IV.3.21), where the α-quantity was called the "Lorentz-asymmetry parameter". As one can see, the Lorentz symmetry is exact for metric (IV.3.31), provided that it is not realized in terms of the simplest conceivable product (IV.4.16), but in terms of our lesser trivial isotopic product (IV.4.23), with

$$T_2 = \text{diag.}\,[(1 - \alpha/3), (1 - \alpha/3), (1 - \alpha/3), (1 + \alpha)] \qquad (5.31)$$

Similar results hold for all possible physically achievable deformations (also called *mutations*) of the Minkowski metric, those of Class I. Theorem IV.5.3 can therefore be called a *technique for reconstructing the exact Lorentz symmetry when believed to be conventionally broken.*

It should be indicated that, while the Lorentz symmetry remains exact for all interior generalizations of the isominkowski of Class I, this is evidently not the case for isospaces of Class II and III.
Other isolorentz transformations can be computed for any given isotopic elements T_2. The combination of the isorotations $\hat{O}(3)$ of Sect. III.3 and the general or special Lorentz-isotopic transformations of this section is here left as an exercise for the interested reader.

It is also instructive to show that the abstract, general and special isotransformations do indeed leave invariant isoseparations (5.1b), (5.26a) and (5.27a), respectively. It is finally suggested to the interested reader to verify that *all general isolorentz transformations in the (3,4) plane can be cast in form (5.29).* We can equivalently say that *the geometrization of the interior dynamical problem characterized by $\hat{O}(3.1)$ on isospaces $\hat{M}^I(x,\hat{g},\hat{\Re})$ can be unified in form (5.29).*

A few comments elaborating the differences between the construction of the $\hat{O}(3.1)$ symmetries of this chapter and the original derivation of Santilli (1983a) are recommendable to prevent possible misinterpretations. To begin, the reader should note that the original derivation used the isounit $\hat{1} = \hat{g}^{-1}$, while in this chapter we used the form $\hat{1} = T_2^{-1}, \hat{g} = T_2 \eta$.

This is due to the fact that the original derivation followed the general theorem of isosymmetries of Sect. II.8, and therefore constructed the Minkowski-isotopic spaces as isotopes of the Euclidean space in four-dimension. Consequently, the Lorentz-isotopic symmetries $\hat{O}(3.1)$ were constructed as isotope of the orthogonal symmetry $O(4)$

$$E(x,\delta,\mathfrak{R}) \Rightarrow \hat{M}(x,\hat{g},\mathfrak{R}), \qquad O(4) \Rightarrow \hat{O}(3.1), \tag{5.32}$$

while in this paper we have considered the simpler liftings

$$M(x,\eta,\mathfrak{R}) \Rightarrow \hat{M}(x,\hat{g},\mathfrak{R}), \qquad O(3.1) \Rightarrow \hat{O}(3.1), \tag{5.33}$$

Equivalently, in the original derivation of the *noncompact* isosymmetries $\hat{O}(3.1)$ as isotopes of $O(4)$ (*loc. cit.*) the generators of $\hat{O}(3.1)$ are the *compact* generators of $O(4)$ (see ref.[28] of (*loc. cit.*), p. 550), while the generators used in this chapter are *noncompact* forms (5.15).

This illustrates the existence of *isotopies of isotopies* (evidently in the same dimensions), which we here submit in the symbolic form

$$E(x,\delta,\mathfrak{R}) \Rightarrow M(x,\eta,\mathfrak{R}) \Rightarrow \hat{M}(x,\hat{g},\mathfrak{R}), \tag{5.34a}$$

$$O(4) \Rightarrow O(3.1) \Rightarrow \hat{O}(3.1) \tag{5.34b}$$

Moreover, the Lie-isotopic brackets of the original derivation exhibit the explicit presence of the isotopic element T,

$$[A \hat{,} B] = ATB - BTA, \tag{5.35}$$

while the Lie-isotopic brackets (4.23) of this chapter exhibit the explicit presence of the isounit $\hat{1}$. This is due to the inversion of role between T and $\hat{1}$ in the universal enveloping associative algebras, from their (covariant), in the transition form for their abstract (matrix) representations, to their classical (contravariant) realization.

The following important property is an evident consequence of the "direct universality" of Birkhhoffian mechanics mentioned earlier, as well as of the arbitrariness of the $\hat{b}$-functions in Lie-isotopic structure (5.28).

COROLLARY IV.5.3.3 (loc. cit.) : The general isolorentz symmetries, and related isotransformations, are "directly universal", in the sense that they admit as particular cases all possible generalizations of the Lorentz symmetry and related transformations characterized by topology-preserving, linear or nonlinear, and local or nonlocal generalizations of the Minkowski metric of type (5.26a) ("universality"), directly in the frame of the observer ("direct universality").

It is best to illstrate the above property with specific examples. Consider first *Bogoslovski's special relativity* (Bogoslovski (1977), (1984)). It is essentially characterized by a *homogeneous but anisotropic generalization of space-time* induced by the lifting of the Minkowski metric of the Finslerian form

$$\eta \Rightarrow \hat{g} = T\eta = \frac{(-\nu^{\alpha}\eta_{\mu\nu}x^{\beta})^2}{-x^{\mu}\eta_{\mu\nu}x^{\nu}}\eta \tag{5.36}$$

where $\nu = (\nu_o, 1)$ is a null vector along the direction of anisotropy, $\nu^2 = \nu^{\alpha}\nu_{\alpha} = 0$, and r is a scalar parameter. *Bogoslovski's transformations* were constructed via the use of the conventional Lie's theory, and, in one of their forms, can be written

$$\begin{cases} x'^1 = x^1, & (5.37a) \\ x'^2 = x^2, & (5.37b) \\ x'^3 = \tilde{\gamma}(x^3 - \beta x^4), & (5.37c) \\ x'^4 = \tilde{\gamma}(x^4 - \beta x^3), & (5.37d) \end{cases}$$

where

$$\beta = v/c_o\,,\quad \tilde{\gamma} = \gamma\,(\frac{1 - \nu_o/c_o}{1 + \nu_o/c_o})^{r/2}{}^{r/2}, \tag{5.38}$$

It is an instructive exercise for the interested reader to prove that, when the isotopic element T of lifting (5.36) are positive-definite,

$$T = \frac{(-\nu^{\mu}\eta_{\mu\nu}x^{\nu})^2}{-x^{\mu}\eta_{\mu\nu}x^{\nu}}^{r/2} > 0. \tag{5.39}$$

Bogoslowski's transformations (5.37) are a particular case of our general Lorentz-isotopic transformations on $\hat{M}^I$. In the general case of an isotopic element T of undefined topology, Bogoslowski's transformations are a particular case of our isolorentz transform,ations on $\hat{M}^{II}$.

Bogoslovski's special relativity is particularly important for our analysis. In fact, it is sufficient to re-interpret the theory as *representing homogeneous but anisotropic physical media.* As such, the theory is particularly useful, physically, whenever only anisotropy is needed. The theory is also useful mathematically, to see how our Lie-isotopic commutation rules (5.16) can be equivalently rewritten via the *conventional* commutation rules, although in a predictably much more complex form.

Note that, conventionally, transformations (5.37) require a combination of Lorentz and scale transformation, while the latter are absorbed in the structure of the brackets in the Lie-isotopic theory.

Above all, Bogoslovski's special relativity is particularly useful in hadron physics along anisotropic lines studied by a number of authors (see, e.g., Preparata (1981)). In fact, the assumption of an internal anisotropy constitutes the first ground for the achievement of a *true quark confinement*, that is a confinement not only with an infinite potential barrier, but also with an identically null, and explicitly computed probability of tunnel effects of free quarks (see the comments of Sect. IV.1). The understanding is that the full differentiation of this type is achieved via an interior geometry of both anisoptropic *and* inhomogeneous type.

Despite these possibilities, Bogoslovski's special relativity has been substantially ignored, with very few exceptions known to this author. This is regrettable because, on phenomenological grounds, Bogoslowski's and Einstein's special relativities have essentially the same predictions for all relativistic speeds currently attainable in particle accelerators, and diverge only when the speeds are sufficiently close to that of light.

Another particular case is given by the transformations identified by Edwards (1963) and Strelt'sov (1990) via the invariant

$$x^{\mu}\,\hat{\eta}_{\mu\nu}\,x^{\nu}$$

$$= x^1x^1 + x^2x^2 + x^3x^3 + x^4[c_{o1}c_{o2} - (\frac{c_{o2}}{c_{o1}} - \frac{c_{o1}}{c_{o2}})]x^4. \quad (5.40)$$

which characterizes the generalized Lorentz transformations

$$\begin{cases} x'^1 = x^1, & (5.41a) \\ x'^2 = x^2, & (5.41b) \\ x'^3 = \gamma\{[1 + \frac{1}{2}(1/c_{o1} - 1/c_{o2})v]x^3 - vx^4\}, & (5.41c) \\ x'^4 = \gamma\{[1 + \frac{1}{2}(1/c_{o1} - 1/c_{o2})v\, x^4 - v\, x^3/(c_{o1}c_{o2}), & (5.41d) \end{cases}$$

As one can see, the *Edward-Strelt'sov transformations* (5.41) are clearly based on an *anisotropy of time*, with different speeds of light in the forward and backward direction, c_{o1} and c_{o2}. The conventional Lorentz transformations are evidently recovered for $c_{o1} = c_{o2} = c_o$. It is evident that, when

$$T_{44} = c_{o1}c_{o2} - \left(\frac{c_{o2}}{c_{o1}} - \frac{c_{o1}}{c_{o2}}\right) \frac{x^3}{x^4} > 0, \qquad (5.42)$$

the Edward-Strelt'sov transformations are a particular case of our general isolorentz transformations on $\hat{M}^I$. Otherwise, the transformations are a particular case of our general Lorentz-isotopic transformations on $\hat{M}^{II}$.

The study of a virtually endless variety of other particular cases is left to the curiosity of the interested reader.

The above two particular cases serve also to illustrate the importance of our distinction between isospaces of Class I, with positive-definite isotopic elements T, and isospaces of Class II, with isotopic elements of undefined topology. In fact, our isotopies of the special relativity (Sect. IV.9) will be restricted to isospaces of Class I.

The reader should remember this point in the use of Bogoslowski's anisotropy for interior physical media, because no known physical event in the transtition from motion in vacuum to motion within physical media can yield a negative-definite isotopic element T.

Nevertheless, the transition from isospaces of Class I to those of Class II is theoretically intriguing and important on a number of counts, e.g., for a better understanding of the contemporary notion of tachyons (see Sect. IV.9), or for the incorporation of all inversions in the isounit of the theory.

At this point it is important to review the so-called *Recami-Mignani (1972) superluminal transformations*

$$x'^{\hat{2}} = x^t \hat{g}\, x = -x^2 = -x^t \eta\, x, \qquad (5.43)$$

whose isometric $\hat{g} = -\eta$ is evidently of Class II. The above transformations are important for our analysis bescause they allow the mapping of all generalized Lorentz transformations with T > 0, into their image for T < 0. Additional applications of Recami-Mignani transformations will be indicated later on in this section and in Sect. IV.9.

We now close this section with the classification (and unification) of all possible simple isolorentz groups on $\hat{M}^{III}(x,\hat{g},\hat{\mathfrak{R}})$. To begin, let us recall from Sect. II.8 that all possible, simple, compact and noncompact isogroups can be constructed, in their most general possible form, as isotopes of the conventional compact group in the same dimension.

In regard to the classification of isogroups $\hat{O}(3.1)$, we shall therefore return to the original derivation of Santilli (1983a) which assumes the Euclidean space $E(x,\delta,\mathfrak{R})$ in four-dimension and the compact orthogonal group O(4) as the fundamental quantities.

Next, we introduce the following isorelativistic extension of the notion of isodual isogalilean symmetries (Sect. III.8).

DEFINITION IV.5.3 (Santilli (loc. cit.)): Let $\hat{O}(3.1)$ *be an isolorentz algebra characterized by the diagonal isometrics* $\hat{g} = diag.\ (\hat{g}_{\mu\mu})$. *Then, the "isodual isolorentz algebra"* $\hat{O}^d(3.1)$ *of* $\hat{O}(3.1)$ *is the algebra characterized by the the isometric*

$$\hat{g}^d = -\hat{g} = (-\hat{g}_{\mu\mu}). \tag{5.44}$$

Recall also that *isodual Lie-isotopic algebras are locally isomorphic* (Sect. II.6). Thus, *all conventional and generalized groups and symmetries admit an isomorphic isodual image.*

Denote now the conventional compact and noncompact orthogonal groups with the symbols $O_I(4)$, $O_I(3.1)$, and $O_I(2.2)$, where I = diag. (1,1,1,1) is the trivial unit of the contemporary formulation of Lie's theory. From Definitioon IV.5.1 and Lemma IV.5.1 we learn that these groups admit the isodual images $O_I{}^d(4)$, $O_I{}^d(3.1)$ and $O_I{}^d(2.2)$. The classification of Figure IV.5.1 then follows.

This illustrates Conjecture III.3.1 for the case of six dimension, following the unification of all simple Lie groups of three dimensions of Sect. III.3.

We assume the reader is familiar with the fact that, in conventional relativistic settings, the Lorentz symmetry is not freely defined in Minkowski space, but rather on the hypersurface of Dirac's subsidiary constraints (1964). Exactly the same situation occurs for the isolorentz symmetry, although in a predictably generalized way (Appendix IV.B).

We finally recall that the isolorentz symmetries of this section are a particular case of the broader *Lorentz-admissible*

$O_O(4)$: $\hat{g}$ = diag. (+1, +1,+1,+1)	$O_O^d(4)$: $\hat{g}$ = diag. [-1,-1,-1,-1)
$O_O(3.1)$: $\hat{g}$ = diag. (+1,+1,+1,-1)	$O_O^d(3.1)$: $\hat{g}$ = diag. (-1,-1,-1,+1)
$O_O(2.2)$: $\hat{g}$ = diag. (+1,+1,-1,-1)	$O_O^d(2.2)$: $\hat{g}$ = diag. (-1,-1,+1,+1)
$\hat{O}_1(4)$: sig. $\hat{g}$ = (+, +, +, +)	$\hat{O}_1^d(4)$: sig. $\hat{g}$ = (-, -, -, -)
$\hat{O}_1(3.1)$: sig. $\hat{g}$ = (+, +, +, -)	$\hat{O}_1^d(3.1)$: sig. $\hat{g}$ = (-, -, -, +)
$\hat{O}_2(3.1)$: sig. $\hat{g}$ = (+, +, -, +)	$\hat{O}_2^d(3.1)$: sig. $\hat{g}$ = (-, -, +, -)
$\hat{O}_3(3.1)$: sig. $\hat{g}$ = (+, -, +, +)	$\hat{O}_3^d(3.1)$: sig. $\hat{g}$ = (-, +, -, -)
$\hat{O}_4(3.1)$: sig. $\hat{g}$ = (-, +, +, +)	$\hat{O}_4^d(3.1)$: sig. $\hat{g}$ = (+, -, -, -)
$\hat{O}_1(2.2)$: sig. $\hat{g}$ = (+, +, -, -)	$\hat{O}_1^d(2.2)$: sig. g = (-, -, +, +)
$\hat{O}_2(2.2)$: sig. $\hat{g}$ = (+, -, +, -)	$\hat{O}_2^d(2.2)$: sig. $\hat{g}$ = (-, +, -, +)
$\hat{O}(4)$: $\hat{g}$ = diag. ($g_{11}, g_{22}, g_{33}, g_{44}$)	

FIGURE IV.5.1. The twenty one most significant isotopes in the classification of all possible isoorthogonal group in four dimensions studied in Santilli (1988c). The most general possible isotope is the last one, denoted with $\hat{O}(4)$, with an arbitrary topology of its isometric, which *unifies all possible six-dimensional simple Lie groups of Cartan's classification. In particular, this isotope is the abstract Lorentz-isotopic group of Theorem IV.5.1.* In fact, depending on the local topology of the isometric, $\hat{O}(4)$ can assume the form of: any one of the six-dimensional simple Lie groups O(4), O(3.1) and O(2.2) (and others locally isomoprhic to the latters); any one of their isodual; as well as any one of their isotopes. The emerging infinite number of possible realizations can be first divided into two classes interconnected by isoduality. Then, among each of these classes, only three essential isogroups emerge, those isomorphic to O(4) or O(3.1) or O(2.2). Finally, note that the above classification essentially identifies all possible signatures that can be reached under isotopy.

symmetries of Santilli (1981a). The former symmetries are recomnmendable, e.g., for the characterization of generalized particles in stable orbits (Sect. IV.7), while the latter symmetries are more effective for the description of generalized particles in nonconservative conditions (Appendix II.E).

IV.6: CONSTRUCTION OF THE POINCARE'-ISOTOPIC SYMMETRIES

We now study the construction of the classical realization of the *inhomogeneous Lorentz-isotopic symmetries,* or *Poincare'-isotopic symmetries* , called *isopoincaré symmetries* for short, and denoted with the symbol $\hat{P}3.1)$, first introduced in Santilli (1988c), which is followed in this section. For additional studies one may consult Santilli (1991c).

Consider a set of N particles denotes with the index a = 1, 2,..., N, in isominkowski spaces $M^{III}{}_2(x,\hat{\eta},\hat{\Re}) = \hat{M}^{III}(x,\hat{g},\hat{\Re})$ with local separation[20]

$$x_{ab}^{\hat{2}} = (x_a^{\mu} - x_b^{\mu})\, \hat{g}_{\mu\nu}(x, v, a, \mu, \tau, n,...)\, (x_a^{\nu} - x_b^{\mu}). \qquad (6.1)$$

DEFINITION IV.6.1 (Santilli (1988c)):The "abstract Poincare'-isotopic (isopoincaré) symmetries" $\hat{P}(3.1)$ are the largest possible, ten-dimensional Lie-isotopic group of isometries of isoseparation (6.1) which are isolinear and isolocal on isominkowski spaces $M^{III}(x,\hat{g},\hat{\Re})$, but nonlinear and nonlocal on the conventional Minkowski space $M(x,\eta,\Re)$. The "general Poincaré-isotopic (isopoincaré) symmetries" are the most general possible, isolinear and isolocal isosymmetries of isoseparation (6.1) on isominkowski spaces $\hat{M}^{I}(x,\hat{g},\hat{\Re})$. The "restricted Poincaré-isotopic (isopoincaré) symmetries" are the most general possible, linear and local isometries of isoseparation (6.1) on isominkowski spaces $\hat{M}^{I}(x,\hat{\eta},\hat{\Re})$, with isometric $\hat{\eta}$ independent from the local coordinates and all their derivatives.

As is well known (see, e.g., Schweber (1962)), the conventional Poincaré group possesses the structure of the semidirect product

$$P(3.1) = O(3.1) \otimes T(3.1), \qquad (6.2)$$

where O(3.1) is the (simple) conventional Lorentz subgroup, and T(3.1) is the (Abelian) invariant subgroup of translations in Minkowski space.

[20] We assume the reader is familiar with the property, from Sect. II.3, that the separation in an isospace is an element of the underlying isofield. Thus, on rigorous mathematical grounds, separation (6.1) should be multiplied by $\hat{1}$ to belong to $\hat{\Re}$, and the same holds true for separation (IV.5.1b). However, the presence of the isounit is redundant from a symmetry viewpoint, trivially, because all symmetries leave invariant their own unit, and it has been omitted for simplicity in notations.

The conventional Poincaré transformations are given by the well known *linear and local transformations* on $M(x,\eta,\Re)$

$$x' = \Lambda x + x^{\circ}, \qquad \Lambda \in O(3.1)\ ,\ x^{\circ} = (x^{\circ\mu}) \in \Re, \tag{6.3}$$

A classical realization of the Poincare' symmetry for the case of N particles with non-null masses is given by the ten (ordered) parameters

$$w = (w) = (\theta, u, x^{\circ}), \quad k = 1, 2, \ldots, 10, \tag{6.4}$$

and generators in $T^{*}M(x,\eta,\Re)$

$$X = (X_k) = (J_{\mu\nu}, P_{\mu}),\ k = 1, 2, \ldots, 10, \tag{6.5a}$$

$$J_{\mu\nu} = \sum_a (x_{a\mu}\, p_{a\nu} - x_{a\nu}\, p_{a\mu}), \qquad P_{\mu} = \sum_a p_{a\mu}, \tag{6.5b}$$

with Lie algebra $\mathbf{P}(3.1)$ characterized by the commutation rules in terms of brackets (IV.4.16)

$$\mathbf{P}(3.1): [J_{\mu\nu}, J_{\alpha\beta}] = \eta_{\nu\alpha} J_{\beta\mu} - \eta_{\mu\alpha} J_{\beta\nu} - \eta_{\nu\beta} J_{\alpha\mu} + \eta_{\nu\beta} J_{\alpha\nu}, \tag{6.6a}$$

$$[J_{\mu\nu}, P_{\alpha}] = \eta_{\mu\alpha} P_{\nu} - \eta_{\nu\alpha} P_{\mu}, \tag{6.6b}$$

$$[P_{\mu}, P_{\nu}] \quad 0, \qquad \mu, \nu, \alpha = 1, 2, 3, 4, \tag{6.6c}$$

with Casimir invariants

$$C^{(o)} = I\ , \quad C^{(1)} = P^2 = P^{\mu} P_{\mu} = P^{\mu} \eta_{\mu\nu} P^{\nu} \tag{6.7a}$$

$$C^{(2)} = W^{\mu} W_{\mu} = W^{\mu} \eta_{\mu\nu} W^{\nu}, \quad W_{\mu} = \epsilon_{\mu\alpha\beta\rho} J^{\alpha\beta} P^{\rho}, \tag{6.7b}$$

Lie group structure

$$P(3.1): \quad a' = \{\textstyle\prod_k e|_{\xi}^{\, w_k\, \omega^{ij} (\partial_j X_k)(\partial_i)}\}\, a, \tag{6.8}$$

and discrete invariant subgoup

$$\phi(3.1): \ Px = (-r, x^4),\ Tx = (r, -x^4),\ PTx = (-r, -x^4), \tag{6.9}$$

The following isotopic liftings then occur.

THEOREM IV.6.1 (Santilli (loc. cit.)): The Poincare' symmetry P(3.1) on the conventional Minkowski space $M(x,\eta,\Re)$ *admits an infinite number of abstract Poincare'-isotopic generalizations* $\hat{P}(3.1)$, *called "isopoincare' symmetries", on the infinite number of corresponding isominkowski spaces* $\hat{M}^{III}(x,\hat{g},\hat{\Re})$ *with diagonal, nonsingular and Hermitean isometrics* $\hat{g} = T_2\,\eta$, *each element* $\hat{P}(3.1)$ *being characterized by an isounit* $\hat{I}_2 = T_2^{-1}$. *All isosymmetries* $\hat{P}(3.1)$ *admit the decomposition into the semidirect product*

$$\hat{P}(3.1) = \hat{O}(3.1) \otimes \hat{T}(3.1), \tag{6.10}$$

where the subgroups $\hat{O}(3.1)$ *are the (simple) isolorentz subgroups of the preceding section,*

$$\hat{O}(3.1): \qquad x' = \hat{\Lambda} * x = \hat{\Lambda}\, T_2\, x, \qquad \hat{\Lambda}^t \hat{g} \hat{\Lambda} = \Lambda \hat{g} \hat{\Lambda}^t = \hat{g}^{-1}, \tag{6.11}$$

and the groups $\hat{T}(3.1)$ *are the (isobelian) invariant subgroups of isotranslations*

$$\hat{T}_a(3.1): \quad x'^{\mu} = \hat{T}(x^\circ) * x^{\mu} = x^{\mu} + x^{\circ\mu}\, \hat{b}^{\mu\,-2}(x, \dot{x}, \mu, \tau, n, ...), \tag{6.12a}$$

$$p'^{\mu} = \hat{T}(x^\circ) * p^{\mu} = 0, \tag{6.12b}$$

where the $\hat{b}$*-functions are generally nonlinear and nonlocal in all their arguments to be identified below.*

All isotopes $\hat{P}(3.1)$ *admit the following classical realization for a system of N particles in the isocotangent bundle* $T^*\hat{M}^{III}(x,\hat{g},\hat{\Re})$ *with local charts* $a = (a^i) = (x, p) = (x^{a\mu}, p^{a\mu})$, $i = 1, 2, ..., 8N$, $\mu = 1, 2, 3, 4$:

1) The same ordered set of generators (6.4) of the conventional symmetry;

2) The same (ordered set of) generators (6.5) of the conventional symmetry;

3) the isocommutation rules in terms of the Lie-isotopic brackets (IV.4.23)

$$\hat{P}(3.1): [J_{\mu\nu} \,\hat{,}\, J_{\alpha\beta}] = \hat{g}_{\nu\alpha} J_{\beta\mu} - \hat{g}_{\mu\alpha} J_{\beta\nu} - \hat{g}_{\nu\beta} J_{\alpha\mu} + \hat{g}_{\mu\beta} J_{\alpha\nu}, \tag{6.13a}$$

$$[J_{\mu\nu} \,\hat{,}\, P_\alpha] = \hat{g}_{\mu\alpha} P_\nu - \hat{g}_{\nu\alpha} P_\mu, \tag{6.13b}$$

$$[J_\mu \hat{,} P_\nu] = 0, \qquad \mu, \nu = 1, 2, 3, 4 \qquad (6.13c)$$

$$[A,B] = \frac{\partial A}{\partial x^\mu} \hat{g}^{\mu\nu} \frac{\partial B}{\partial p^\nu} - \frac{\partial B}{\partial x^\mu} \hat{g}^{\mu\nu} \frac{\partial A}{\partial p^\nu}, \qquad (6.13d)$$

$$(\hat{g}^{\mu\nu}) = (\hat{1}_2{}^\mu{}_\alpha \eta^{\alpha\nu}) = (\hat{g}_{\mu\nu})^{-1} = (\eta_{\mu\alpha} T_2{}^\alpha{}_\nu)^{-1}, \qquad (6.13d)$$

4) the local isocasimir invariants

$$\hat{C}^{(o)} = \hat{1}_2 = \hat{T}_2{}^{-1}, \qquad (6.14a)$$

$$\hat{C}^{(1)} = P^{\hat{2}} = (P^\mu \hat{g}_{\mu\nu} P^\nu) \hat{1}_2 = (P^\mu P_\mu) \hat{1}_2, \qquad (6.14b)$$

$$\hat{C}^{(2)} = W^{\hat{2}} = (W^\mu \hat{g}_{\mu\nu} W^\mu) \hat{1}_2, \quad W_\mu = \epsilon_{\mu\alpha\beta\rho} J^{\alpha\beta} P^\rho, \qquad (6.14c)$$

5) The Lie-isotopic group structure for the connected part $\hat{SO}(3.1) \otimes \hat{T}(3.1)$ on $T^*\hat{M}^{III}(x,\hat{g},\hat{\Re})$

$$a' = \{\hat{\Lambda}, \hat{T}\} * a = \{ [e_{|\xi}^{w_k \omega^{ir} I_{2r}{}^j (\partial_j X_k)(\partial_i)}] \hat{1}_2 \} * a =$$

$$= \{S_{\hat{g}(\theta,u)}, T_{\hat{g}}(x^\circ)\} a, \qquad (6.15)$$

6) the $\hat{b}$-functions are explicitly given by

$$\hat{b}_\mu{}^{-2} = \hat{b}_\mu{}^{-2} + a^\alpha [\hat{b}_\mu{}^{-2} \hat{,} P_\alpha]/2! + x^{\circ\alpha} x^{\circ\beta} [\hat{b}_\mu{}^{-2} \hat{,} P_\alpha] \hat{,} P_\beta]/3! + \ldots\ldots \qquad (6.16)$$

and

7) the invariant discrete component is the same as that in $\hat{O}(3.1)$, i.e.,

$$\hat{\phi}(3.1): \hat{P} * x = (-r, x^4), \quad \hat{T} * x = (r, -x^4), \quad (\hat{P} * \hat{T}) * x = (-r, -x^4). \qquad (6.17)$$

The proof of the above properties is just a reformulation of the proof of Theorem III.4.1 on the *isoeuclidean symmetries* $\hat{E}(3)$ for the case of isometrics with signature (+, +, −). As such, it will be left as an exercise for the interested reader.

Despite the manifest similarities between the conventional Poincare' symmetries and all its isotopes, the latter have non-trivial physical implications, as illustrated by the application of isosymmetries $\hat{P}(3.1)$ for the characterization of relativistic, closed, nonhamiltonian systems (Sect. IV.7), the consequential, necessary isotopic liftings of Einstein's special relativity (Sect. IV.8), or the characterization of a generalized notion of isoparticle (Sect. IV.7).

The following property is by now evident.

COROLLARY IV.6.1.1 (loc. cit.): All infinitely possible general and restricted isopoincaré symmetries $\hat{P}(3.1)$ on isominkowski spaces $M^I(x,\hat{g},\hat{\Re})$, are locally isomorphic to the conventional Poincaré symmetry $P(3.1)$.

It is understood that on isospaces of Class II or III the above isomorphisms are no longer guaranteed. As an example, the use of the classification of $\hat{O}(3.1)$ shows that some of the isotopes $\hat{P}(3.1)$ on isospaces of Class II are locally isomorphic to $O(4)\times T(4)$ or $O(2.2)\times T(2.2)$. The classification of all possible isotopes $\hat{P}(3.1)$ on isospaces of Class III is left to the interested reader.

It is evident that the general and restricted isopoincaré symmetries constitute isotopic coverings of the conventional symmetry in the sense indicated in Sect. IV.5 for the Lorentz symmetry.

As concluding remarks, note the explicit dependence of the isometric from the local coordinates x. Also, the composition law of two isopoincaré transformations $\{\hat{\Lambda}_k , \hat{T}_k\}$, k = 1, 2, is given by

$$\{\hat{\Lambda}_1 , \hat{T}_1\} * \{\hat{\Lambda}_2 , \hat{T}_2\} = \{\hat{\Lambda}_1 * \hat{\Lambda}_2 , \hat{T}_1 + \hat{\Lambda}_2 * \hat{T}_2\}. \qquad (6.18)$$

Finally, note for future use the isotopy of the group of translations

$$T(3.1) = e_{|\xi}^{x^{\circ\mu}\eta_{\mu\nu}p^{\nu}} \Rightarrow \hat{T}(3.1) = e_{|\xi}^{x^{\circ\mu}\hat{g}_{\mu\nu}p^{\nu}}, \qquad (6.19)$$

which is at the foundation of our representation of electromagnetic wave propagating within an inhomogeneous and anisotropic physical medium (Sect. IV.9).

Again, as it was the case for the conventional and isotopic Lorentz symmetries, the isopoincaré symmetries are not freely defined in isospaces $\hat{M}^{III}(x,\hat{g},\hat{\Re})$, but rather on the hypersurface of the constraints.

Finally, the Poincaré-isotopic symmetries of this section are a particular case of the *Poincaré-admissible symmetries* of Santilli (1981a), the latter being the most general possible symmetries for extended-deformable particles under the most general known nonconservative dynamical conditions.

IV.7: RELATIVISTIC ISOPARTICLES AND ISO-QUARKS

The best way to illustrate the physical implications of the isotopic liftings of the Lorentz and Poincaré symmetries is by identifying their implications for the characterization of particles. In turn, this will predictably provide a number of possibilities for experimental verifications, first, for the identification of the physical conditions under which the isotopies are applicable and, second, for the test of the quantitative predictions of the novel theory in the arena of its applicability.

As well known, Einstein's special relativity characterizes particles as *massive points.* But point are perennial and immutable geometric objects. Thus, *according to contemporary relativistic theories, elementary particles preserve their intrinsic characteristics for all conceivable physical conditions existing in the Universe .*

In his limpid writings, Einstein (1905) avoided such a manifestly excessive assumption, because he identified quite clearly the arena of applicability of his views. In this volume we therefore assume as exact Einstein's views, but not necessarily those of his contemporary followers. In particular, *throughout our analysis we shall assume that elementary particles preserve their intrinsic characteristics under the following Einstenian conditions:*

1) Particles can be well approximated as being massive points;

2) when moving in the homogeneous and isotropic vacuum (empty space);

3) while experiencing only action-at-a-distance, local-potential (selfadjoint) forces.

In this volume we are interested in studying particles in *noneinsteinian conditions,* i.e., we shall study

$\hat{1}$) Particles (and/or their wavepackets) which cannot be approximated as being point-like, but require a representation of their actual extended, and therefore deformable shape;

$\hat{2}$) when moving within generally inhomogeneous and anisotropic physical media;

$\hat{3}$) while experiencing conventional, action-at-a-distance, local-potential forces, as well as contact, instantaneous (classically range), nonlinear, nonlocal and nonhamiltonian interactions with the medium itself.

In Sect. III.7 we have introduced the notion of *nonrelativistic isoparticle* as a representation of one of the infinitely possible isogalilean symmetries $\hat{G}(3.1)$. The preceding results of this section then allow us to introduce the following

DEFINITION IV.7.1 (Santilli (1988c), (1989), (1991d)): A classical relativistic isoparticle is a representation of one of the infinitely possible isopoincaré symmetries in the isominkowski spaces $\hat{M}^I(x,\hat{g},\hat{\Re})$,

$$\hat{P}(3.1):\ x' = \{\hat{\Lambda}(\theta,u)\ ,\ \hat{T}(x^\circ)\} * x = (\hat{\Lambda}(\theta,u)\ ,\ \hat{T}(x^\circ)\}\, T_2\, x$$

$$= \{\Pi_k\, e|_\xi^{\ w_k\, \omega^{ir}\, \hat{1}_{2r}{}^{j}\, (\partial_\nu X_k)\, (\partial_j)}\, \hat{1}_2\} * x, \qquad (7.1a)$$

$$w = (u,x^\circ),\quad X = (J_{\mu\nu}\ , P_\mu), \qquad (7.1b)$$

$$\hat{M}^I(x,\hat{g},\hat{\Re}):\quad \hat{g} = T_2\, \eta,\quad \hat{\Re} = \Re\hat{1},\ \hat{1}_2 = \hat{T}_2^{-1} > 0, \qquad (7.1c)$$

Equivalently, a classical relativistic isoparticle is the generalization of the classical Einsteinian particle characterized by the isotopic lifting of the trivial unit I of the conventional Poincaré symmetry, into one of the infinitely possible isounits $\hat{I}_2 > 0$ of the isopoincaré symmetries $\hat{P}(3.1)$.

In the nonrelativistic case, the central consequences of the isotopies $G(3.1) \Rightarrow \hat{G}(3.1)$ are given by the characterization of particles with an infinite number of *different* intrinsic characteristics (e.g., infinitely many possible deformed shapes), depending on the local physical conditions considered.

The alteration of the intrinsic characteristics of a particle in the transition, from motion in vacuum (Einsteinian conditions), to motion within physical media (noneinsteinian conditions), was called *mutation* (Santilli (1978b)), and the same terminology has been adopted in this volume.

As an example, we pointed out in Sect. III.7 that a *deformation of the charge distribution of a neutron under sufficiently intense external fields or collisions, must necessarily produce a mutation of its intrinsic magnetic moment, as preliminarily measured by Rauch and his collaborators (1981), (1983)* (see Chapter VII for details).

It is evident that, once *one* intrinsic characteristics is mutated, the mutation of all remaining characteristics follows on a number of independent arguments known since the original proposal ((*loc. cit*). Sect. 4.11), where it was shown that the addition to Dirac's equations of a nonselfadjoint coupling implies the necessary mutation of all intrinsic characteristics.

The relativistic extension of the notion of isoparticles is intended precisely to provide the technical means for the study of the inter-relationship among seemingly different mutations.

The existence of an infinite number of possible different isosymmetries $\hat{P}(3.1)$ is intended to represent precisely the possible existence of an infinite number of different intrinsic characteristics for the same particle, evidently depending on the infinite number of possible local conditions for the interior problem.

It may be of some value to indicate already in these introductory words the expected physical relevance of the notion of mutation.

First, the reader should be aware that the notion considered is inapplicable in the atomic structure, because of its exact Einsteinian character due to the large mutual distances among the constituents.

Second, the notion of mutation begins to acquire possible physical relevance in nuclear physics because a small nuclear force of short range, nonlocal and nonhamiltonian type is expected from the experimentally established mutual penetration of the nucleon constituents for an average of 10^{-3} parts of their volume.

In fact, the possible mutation of the magnetic moment of nucleons, when inside a nuclear structure, offers some intriguing possibilities for attempting a final understanding of the still elusive, total, nuclear magnetic moments.

Also, the notion of *relativistic isoneutron* as per Definition IV.7.1 appears to be useful for a deper understanding of the nuclear stability while one of its constituents is unstable. In fact, the isoneutron appears to have a much longer meanlife than that of the conventional neutron when a member of a nuclear structure (Sect. IV.9).

More generally, the notion of *isonucleon* appears to be useful in a number of other aspects of nuclear physics, with the understanding that, again, conceivable mutations can at best be small in nuclear physics.

Within the context of hadron physics, for which the notion was conceived, we expect mutations quantitatively much bigger than those of the nuclear structure. This is due to the fact that all massive particles have a wavelength of the order of the size of hadrons, as well as of the range of the strong interactions themselves.

Then, the hadronic constituents, to be physical particles, are expected to be in conditions of total mutual immersion within the hyperdense media in the interior of hadrons (called *hadronic media*) This activates precisely the nonlinear, nonlocal and nonhamiltonian forces studied in these volumes, and justifies the introduction of the following

DEFINITION IV.7.2 (Santilli (1988c): A "relativistic isoquark" is a representation of one of the infinitely possible isosymmetries $\hat{P}(3.1)\times S\hat{U}(3)$, where $\hat{P}(3.1)$ characterizes the space-time isosymmetries on $\hat{M}(x,\hat{g},\hat{\mathfrak{R}})$, and $S\hat{U}(3)$ represents the isotopes of the unitary symmetry $S\hat{U}(3)$ (Mignani (1984), Mignani and Santilli (1991)).

As we shall see, the above notion of isoquark with its mutated intrinsic characteristics, offers genuine possibilities of resolving the now vexing problems of hadron physics, such as

a) attempting a *true quark confinement*, i.e., a confinement not only with an infinite potential barrier, but also with an identically null probability of tunnel effect for free quarks (see Sect. IV.9 for preliminary possibilities);

b) the possibile identification of quarks with ordinary massive particles, although in suitably altered (mutated) conditions;

c) the possibility of achieving under isotopy convergent perturbative expansions for strong interactions, prior to renormalization techniques, as preliminarily studied in Santilli (1989)).[21]

[21] The main ideas are so simple to appear trivial. Consider a classical, relativistic canonical series expressed in terms of brackets (IV.4.16) which is *divergent*. Then, there exist an infinite number of isotopies $\eta \Rightarrow \hat{\eta} = T\eta$ under which the same series expressed in terms of the Lie-isotopic brackets (IV.4.23) becomes *convergent*, as one can easily prove, e.g., for $|\hat{I}| < 1$. The isotopic regeneration of convergence at the operator level is equally simple. Consider a perturbative series expressed in terms of the trivial Lie prouct of quantum mechanics $[A,B] = AB - BA$ which is *divergent*, e.g., because of the high value of the coupling constant. Then, there exist infinitely possible isotopies $I \Rightarrow \hat{I} = T^{-1}$ under which the same series expressed in term of the isotopic product of the covering hadronic mechanics, $[A \hat{,} B] = ATB - BTA$, becomes *convergent*. These aspects are here mentioned to indicate the truly novel

A quantitative appraisal of these possibilities will predictably require a long chain of studies, beginning with the operator formulation of the theory, then passing to the construction of the $\hat{P}(3.1)$ invariant field equations, and finally formulating specific experiments in particle physics (for preliminary proposals see, Santilli (1990) and (1991d)). Evidently, in this volume we can provide only the classical, relativistic and spinless background.

To begin, the necessary conditions for the existence of a mutation are readily given by the following proposition.

PROPOSITION IV.7.1 (Santilli (1988c): A necessary condition for the mutation of the intrinsic characteristics of particles is that the local physical conditions imply a violation of the conventional Lorentz symmetry, of nonlinear, or nonlocal or nonhamiltonian type.

We are now interested in the necessary and sufficient conditions for a mutations. An inspection of the invariants of the isotopies $\hat{P}(3.1) \Rightarrow \hat{P}(3.1)$,

$$P^2 = P^\mu \eta_{\mu\nu} P^\nu \quad \Rightarrow \quad P^{\hat{2}} = P^\mu \hat{g}_{\mu\nu} P^\nu , \tag{7.2a}$$

$$W^2 = W^\mu \eta_{\mu\nu} W^\nu \quad \Rightarrow \quad W^{\hat{2}} = W^\mu \hat{g}_{\mu\nu} W^\nu, \quad W_\mu = \epsilon_{\mu\alpha\beta\gamma} J^{\alpha\beta} P^\gamma, \tag{7..2b}$$

and prior to any study of the isorepresentation theory, yields the following:

PROPOSITION IV.7.1 (loc.. cit.): A necessary and sufficient condition for the mutation of the intrinsic characteristics of elementary particles is that the isometric $\hat{g}$ is a nontrivial isotopy of the Minkowski metric η, $\hat{g} = T_2 \eta$, $T_2 \neq I$, $T_2 > 0$.

Finally, the reader should not assume that, under mutation, we lose fundamental space-time symmetries. This leads to the followiong restriction on the mutations evidently imposed by the local

---

possibilities of our isotopic relativities. In fact, the isotopic regeneration of convergence is structurally equivalent to the isotopic regeneration of exact space-time symmetries when believed to be conventionally broken. More generally, we can state that *a divergent, Poincaré-invariant, perturbative, classical or quantum mechanical series, can always be turned into a convergent series via its reformulation into a form invariant under Poincare-isotopic symmetries* (Santilli (1989c, d). In different terms, the current divergence of perturbative series for strong interactions may eventually result to be a manifestation of the current, excessively simplistic realization of the Poincaré symmetry for the physical context considered.

isomorphisms $\hat{P}(3.1) \approx P(3.1)$.

PROPOSITION IV.7.3 (Santilli (loc. cit.)): All infinitely possible configurations of a relativistic isoparticle must coincide with the conventional Einsteinian form of the same particle at the abstract, realization-free level.

In different terms, mutations are not arbitrary, but geometrically restricted to such a class that conventional and mutated intrinsic characteristics can be unified at the abstract, realization-free level, as we shall see in detail later on.

Further advances in the topics of the above propositions require the study the isorepresentation theory, which we hope to present at some future time.

We now outline the three methodological tools needed for the quantitative characterization of a relativistic isoparticle.

ISORELATIVISTIC KINEMATICS (Santilli (1988c)). The first tools are given by the generalization of the conventional relativistic kinematic caused by the lifting of the carrier space $M(x,\eta,\mathfrak{R}) \Rightarrow \hat{M}^{I}_{2}(x,\hat{g},\hat{\mathfrak{R}})$, hereon considered with the diagonal isometric

$$\hat{g} = \text{diag.}\,(\hat{b}_1^2, \hat{b}_2^2, \hat{b}_3^2, -\hat{b}_4^2), \quad \hat{b}_\mu = \hat{b}_\mu(s, x, p, ..) > 0, \tag{7.3}$$

Isorelativistic kinematics is based on the isoinvariant

$$ds^{\hat{2}} = -dx^\mu \hat{g}_{\mu\nu} dx^\nu = = dr^k b_k^2 dr^k - dt\, c^2 dt = -1, \tag{7.4a}$$

$$c = c_o \hat{b}_4, \tag{7.4b}$$

from which we can write the conditions on the *isofourvelocity* $u^\mu = dx^\mu/ds$

$$u^{\hat{2}} = u^\mu \hat{g}_{\mu\nu} u^\nu = -1, \tag{7.5}$$

But

$$u^\mu = dx^\mu / ds, \tag{7.6}$$

Thus the components of the isofourvelocity are given by

$$u^4 = \frac{dx^4}{ds} = \frac{dt}{ds} = \hat{\gamma}\, c = \hat{\gamma} c_o b_4, \tag{7.7a}$$

$$u^k = \frac{du^k}{ds} = \frac{dx^4}{ds}\frac{dx^k}{dx^4} = \hat{\gamma}\, c\, v^k = \hat{\gamma} c_o b_4 v^k, \quad (7.\overline{7}b)$$

$$\hat{\gamma} = (1 - \hat{\beta^2})^{-\frac{1}{2}}, \quad \hat{\beta}^2 = (v^k b_k{}^2 v_k) / (c_o b_4{}^2 c_o), \quad (7.7c)$$

where v is the velocity in Euclidean isospace $\hat{E}_2(r,\hat{\delta},\mathfrak{R}) = \hat{E}(r,\hat{G},\mathfrak{R})$ of Sect. III.3.

We now introduce the *isofourmoMentum* as the isofourvector in $\hat{M}^{l}(x,\hat{g},\mathfrak{R})$

$$p = (p^\mu) = (\hat{m} u^\mu) = (m_o \hat{\gamma}\, c\, v^k, m_o\, \hat{\gamma}\, c), \quad (7.8a)$$

$$\hat{m} = m_o\, \hat{\gamma}, \quad (7.8b)$$

The isocasimirs (IV.6.14) then imply the following *fundamental isoinvariant of the Poincaré-isotopic symmetries*

$$p^{\hat{2}} = p^\mu\, \hat{g}_{\mu\nu}\, p^\nu = p^k\, b_k{}^2\, p^k - p^4 c^2\, p^4$$

$$= m_o{}^2\, \hat{\gamma}^2\, c^2\, v^k\, b_k{}^2\, v^k - m_o{}^2\, \hat{\gamma}^2\, c^4 =$$

$$= - m_o{}^2 \hat{\gamma}^2\, c^4\, (1 - \beta^{\hat{2}}) = - m_o{}^2\, c^4 = - m_o{}^2\, c_o{}^4\, b_4{}^4, \quad (7.9)$$

or, equivalently,

$$(p^\mu\, \hat{g}_{\mu\nu}\, p^\nu) / m_o{}^2 c^4 = -1, \quad (7.10)$$

Note that in the above derivation we have ignored the multiplication of the isotopic square by the isounit because inessential for practical applications (Sect.s II.3 and II.6).

RELATIVISTIC, CONSTRAINED BIRKHOFFIAN-ISOTOPIC MECHA-NICS (*loc. cit.*). The next tool needed for the characterization of a relativistic isoparticle is the applicable analytic dynamics. A preliminary outline of the mechanics was presented in Sect. IV.4 to identify the underlying geometry and algebra, and without constraints. We shall now reconsider the mechanics in its more general constrained form.

Recall from Sect. IV.4 that relativistic Hamilton-isotopic mechanics is *nondegenerate* (also called *nonsingular*), in the sense that it admits a unique solution $\dot{x}(x,p)$ expressing the velocities in terms of the coordinates and moments (i.e., det $(\partial^2 L / \partial x^\mu \partial x^\nu) \neq 0$).

Nevertheless, the theory is necessarily *constrained* in the sense of requiring as subsidiary constraint the fundamental isoinvariant (7.4), as well as all needed subsidiary constraints (IV.2.5b)–(IV.2.5d) for total conservation laws.

As a result of this occurrence, the subsidiary constraints can be first formulated via *Lagrange's multipliers* as done in this section (for a rigorous treatment within the context of the problem of Bolza, see Bliss (1946)). The theory then requires *Dirac's (1964) method for subsidiary constraints* to identify the hypersurface of the definition of the system and of its isosymmetries, as studied in Appendix IV.B.

Consider first the conventional space $\Re_s \times T^*M(x,\eta,\Re)$ with local coordinates $a = (a^i) = (x,p) = (x^\mu, p^\mu)$. The most general possible *nondegenerate, constrained, relativistic, Pfaffian variational principle* can be written

$$\delta\hat{A} = \delta\int_1^2 ds[R_i(s, a, ...)\, \dot{a}^i + \dot{\Phi}_\chi \lambda_\chi - B(s, a, \lambda, \Phi, ...)] = 0, \quad (7.11a)$$

$$\dot{a} = da/ds = (\dot{x}, \dot{p}), \ \dot{x} = dx/ds, \ \dot{p} = dp/ds, \quad (7.11b)$$

$$i = 1, 2, ..., 8N, \quad \mu = 1, 2, 3, 4, \quad \chi = 1, 2, ..., n$$

where the λ_χ are the Lagrange multipliers, the Φ_χ are the subsidiary constraints, B is the *relativistic Birkhoffian*, and the R_i are the *relativistic Pfaffian functions*.

The functions R_i and B can be computed from the given equations of motion via one of the various methods of Santilli (1982a), while the subsidiary constraints are assumed to be known from the problem at hand.

These methods generally result in the *nonautonomous* form of principle (7.11) with R = R(s, a,...) and B = B(s, a,...). As such, the related Birkhoff's equations do not admit a consistent algebraic structure (Appendix II.A).

The above principle must therefore be reduced to the equivalent *semiautonomous form* with R = R(a) and B = B(s, a,...). This can be done via the use of the *relativistic Birkhoffian gauge transformations* (IV.4.28), i.e.,

$$R'_i(a, ...) = R_i(s, a, ...) + \partial_i G(s, a), \quad \partial_i = \partial/\partial a^i, \quad (7.12a)$$

$$B'(s, a, ...) = B(s, a, ...) - \partial_s G(s, a). \quad (7.12b)$$

As one can see, the above transformations do redefine the R and B functions, but leave completely unaffected the multipliers and subsidiary constraints and, thus, they remain fully applicable in the constrained version of the theory.

To achieve a $\hat{P}(3.1)$-invariant mechanics, we have to consider essentially the same *restrictions* introduced in Sect. IV.4. The first restriction is that the semiautonomous Pfaffian principle must be of a form representable in the isospace $\hat{\mathfrak{R}}_s \times T^*\hat{M}^I_1(x,g,\hat{\mathfrak{R}})$, i.e., of the type

$$\delta\hat{A}^\circ = \delta\int_1^2 [\hat{R}^\circ T_1(a, ...)\, da^i + \Phi\,\lambda - B(s, a, \lambda, \Phi, ...)T_t ds\,] =$$

$$= \delta\int_1^2 ds\{\, p^{a\mu}\hat{\eta}_{\mu\nu}(x, p, ...)\, \dot{x}^{a\nu} + \Phi_\chi\lambda_\chi - B(s, x, p, \lambda, \Phi, ...)] = 0, \qquad (7.13a)$$

$$\hat{R}^\circ = (p, 0), \qquad \hat{\eta} = T_1\eta, \ \ T_1 > 0, \qquad (7.13b)$$

As second restriction, the isometric $\hat{\eta}$ must be such to induce an exact, isosymplectic two-isoform (IV.4.18), i.e.,

$$\hat{\Omega}^\circ_2 = d\hat{\Theta}^\circ_1 = = d(\hat{R}^\circ_i T_1{}^i{}_j\, da^j) =$$

$$= \tfrac{1}{2}\omega_{ir}\, T_2{}^r{}_j\,(a)\, da^i \wedge da^j \qquad (7.14)$$

As one can see, the above structure is unaffected by the subsidiary constraints. We reach in this way the important conclusion that the constrained version of relativistic Hamilton-isotopic mechanics preserves its Lie-isotopic structure in full.

Under the above conditions, principle (7.13) implies the following *relativistic, nondegenerate, constrained, Hamilton-isotopic equations*

$$\frac{da^i}{ds} = \Omega^{ij}(a,..)\,\frac{\partial B(s,a,\lambda,...)}{\partial a^j} = \begin{cases} \dfrac{dx^{a\mu}}{ds} = \hat{g}^{\mu\nu}(x,p,..)\,\dfrac{\partial B(s, a, \lambda, ...)}{\partial p^{a\nu}}, \\[2ex] \dfrac{dp^{a\mu}}{ds} = -\,\hat{g}^{\mu\nu}(x,p,...)\,\dfrac{\partial B(s,a,\lambda,..)}{\partial x^{a\nu}}, \end{cases} \qquad (7.15a)$$

$$\frac{d\Phi_\chi}{ds} = \frac{\partial B(s,a,\lambda,...)}{\partial\lambda_\chi} = 0, \qquad (7.15b)$$

$$(\hat{g}^{\mu\nu}) = (\hat{g}_{\alpha\beta})^{-1}, \tag{7.15c}$$

The underlying algebraic brackets are then given by Eq.s (IV.4.23), plus additional degrees of freedom for the constraints which can be ignored for the study of the isosymmetries.

We have reached in this way an analytic formulation of relativistic Hamilton-isotopic mechanics whose algebraic structure *coincides* with that of the Poincaré-isotopic symmetries of the preceding section.

Finally, to achieve a $\hat{P}(3,1)$-invariant description, we remain with the third restriction that the Birkhoffian B is the Hamiltonian H properly written in $\hat{M}^4(x,\hat{g},\hat{\mathfrak{R}})$ which, in the general case of N particles, can be written

$$H = [p^{a\mu}\, \hat{g}_{\mu\nu}(s,x,p,)\, p^{a\nu}]\, /2\lambda_a \; - \; \tfrac{1}{2}c^4\lambda_a - C_k(x,p)\lambda_k + V(x^{ab}),$$

$$x^{ab} = [(x^{a\mu} - x^{b\mu})\, \hat{g}_{\mu\nu}\, (x^{a\nu} - x^{b\nu})]^{\frac{1}{2}} \tag{5.16}$$

$$i = 1, 2,..., 8N, \quad a = 1, 2,..., N, \quad \mu, \nu, = 1, 2, 3, 4, \quad k = i, 2,..., n - N,$$

where Lagrange's multiplier λ_a are given by

$$\lambda_a = \hat{m}_a = m_{ao}\, \hat{\gamma}, \tag{7.17}$$

the C's are the subsidiary constraints of systems (IV.2.5), and V is a conventional potential, only properly written in the isospace. The *isorelativistic equations of motion* are then given by

$$\frac{dx^{a\mu}}{ds} = \hat{g}^{\mu\nu}(x,p,..)\,\frac{\partial H(s,a,\lambda,..)}{\partial p^{a\nu}}, \tag{7.18a}$$

$$\frac{dp^{a\mu}}{ds} = -\,\hat{g}^{\mu\nu}(x,p,..)\,\frac{\partial H(s,a,\lambda,..)}{\partial x^{a\nu}}, \tag{7.18b}$$

$$\frac{d\Phi_a}{ds} = \frac{\partial H(s,a,\lambda,..)}{\partial \lambda_a} = -\tfrac{1}{2}\Big(\frac{\hat{p}_a^2}{2\lambda_a^2} + c^4\Big) = 0, \tag{7.18c}$$

$$\frac{d\Phi_k}{ds} = \backslash\,\frac{\partial H(s,a,\lambda,..)}{\partial \lambda_k} = C_k = 0, \tag{7.18d}$$

$$a = 1, 2, ..., N, \quad \mu, \nu = 1, 2, 3, 4, \quad k = 1, 2, ..., n$$

where Eq.s (7.18c) represents isoinvariants (7.14), and Eq.s (7.18d) represent all needed, additional subsidiary constraints (see also next section).

Again, we have presented here the simpler Hamilton-isotopic formulation, while leaving the most general possible Birkhoff-isotopic formulation to the interested reader.

RELATIVISTIC ISOSYMMETRIES AND CONSERVATION LAWS (*loc. cit.*). The third final tools needed (for the rudimentary level of this analysis) are the *symmetries and conservation laws for relativistic constrained Birkhoffian mechanics.*

A relativistic extension of Theorem 6.3.3, p. 240 of Santilli (1982a) can be done, quite simply, by interpreting the integrand of Pfaffian principle (7.13) as being the Lagrangian

$$L(s, a, \dot{a}, \lambda, ...) = \hat{R}_i(a)\, \dot{a}^i + \Phi_\chi\, \lambda_\chi - B(s, a, \lambda, ...). \qquad (7.19a)$$

$$R = (pT_1, 0) \qquad p7.19b)$$

Then, the relativistic constrained Birkhoff's equations (7.15a) coincide (in their covariant form) with the conventional Lagrange equations in the above Lagrangian, i.e.

$$\frac{d}{ds}\frac{\partial L}{\partial \dot{a}^i} - \frac{\partial L}{\partial a^i} \equiv \omega_{ir}\, T_2{}^r{}_j(a)\, \dot{a}^j - \frac{\partial B}{\partial a^i} = 0, \qquad (7.20)$$

with similar equations holding for the subsidiary constraints.

The application of the conventional Noether's theorem to the above Lagrangian formulation, plus the use of the Birkhoffian gauges (7.12), yield the following:

THEOREM IV.7.1 (Relativistic Birkhoffian Noether's Theorem, Santilli (1988c)) : If relativistic Hamilton-isotopic equations (7.18) admit a symmetry under an n-dimensional Lie group S_r of infinitesimal transformations

$$a^i \Rightarrow a'^i = a^i + w_k\, \alpha^i_k(a), \quad a = (x, p), \qquad (7.21)$$

then there exist r quantities

$$Q_k = \hat{R}_i(a)\,\alpha^i_k(a,..) + G_k(s,a,..), \quad i = 1, 2, ..., 8, \quad k = 1, 2, ..., r, \quad (7.22)$$

which are first integrals along a possible path (conserved quantities).

The attentive reader may have noted that the symmetry S_r of Theorem IV.7.1 has an undefined Lie structure, that is, it can be either a conventional Lie or a Lie-isotopic symmetry. In order to narrow the possibilities to those directly applicable to Eq.s (7.18), we introduce the following

DEFINITION IV.7.3 (loc. cit.): A symmetry S_r of relativistic Hamilton-isotopic equations (7.18) is called an "isosymmetry" and denoted with the symbol $\hat{S}_r$ when it admits the infinitesimal form on $T^*\hat{M}(x,\hat{g},\hat{\Re})$

$$a'^i = a^i + w_k\,\omega^{iq}\,\hat{1}_{2q}{}^j\,(\partial_j X_k), \quad (7.23)$$

where the X's are the generators of $\hat{S}_r$ with isocommutation rules in terms of brackets (IV.4.23)

$$[X_i \hat{,} X_j] = (\partial_p X_i)\,\omega^{pt}\,\hat{1}_{2t}{}^q\,(\partial_q X_j) = \hat{C}_{ij}{}^k(a,..)\,X_k, \quad (7.24)$$

and the Ĉ's are the structure functions.

In essence the above definition assures that the algebraic structure of the isosymmetry $\hat{S}_r$ coincides with that of Eq.s (7.18). The necessary and sufficient conditions for an isosymmetry are then given by a simple relativistic extension of Theorem II.8.3

THEOREM IV.7.2 (Integrability conditions for the existence of an isosymmetry (loc. cit.)): Necessary and sufficient conditions for infinitesimal isotransformations (7.23) to be an isosymmetry of relativistic Hamilton-isotopic equations (7.18) are that all generators X_k of the isosymmetry isocommute with the Hamiltonian (7.16), i.e.,

$$[X_k \hat{,} H] \equiv 0, \quad k = 1, 2, ..., r. \quad (7.25)$$

But the isopoincaré symmetries have the same algebraic structure of relativistic Birkhoff's equations (7.18) by constructions. We therefore have from Theorems IV.7.1 and IV.7.2 the following:

THEOREM IV.7.3 (loc. cit.): A necessary and sufficient condition for relativistic Hamilton-isotopic equations (7.18) to be invariant under a general isopoincaré symmetry $\hat{P}(3.1)$ in the same isospace $\Re_s \times T^*\hat{M}^I(x,\hat{g},\Re)$, is that Hamiltonian (7.16), and more particularly, its potential V, are $\hat{P}(3.1)$-invariant, in which case its conventional ten generators $X_k = (J^{\mu\nu}, P^{\mu})$ are automatically conserved.

As an incidental note we have used here the term "invariant" because referred to the invariasnce property of the underlying vector-field, $\Gamma'(s',a',..) \equiv \Gamma(s', a',...)$. When considering the equations of motion, a more appropriate term is that of "covariance", as in the conventional case.

Note also that, exactly as in the conventional case, Hamiltonian (7.16)) does not represent the total energy of the particle, but only its generator of time isoevolution.

Nevertheless, while in conventional treatments the *constant* $\sum_a m_{ao} c_o$ is usually added to render the numerical value of the Hamiltonian null, this is not the case in isorelativistic mechanics, because it would imply adding the *functions* $\sum_a m_{ao} c_o \hat{b}_4(x,p,...)$ with consequential alteration of the equations of motion.

We are now sufficiently equipped to study the following illustrative cases of *one* isoparticle in an *external* medium. The system is therefore open by conception and, as such, it requires only the fundamental constraint (7.4). Throughout the rest of this section we therefore have a = 1, $\lambda_X \equiv \lambda$, and $\Phi_X = \Phi$.

"FREE" RELATIVISTIC ISOPARTICLE (*loc. cit.*). In this case we have a null potential in Hamiltonian (7.16), the isometric has the constant form

$$\hat{\eta} = \hat{g} = \text{diag.}\,(b_1^2, b_2^2, b_3^2, -b_4^2), \qquad b_\mu = \text{constants} > 0, \tag{7.26}$$

and the Hamiltonian is given by

$$H = p^\mu \hat{\eta}_{\mu\nu} p^\nu / 2\lambda \;-\; \tfrac{1}{2} c^4 \lambda \tag{7.27}$$

resulting in the equations of motion

$$\dot{x}_\mu = b_\mu^{-2}\, \partial H/\partial p^\mu = p_\mu/\hat{m}, \quad \text{(no sum)} \tag{7.28a}$$

$$\dot{p}_\mu = -\,b_\mu^{-2}\, \partial H/\partial x^\mu = 0, \quad \text{(no sum)} \tag{7.28b}$$

$$\dot{x}^{\mu}\,\hat{\eta}_{\mu\nu}\,\dot{x}^{\nu} = -1. \qquad (7.28c)$$

which are manifestly $\hat{P}(3.1)$-covariant, nevertheless, they coincide with the conventional equations for a free particle.

The first possible interpretation of system (7.28) is that along the main line of research on these volumes, namely, the representation of *an extended particle moving in vacuum whose caracteristic $\hat{b}$-functions have been averaged into the b-constants.*

In particular, the above case is the relativistic extension of the corresponding nonrelativistic case of Sect. III.7.

A simple example of the isospace under consideration is that characterized by the Nielsen-Picek metric (IV.3.21) for which

$$b_1 = b_2 = b_3 = b = 1 - \alpha/3, \qquad b_4 = 1 + \alpha, \qquad (7.29a)$$

$$\alpha \cong -3.79\times10^{-3} \text{ for pions}, \qquad \alpha \cong +0.61\times10^{-3} \text{ for kaons}. \qquad (7.29b)$$

Note that isoparticle (7.128) is free and, as such, it cannot represent an isoquark.

At the rudimentary level of this example, mutations are expectedly minimal, and mainly restricted to the behavior of physical quantities with speed (see Sect. IV.9). As an example, the rest mass m_o is an independent parameter of the theory which, as such, has to be conventionally assigned and cannot be mutated (at this stage). Nevertheless, its behavior with speed and the corresponding rest energy are mutated (Sect. IV.9). A similar situation occurs for other quantities.

In regard tio the representation of rthe three-dimensional shape of the particle, recall that the conventional Minkowski space $M(x,\eta,\Re)$ can be interpreted as a geometrical space whose space component $E(r,\delta,\Re)$ characterizes a rotationally invariant, rigid and perfect sphere of unit radius, δ = diag. (1,1,1). In the transition to the isominkowski space $\hat{M}^I(x,\hat{\eta},\hat{\Re})$, we can therefore represent the actual shape of the particle considered via the space isotopy

$$\delta = \text{diag.}\,(1,1,1) > 0 \;\Rightarrow\; \hat{\delta} = \text{diag.}\,(b_1^2, b_2^2, b_3^2) > 0, \qquad (7.30)$$

which is merely embedded in the larger isorelativistic context of the full isometric $\hat{\eta}$.

Moreover, one can have an *isotopy of an isotopy* (Sect. IV.5) of the type

$$\hat{M}^{I}(x,\hat{\eta},\hat{\mathfrak{R}}) \Rightarrow \hat{M}^{I}(x,\hat{\eta}',\hat{\mathfrak{R}}), \tag{7.31a}$$

$$\hat{\eta} = \text{diag.}(b_1^{\,2}, b_2^{\,2}, b_3^{\,2}, -b_4^{\,2}) \Rightarrow \hat{\eta}' = \text{diag.}(b'^{\,2}_1, b'^{\,2}_2, b'^{\,2}_3, -b'^{\,2}_4), \tag{7.31b}$$

which evidently represents the deformations of the original shape, and holds, e.g., when the isometric is dependent, say, on an external pressure π, $b_\mu = b_\mu(\pi)$. The infinite number of possible $\hat{P}(3.1)$ symmetries then represent the infinite number of possible shapes.

It should be noted again, as we did it at the nonrelativistic level of Sect. III.7, that *the representation of the actual shape of the particle and of all its infinitely possible deformations is permitted by our formulations already at the classical level of this analysis.*

By comparison, conventional Einstenian theories, first of all, cannot classically represent any extended character of the particles and, secondly, they can do it only after the rather laborious second quantization. Even at that level, one can obtain only some remnants of the shape via the form factors, and not the actual shape itself (say, an oblate spheroidal ellispoid, a quite probable shape of the proton, as well as of all spinning and extended charge distributions), evidently because such a shape is not rotationally invariant. Finally, conventional theories in second quantization cannot possibly represent the deformation of a given shape, evidently because it would imply a direct violation of the conventional rotational and Lorentz symmetries.

Moreover, we should recall that our isorotational group $\hat{O}(3)$ restores the exact rotational invariance for all deformed shapes of type (7.30) or (7.31), of course, at the higher isotopic level.

Note that, in the conventional case of a free relativistic particle in Minkowski space, the translations are given by

$$x \Rightarrow x' = x + x^\circ, \qquad p \Rightarrow p' = p, \tag{7.32}$$

and motion is along a straight line.

For the case of isoparticle (7.28), the isotranslations are given by

$$x \Rightarrow \hat{x} = x + x^\circ b, \qquad p \Rightarrow p'. \tag{7.33}$$

Motion is evidently still in a straight line (from the constancy of the characterictic b-quantities). Nevertheless, the reader should keep in mind the redefinition of the x°-parameters when needed to be referred to translation of the center-of-mass of the isoparticle in vacuum

$$x^\circ \Rightarrow \hat{x}^\circ = x^\circ b. \tag{7.34}$$

ISOPARTICLE UNDER AN EXTERNAL POTENTIAL FORCE (*loc. cit.*). In this case the isometric can also be assumed to be constant, but the potential V in Hamiltonian (7.16) is not null. The best case is given by a charged isoparticle in interaction with an external electromagnetic field with potentials $A^\mu(x)$, see, e.g., Mann (1974). The Hamiltonian can then be written in $T^*\hat{M}^I(x,\hat{\eta}',\hat{\Re})$

$$H = (p^\mu - eA^\mu / c)\, \hat{\eta}'_{\mu\nu}\, (p^\nu - eA^\nu) / 2\lambda \; - \tfrac{1}{2}c^4\lambda\,, \tag{7.35}$$

with equations of motion

$$\begin{cases} \dot{x}_\mu = b'^{-2}_\mu \dfrac{\partial H}{\partial p^\mu} = (p_\mu - \dfrac{e}{c} A_\mu) / m, & (7.36a) \\ \dot{p}_\mu = - b'^{-2}_\mu \dfrac{\partial H}{\partial x^\mu} = \dfrac{e}{cm} (\partial_\mu A_\alpha)(p^\alpha - \dfrac{e}{c} A^\alpha), & (7.36b) \\ \dot{\Phi} = \dfrac{\partial H}{\partial \lambda} = - \dfrac{1}{2\lambda^2} (p - eA/c)^{\hat{2}} - \tfrac{1}{2} c^4 = 0, \; \lambda = m, & (7.36c) \end{cases}$$

which are also manifestly $\hat{P}(3.1)$-covariant (and which admit systems (7.28) when $A^\mu = 0$)

The most significant implication of system (7.36) is the possibility that the external field produces the mutation of shape of the original charge distribution. We are referring to the possibility that, starting with isoparticle (7.28) with given shape (7.30), the addition of a sufficiently intense, external electromagnetic field produce deformations (7.29), i.e,

$$H = p^\mu \hat{\eta}_{\mu\nu} p^\nu / 2\lambda - \tfrac{1}{2}c^4\lambda \;\Rightarrow\; H' = (p^\mu - eA^\mu/c)\; \hat{\eta}'_{\mu\nu}(p^\nu - eA^\nu/c)/2\lambda \; - \; \tfrac{1}{2}c^4\lambda$$

$$\hat{\delta} = \text{diag.}\,(b_1^2, b_2^2, b_3^2) \;\Rightarrow\; \hat{\delta}' = \text{diag.}\,(b'_1{}^2, b'_2{}^2, b'_3{}^2). \tag{7.37}$$

which is precisely a classical relativistic extension of the deformation of an extended charge distribution studied in Sect. III.7, which is applied there to a purely classical, but quantitative interpretation of

Rauch's experiment on the rotational symmetry of neutrons under external nuclear fields in Chapter VII.

For this and other aspects, we refer the reader to Chapter VII. The $\hat{P}(3.1)$-invariant, field theoretical, operator interpretation of Rauch's experiments is presented in Santilli (1991d).

Note that in this example that *the isoparticle is mutated, but the electromagnetic field, being external, is conventional*.

It should be finally noted that system (7.36) remains virtually unchanged by adding a dependence of the isometric on local density, pressure, temperature, etc., thus providing a class of additional examples.

ISOPARTICLE UNDER EXTERNAL NONLINEAR, NONLOCAL AND NONHAMILTONIAN FORCES (*loc. cit.*). As well known, Einstein's special relativity can only represent particles under external local-potential forces. The best way to illustrate the generalized character of our formulations is by showing that they permit the representation of the most general possible combination of linear and nonlinear, local and nonlocal, and potential or nonhamiltonian forces.

As a simple, local, but nonlinear and nonhamiltonian example, consider the case in which the conventional potential V is null, the isometric is given by

$$\hat{g} = e^{-\frac{1}{2}kx^2}\eta, \qquad x^2 \in M(x,\eta,\Re), \tag{7.38}$$

and the Hamiltonian is that for the "free" isoparticle, i.e.,

$$H = p^\mu \hat{g}_{\mu\nu}(x)\, p^\nu / 2\lambda - \tfrac{1}{2} c^4 \lambda, \tag{7.39}$$

Then the equations of motion are given by

$$\begin{cases} \dot{x}^\mu = p / \hat{m} & (7.40a), \\ \dot{p}^\mu = k\, x^\mu (p^\mu \hat{g}_{\alpha\beta} p^\beta)/2m - 4\, k\, x^\mu\, \hat{m}\, c^4, & (7.40b) \\ \dot{\Phi} = -\tfrac{1}{2}(p^\alpha \hat{g}_{\alpha\beta} p^\beta)/\lambda^2 - \tfrac{1}{2}c^4 = 0, \ \lambda = \hat{m}, & (7.40c) \end{cases}$$

which are also manifestly $\hat{P}(3.1)$-covariant.

The above example illustrate the nonlinearity of the theory. A simple example of nonlocal interactions are provided by isotopies characterized by the surface integral

$$\eta \;\Rightarrow\; \hat{g} = e^{K\int_{\sigma} d\sigma F(x,\, p,\, \mu,\, \tau,\, n, ...)}\, \eta, \quad K \in \Re, \qquad (7.41)$$

which is nothing but a relativistic extension of the nonlocal-integral forces experienced by extended bodies with surface σ moving with a resistive medium, as familiar in mechanics.

A virtually endless number of generalizations are then conceivable, via local and nonlocal isometric and with or without conventional potential forces. Their study is left as an instructive exercise to the interested reader.

The above nontrivial forms of isoparticles share the following properties:

1) they provide a classical, relativistic description of interior trajectories, such as a high speed, extended test particle during penetration in the Jovian atmosphere considered as external, or, along similar lines, a first, classical representation of a proton moving within the hyperdense medium in the core of a star;

2) they evidently admit possible mutations of their characteristics; and, last but not least;

3) they all restore the exact Poincaré symmetry, of course, at our isotopic level, by therefore coinciding with the conventional Einsteinian particles at the abstract, realization-free level.

IV.8: CLOSED SYSTEMS OF ISOPARTICLES.

An inspection of the historical successes of the Poincare′ symmetry and of the special relativity reveals the restriction of their exact applicability to closed systems of point-like particles with only local-potential forces, such as the planetary or atomic systems.

In an attempt to enlarge these physical conditions, while preserving the underlying abstract axioms, in the preceding chapter we have studied closed-isolated systems of extended particles with conventional selfadjoint, as well contact nonselfadjoint) internal forces, which nevertheless verify the ten total conservation laws imposed as subsidiary constraints.

Moreover, we have constructed a classical, nonlinear and nonlocal realization of the isogalilean symmetries $\hat{G}(3.1)$ and shown that $\hat{G}(3.1)$-invariant, closed, nonhamiltonian systems verify the ten, total, conventional conservation laws as a result of the isosymmetry, and without subsidiary constraints.

In this way, the isogalilean symmetries essentially select the unconstrained subclass of closed nonhamiltonian systems. In fact, the preservation of the ten total conservation laws is ensured by the preservation of the conventional generators of the Galilei symmetry, while the generalized, symplectic-isotopic/Lie-isotopic structure of $\hat{G}(3.1)$ ensures the existence of the most general known internal forces.

The generalization of these results to a relativistic setting is straighforward.

First, among the infinite class of systems (IV.2.5) of N isoparticles, we *restrict* our attention to the systems whose Birkhoffian vector-field Γ can be consistently written in the isosymplectic form (IV.7.15) in isospaces $T^*\hat{M}^I(x,\hat{g},\hat{\Re})$. More explicitly, consider system (IV.2.3a) in the vector field form

$$\dot{a} = (\dot{a}^i) = \begin{pmatrix} \dot{x}^{a\mu} \\ \dot{p}^{a\mu} \end{pmatrix} = \Gamma(s,a,..) = \begin{pmatrix} p^{a\mu}/m_a \\ K^{a\mu}{}_{SA} + K^{a\mu}{}_{NSA} \end{pmatrix}, \quad (8.1)$$

where we have absorbed the nonlocal forces in the nonselfadjoint ones. The above restriction implies the representation

$$\Gamma^i(s,a,..) = \omega^{iq}\,\hat{1}_{2q}{}^j\,\partial_j B(s,a,...), \quad (8.2)$$

Finally, the Birkhoffian B must be restricted, via the use of the various degrees of freedom of the theory, to represent the conventional Hamiltonian, only properly written in $T^*\hat{M}^I(x,\hat{g},\hat{\Re})$, i.e., Eq. (IV.7.16). The latter requirement has a number of consequences, such as:

1) it implies that all nonselfadjoint forces are represented with the Lie-isotopic tensor of the theory;

2) the Hamiltonian represents only potential forces; and, last but not least

3) the conventional relativistic setting is recovered identically at the limit of null nonselfadjoint forces.

Once the considered closed nonhamiltonian system admits the Lie-isotopic form (IV.7.15) and Hamiltonian (IV.7.16), Theorems IV.7.1, IV.7.2 and IV.7.3 apply.

The isopoincaré invariance then follows when all isocommutators (IV.7.25) hold, namely, when H itself is invariant under $\hat{P}(3.1)$. The conservation of all the *conventional* generators (IV.6.5) then follows.

THEOREM IV.8.1 (Santilli (1988c)): The covariance under a general, isopoincaré symmetry on $T^*\hat{M}^I(x,\hat{g},\hat{\mathfrak{R}})$ characterizes closed-isolated systems of extended isoparticles with conventional, local, action-at-a-distance interactions represented by the Hamiltonian, plus contact, nonlinear, nonlocal and nonhamiltonian interactions represented by the isounit.

In different terms, the generally open, nonconservative, relativistic, Birkhoffian systems (IV.2.5a) are "closed" by the imposition of the isopoincaré symmetries $\hat{P}(3.1)$, exactly as expected from the corresponding nonrelativistic counterpart of the preceding chapter.

The above results can be summarized as follows. The subclass of $\hat{P}(3.1)$-invariant systems (IV.2.3) are those admitting the following representation

$$\begin{cases} \dot{x}^{a\mu} = = p^{a\mu}/m_{ao} = \hat{g}^{\mu\nu}(x, p)\, \partial_{a\nu} H(x, p, \lambda, ...)\,, & (8.3a) \\ \dot{p}^{a\mu} = K^{a\mu}{}_{SA} + K^{a\mu}{}_{NSA} = - \hat{g}^{\mu\nu}(x, p)\, \partial_{a\nu} H(x,p,\lambda,...), & (8.3b) \\ \dot{\Phi}_a = \partial H(x,p,\lambda,...)/\partial\lambda_a = -\tfrac{1}{2} c^4 (\dot{x}^{a\alpha} \hat{g}_{\alpha\beta} \dot{x}^{a\beta} + 1) = 0, & (8.3c) \end{cases}$$

$$H = \sum_a [\, (p^{a\mu} \hat{g}_{\mu\nu}(x, p)\, x^{a\nu})/2\lambda_a - \tfrac{1}{2} c^4 \lambda_a + V(x_{ab}\hat{^2}), \qquad (8.3d)$$

$$x_{ab}\hat{^2} = (x_a{}^{\mu} - x_b{}^{\mu})\, \hat{g}_{\mu\nu} (x_a{}^{\nu} - x_a{}^{\nu}), \qquad (8.3e)$$

where one should note the *lack of the subsidiary constraints (IV.2.3a) and (IV.2.3b) because they are now automatically verified by the $\hat{P}(3.1)$-invariance.*

In case the more general systems (IV.2.5) are desired, their $\hat{P}(3.1)$-invariant subclass are characterized by the Hamiltonian (IV.7.16) for the systems considered, i.e.,

$$H = \sum_a [\, (p^{a\mu} \hat{g}_{\mu\nu} p^{a\nu})/2\lambda_a - \tfrac{1}{2} c^4 \lambda_a + V(x^{ab\hat{2}})] + \sum_k c_k \lambda_k, \quad (8.4a)$$

$$C_1 = F_1(d\hat{P}^2/ds), \quad C_2 = F_2(d\hat{W}^2/ds), \quad C_3 = F_3(\dot{X}^{\alpha}\eta_{\mu\nu}\dot{X}^{\nu} + 1), \quad (8.4b)$$

where one recognizes the three subsidiary constraints (IV.2.5c)-(IV.2.5e) multiplied by arbitrary (sufficiently smooth and regular) functions of the local variables F_k, k = 1, 2, 3.

The $\hat{P}$(3.1)-invariant subclass of closed nonselfadjoint systems (IV.2.5) are then represented by Eq.s (8.3) plus the additional subsidiary constraints

$$\dot{\Phi}_1 = \partial B/\partial\lambda_1 = F_1(d\hat{P}^2/ds) = 0, \quad (8.5a)$$

$$\dot{\Phi}_2 = \partial B/\partial\lambda_2 = F_2(d\hat{W}^2/ds) = 0, \quad (8.5b)$$

$$\dot{\Phi}_3 = \partial B/\partial\lambda_3 = F_3(\dot{X}^{\mu}\eta_{\mu\nu}\dot{X}^{\nu} + 1) = 0, \quad (8.5c)$$

where one should note, again, the lack of subsidiary constraints (2.5b) because ensured by the $\hat{P}$(3,1) symmetry.

This completes the treatment of the systems considered with Lagrange's multipliers. To study their consistency and identify their hypersurface of the constraints, one has to use Dirac's method (Appendix IV.B).

As one recalls from Sect. IV.2, stricter class (8.5) is requested when one wants no remnant whatever of the generalized interior structure in the conventional exterior setting. In fact, *the $\hat{P}$(3.1) isosymmetry ensures the validity of the ten, total, conventional, conservation laws, while the additional constraints (8.5) ensure that all total quantities are defined in the conventional Minkowski space.*

In the derivation of systems (8.3) and (8.4), we assume the reader is familiar with the techniques underlying analytic equations with Lagrange's multipliers, such as the fact that the equations must be computed first along an arbitrary path which is *not* the solution of the system, while the multipliyers are originally considered as new independent variables. The systems are then computed along an actual path. In particular, the subsidiary constraints hold only along the actual path. For additional technical details, we refer the interested reader to Bliss (1946).

Note, for use in the next section, that the center-of-mass of systems (8.4) strictly verifies all Einstenian laws, e.g., time dilation, space contraction, etc.

The relativistic generalizations of the two-body and three-body closed nonhamiltonian systems of Appendix III.A is left as an instructive exercise for the interested reader. Note that all examples of the preceding section for one isoparticle with local coordinate x can be reinterpreted as two-particle nonhamiltonian systems where x represents the relative coordinate.

The system characterized Hamiltonian (8.4) for a sufficiently large number of constituents N constitutes our *relativistic model of the structure of Jupiter* as a closed nonhamiltonian system in an isotopic, yet fully flat space $M^I(x,\hat{g},\hat{\Re})$, and prior to our gravitational consideration on isospace M^{III} of the subsequent chapter.

It is another instructive exercise for the interested reader to see that the model does indeed achieve our objectives, that is, the representation of the local, internal, *nonconservative* structure, thus eliminating the existence of the perpetual motion in a physical environment (Sect. IV.1).

Systems (8.4) for N = 2 and 3 also constitute *classical relativistic, closed, nonhamiltonian structure models of hadrons as bound states of isoquarks* conceived precisely as an operator image of Jupiter.

IV.9: ISOTOPIC LIFTINGS OF EINSTEIN'S SPECIAL RELATIVITY

In the preceding chapters we have first reviewed the impossibility of reducing the Universe to a finite number of point-like particles with only local-potential interactions, because of the impossibility of reducing the nonlinear, nonlocal and nonconservative systems of our macroscopic reality to a finite collection of stable elementary orbits.

In this way, we have confirmed the historical distinction by the Founding Fathers of analytic dynamics between the *exterior* and the *interior dynamical problem* (Chapter I).

We have then identified the differentiation between the local and potential character of the exterior center-of-mass motion of particles in vacuum, and the generally nonlinear, nonlocal and nonhamiltonian character of the interior trajectories, such as for a spaceship penetrating within the Jovian atmosphere or, along conceptually similar lines, a hadronic constituent with extended wavepackets while moving within the medium composed by the wavepackets of the remaining constituents.

We have finally achieved the compatibility between the above exterior and interior dynamics via our nonrelativistic and relativistic, closed nonhamiltonian systems and their isosymmetries, as well as the abstract geometrical identity of our generalized systems and symmetries with the conventional ones.

The above results were reached via a *necessary* isotopy of all structural features of conventional formulations, such as: fields; metric spaces, analytic mechanics, Lie's theory; symplectic geometry; symmetries, etc.

A primary result of our analysis is therefore that the fundamental symmetries of contemporary physics, Galilei, Lorentz and Poincare' symmetries, are not violated in the transition from the exterior to the interior dynamics, but preserved in full, because we merely perform the transition from their simplest conceivable, to their most general possible realizations.

It is evidently necessary to complement these studies with the identification of the generalized relativities emerging from these techniques, as well as of their primary consequences.

On mathematical grounds, it is easy to see that *the isotopic liftings $P(3.1) \Rightarrow \hat{P}(3.1)$ imply necessary, corresponding, isotopies of Einstein's special relativity.*

The main objective of this section is therefore that of identifying the nonlinear, nonlocal and nonhamiltonian generalizations of the basic postulates of the special relativity which are implied by our isosymmetries $\hat{O}(3.1)$ and $\hat{P}(3.1)$ on isospaces $M^I_2(x,\hat{g},\hat{\mathfrak{R}})$.

The foundations of the generalized relativities, under the name of *Lorentz-isotopic relativities*, or *isospecial relativities* for short, were first achieved in Santilli (1983a), and subsequently expanded in Santilli (1988c), which is followed in this review. Additional studies can be found in Santilli (1991c).

The analysis of this section can also be considered to be a relativistic generalization of the *isogalilean relativities* for the interior problem of ref. [10]. As a matter of fact, *the relativities studied in this section were constructed in such a way to admit the isogalilean relativities under conventional nonrelativistic limits* (see Chapter VI for the contraction of the former into the latter).

Finally, the content of this sectin should be considered as a basis for the gravitational studies on the interior gravitational problem of the next chapter. In fact, *the isotopic liftings of Einstein's gravitation of the next chapter were also constructed in such a way to admit, locally, the isospecial or isogalilean relativities.*

The problem of the experimental verification of the deviations from the special relativity predicted by the Lorentz-isotopic relativities in the interior problem, will be considered in Chapter VII.

Let us begin with the following:

DEFINITION IV.9.1 (Santilli (1983a), (1988c)): The "general isotopic liftings" of Einstein's special relativity, herein called "general isospecial relativities," are given by the generalizations characterized by the most general possible, nonlinear, nonlocal and nonhamiltonian isopoincaré symmetries $\hat{P}(3.1)$ on isominkowski spaces $M^I(x,\hat{g},\hat{\Re})$ (Theorem IV.6.1)

$$\hat{O}(3.1): \qquad x' = \hat{\Lambda} * x = \hat{\Lambda} T_2 x, \quad T_2 > 0, \tag{9.1a}$$

$$\hat{\Lambda}^t T_2 \eta \hat{\Lambda} = \hat{\Lambda} T_2 \eta \hat{\Lambda}^t = \hat{1}_2 \eta, \quad \hat{1}_2 = T_2^{-1}, \tag{9.1b}$$

$$\hat{T}(3.1): \qquad x' = x + x^\circ \hat{b}(x, p, ...), \tag{9.1c}$$

or, equivalently, by the most general possible, nonlinear and nonlocal realizations of the isounits $\hat{I}_2 > 0$ of the theory. The "restricted isotopic liftings" of Einstein's special relativity, or "restricted isospecial relativities", are characterized instead by the most general possible linear and local isopoincaré transformations on isospaces $\hat{M}^I(x,g,\hat{\Re})$ with isometric independent from the local variables and their derivatives, but dependent on other physical characteristics of the medium considered.

A few comments are here in order. The above definitions have been conceived to express the construction of the isospecial relativities with the same spirit of the remaining isotopic formulations, i.e., in such a way to coincide with the conventional relativity at the abstract, realization-free level (see later on Theorem IV.9.1).

The reader should therefore be aware from the outset that all the deviations predictaed by the isospecial from the conventional relativity are directly permitted by Einstein's basic postulates, only realized in their most general possible form.

Such an ultimate abstract unity is ensured by the isominkowski spaces of Class I (Sect. IV.3). In fact, as the reader will recall, all the isospaces considered imply the local isomorphisms $\hat{O}(3.1) \approx O(3.1)$. Moreover, the use of isospaces of Class I eliminates *ab inition* all gravitational effects due to curvature, thus restrictuing all isotopies of this section to flat isospaces.

The situation becomes fundamentally different if one allows more general isospaces. In fact, as we shall study in more details in Chapter VI, *Einstein's gravitation is a form of isotopy of the special relativity.*

The understanding is that our most general possible isotopies of the special relativity will characterize certain *generalizations* of Einstein's gravitation in isospaces $M^{III}(x,\hat{g},\hat{\Re})$.

In different terms, our isotopies identify the ultimate geometrical axioms of a given theory and realize them in the most general possible forms. In this process one recovers conventional theories, but also identifies an infinite number of possible generalized theories.

Throughout the analysis of this section we shall restrict our attention, for simplicity but without loss of generality, to general isospaces of Class I of the diagonalizable form

$$\hat{M}^I(x,\hat{g},\hat{\Re}): \quad x^2 = x^\mu \, \hat{g}_{\mu\nu} \, x^\nu =$$

$$= x^1 \, \hat{b}_1^2 \, x^1 + x^2 \, \hat{b}_2^2 \, x^2 + x^3 \, \hat{b}_3^2 \, x^3 - x^4 \, \hat{b}_4^2 \, x^4,$$

$$= x^1 \frac{1}{b_1^2} x^1 + x^2 \frac{1}{n_2^2} x^2 + x^3 \frac{1}{n_3^2} x^3 - t \frac{c_o^2}{n_4^2} t, \tag{9.2a}$$

$$x = (r, x^4) = (r, c_o t), \quad r \in \hat{E}_2(r,\hat{G},\hat{\Re}) \tag{9.2b}$$

$$\hat{g} = T_2 \, \eta, \tag{9.2c}$$

$$\eta = \text{diag.}\,(1,1,1,-1) \in M(x,\eta,\Re), \tag{9.2d}$$

$$T_2 = \text{diag.}\,(\hat{b}_1^2, \hat{b}_2^2, \hat{b}_3^2, \hat{b}_4^2) > 0, \quad \hat{\Re} = \Re \, \hat{I}_2, \quad \hat{I}_2 = T_2^{-1}, \tag{9.2e}$$

$$\hat{b}_\alpha = 1 / n_\alpha = \hat{b}_\alpha(s, x, u, a, \mu, \tau, n, ...) > 0, \quad \alpha = 1, 2, 3, 4, \tag{9.2f}$$

$$\hat{b}_1 = \hat{b}_2 = \hat{b}_3 = 1/n_1 = 1/n_2 = 1/n_3 = \hat{b} = 1/n_3, \tag{9.2g}$$

$$c = c_o \hat{b}_4 = c_o / n_4, \tag{9.2h}$$

where: conditions (9.2g) are assumed for the specific purpose of identifying the relativistic effects of the interior dynamical problem, and separate them from the effects due to isorotations $\hat{O}(3)$ studied in the preceding chapter; conditions (8.1f) are assumed, in addition to the conditions $\hat{b}_\mu^2 > 0$, to permit the identification of the b-functions with physical quantities (see also next section); the quantities $\hat{b}_\mu$ have the most general possible nonlinear and nonlocal depedence in all permitted variables and quantities; the metric $\hat{g}$ is that of the isocotangent bundle $T^*\hat{M}^I_2(x,\hat{\eta},\hat{\Re})$, i.e., it is such that brackets (IV.4.23) characterized by the inverse $\hat{g}^{-1}$ verify the classical Lie-isotopic axioms (IV.4.12); and Eq. (9.2h) represents the geometrization of the conventional speed of light in vacuum c_o, characterized by our isotopies.

The analysis for the restricted isotopies will be conducted in the isospaces of Class I of the particular form

$$\hat{M}^I(x,\hat{\eta},\hat{\mathfrak{R}}) \equiv \hat{M}^I(x,\hat{g},\hat{\mathfrak{R}}), \tag{9.3a}$$

$$\hat{\eta} = \text{diag.}\,(b_1^2, b_2^2, b_3^2, -b_4^2) = \text{local constant} > 0, \tag{9.3b}$$

$$b_\alpha = \hat{b}_\alpha = 1/n_\alpha = \text{constants} > 0, \tag{9.3c}$$

$$b_1 = b_2 = b_3 = b = 1/n_1 = 1/n_2 = 1/n_3, \quad c = c_o b_4 = c_o/n_4. \tag{9.3c}$$

We assume the reader is familiar with the primary differences between the "general" and "restricted" isotopies of Definition IV.9.1 identified in the preceding analysis.

It should be recalled from Sect. IV.6 that the generally nonlinear characteristic $\hat{b}$-functions of a given interior medium can always be averaged into b-constants. This permits the regaining of the linearity and locality for the isospecial relativities, although their predictions remain different than those of Einstein's special relativity, as it will be evident in a moment.

We are now equipped to review the most general possible formulation of the main postulates of the isospecial relativities.

Let us begin by studying the invariant speed under isotopies. For this purpose we need first the maximal possible causal speed, which, as in the conventional case, is characterized by the null isovector

$$ds^{\hat{2}} = dr^k \,\hat{b}_k^{\,2}\, dr^k - dt\, c_o^{\,2}\hat{b}_4^{\,2}\, dt = 0, \tag{9.4}$$

yielding the following

DEFINITION IV.9.2 (Santilli (1983a), (1988c)): The maximal, local, causal speed of the general and restricted isospecial relativities is the maximal speed of a massive particle and/or of a signal verifying the principle of cause and effects, given explicitly by

$$V_{Max} = \left|\frac{dr}{dt}\right|_{Max} = c_o \frac{\hat{b}_4}{b_3} = c_o \frac{n_3}{n_4} \tag{9.5}$$

The first fundamental postulate of the isospecial relativities can then be formulated as follows.

POSTULATE I *(loc. cit.): The invariant speeds of the general and restricted Lisospecial relativities are not given, in general, by the speed of light, but by the local, maximal, causal speed for each given physical medium.*

Consider two identical speeds $v_1 = v_2 = c = c_o \hat{b}_4$ Then, their isorelativistic composition yields

$$V_{Tot} = \frac{2c}{1 + \hat{b}_3} \neq c, \tag{9.6}$$

Consider, on the contrary, two maximal causal speeds, $v_1 = v_2 = V_{Max} = c/\hat{b}_3$. Then their isorelativistiv sum is given by

$$V_{Tot} = \frac{2c_o b_4/\hat{b}_3}{2} = V_{Max}, \tag{9.7}$$

thus illustrating the postulate.

Einstein's special relativity is a trivial particular case of Postulate I because it implies $V_{Max} = c = c_o$. Nevertheless, it is remarkable to note that the invariant quantity of the special relativity, strictly speaking, is not c_o, but the maximal causal speed.

Another special case is that of Bogoslovski's special relativity (1974), (1994) (Sect. IV.5) which does indeed verify Postulate I, as the reader can verify.

The plausibility of the above postulate is readily illustrated by the case of the *Cherenkov light. Since $c = c_o/n_4 < c_o$ in water (and the speed of light varies from transparent medium to transparent medium), the speed of light c cannot possibly be an invariant of any relativity, irrespective of wether conventional or Lorentz-isotopic.*

On the contrary, it is important for consistency that the invariant speed is given by the local, maximal, causal speed, as correctly identified by the isospecial relativities.

The most visible departure of the isospecial from the conventional relativity is given by the following

POSTULATE II *(loc. cit.): The maximal possible, causal speed under the general and restricted isospecial relativities*

can be smaller, equal or bigger than the speed of light in vacuum c_o

$$V_{Max} = c_o \frac{b_4}{\hat{b}} \gtreqless c_o, \tag{9.8}$$

depending on the local physical conditions of the medium considered.

The most intriguing application of the above postulate is for the possible achievement of a *true quark confinement,* i.e. (Sect. IV.7) a confinement with an infinite potential barrier, as well as an identically null probability of tunnel effects into free quarks. Evidently, we can outline here only the main idea, and work out the technical details at some future time when dealing with the operator formulation of the tnew relativities.

According to the original conjecture by Santilli (1982b), *the maximal causal speed in the interior of hadrons (e.g., the speed of the hadronic constituents) is bigger than* c_o. If verified, such a conjecture would evidently permit a *"true isoquark confinement"* because of the incoherence of the interior and exterior Hilbert spaces. By keeping in mind that the maximal causal speed in the Einsteinian exterior world is c_o, an isoquark at speeds $V_{Max} > c_o$ has a necessarily null probability of tunneling into a *free* state with speed $\leqq c_o$. In fact, in order for an isoquark to tunnel into a free state it must first decompose into ordinary particles, and this could explain the impossibility of observing free quarks.

Intriguingly, all modifications of the interior Minkowski metric derived from the phenomenological studies on the behavior of the meanlife with speed (Sect. IV.3) appear to confirm that the maximal causal speed in the interior of hadrons is bigger than c_o.

As an example, Nielsen-Picek isometric (IV.3.21) for the interior of light mesons yields for Eq. (9.5)

$$V_{Max} = c_o \frac{1 + \alpha}{1 - \alpha/3} \tag{9.9}$$

thus characterizing the maximal speed for the interior of kaons

$$\alpha \cong + 0.61\times10^{-3}, \qquad V_{Max} \cong (1 + 0.8\times10^{-3})\, c_o > c_o, \tag{9.10}$$

We can then expect maximal causal speeds bigger than c_o for all remaining hadrons, evidently owing to the increase of the density with mass.

In fact, following the original proposal in Santilli (*loc. cit.*), De Sabbata and Gasperini (1982) computed the maximal possible causal speed in the interior of hadrons via the use of the conventional gauge theory, and found in the average

$$V_{Max} \cong 75\, c_o. \tag{9.11}$$

It is evident that, if the contact nonhamiltonian interactions caused by total mutual overlapping of the wavepackets do indeed permit the isoquarks of attaining speeds bigger than c_o, their true confinement is expected to be consequential.

In the transition from the hadronic to the nuclear structure we expect a necessarily different setting, because of the transition from total mutual immersion of the wavepackets of the constituents, to mutual penetration of the (average) order of 10^{-3} parts of a nucleon's volume. In fact, also according to the conjecture in Santilli (1982b), *the maximal causal speed for nuclear constituents is lower than* c_o .

Note that the conjecture implies the impossibility for the nuclear constituents of attaining the speed c_o even under the availability of infinite energies.

This latter conjecture was formulated on the basis of the fact that nonrelativistic quantum mechanics provides a suprisingly good approximation of the nuclear phenomenology, when compared to the necessity of relativistic formulations for the atomic structure.

The quantitative study of the conjecture herein considered for the nuclear setting will evidently require the calculation of the modification of the space-time metric caused by the nuclear matter, and their averaging to constants per each nucleus, much along the phenomenological papers on the behavior of the meanlife with speed.

At this point we merely illustrate the possibility via Nielsen-Picek metric (IV.3.21) for pions under which the maximal causal speed (9.5) becomes

$$\alpha \cong -3.8\times 10^{-3}, \qquad V_{Ma} = 0.9995\, c_{o_x} < c_o, \tag{9.12}$$

In summary, the phenomenological studies on the behavior of the meanfile with speed of Sect. IV.3 indicate the possibility that the maximal causal speed increases with the density of hadrons, that is, it increases with the short, range nonlocal, nonhamiltonian interactions, and are expected to become Newtonian under extreme, limiting conditions. In fact, in the core of a star undergoing gravitational collapse, unlimited speeds are locally and instantaneously permitted by the isospecial relativities. By the same token, it is possible to assume that the same maximal causal speed decreases with the density, that is, it decreases with the short range, nonlocal, nonhamiltonian forces, by therefore assuming values smaller than c_o in the nuclear structure.

In this way, we have proved the following

PROPOSITION IV.9.1 (loc. cit.): Any (topology preserving[22]*) modification of the Minkowski metric implies a necessary alteration of the maximal causal speed which can be smaller, equal or bigger than* c_o *depending on the conditions at hand.*

The reader should be aware that *the quantity* c_o *is a universal constant for Einstein's special relativity, while the quantity* V_{Max} *of the isolorentz relativities is a local invariant*, evidently because it can be only defined in the neighborhood of a given point, and it varies from point to point of each given interior medium.

Notice also that the quantity $c = c_o b_4$ is, in general, a geometric quantity and does not necessarily represent a physical speed, evidently because the medium considered can be opaque to all electromagnetic waves, yet permits the motion of particles (see below).

This occurrence can be illustrated via the use, again, of Nielsen-Picek metric (IV.3.21). In fact, we have for kaons

$$c = c_o b_4 = c_o(1 + \alpha) > c_o , \quad \alpha > 0, \tag{9.13}$$

The point is that the above value does not necessarily represent the speed of light, because in our classical approximation light cannot propagate inside a hadron (fotons and neutrinos do not exist at this primitive classical level). Also note that a given value of $c > c_o$ is not sufficient, per se, to imply that the actual causal speed can indeed be bigger than c_o, because that speed must be computed via quantity (9.5).

Needless to say, the quantity $c = c_o b_4$ can indeed represent the speed of light in particular transparent media according to our now familiar notation

$$b_4 = 1/n_4, \quad c = c_o/n_4, \tag{9.14}$$

---

[22] By "topology preserving" we mean any sufficiently smooth and nonsingular modification of the Minkowski metric η which preserves its signature (+1, +1, +1, −1).

where n is the index of refraction.

The above results then imply the following additional property:

PROPOSITION IV.9.2 (loc. cit.): Any (topology preserving) modification of the Minkowski metric implies that the maximal causal speed can be smaller, equal or bigger than the local value $c = c_o b_4 = 1/n_4$

In different terms, besides postulate (9.8) we have the following more general postulate

$$V_{Max} = \left| \frac{dr}{dt} \right|_{Max} \begin{matrix} > \\ = \\ > \end{matrix} c = c_o \hat{b}_4 = c_o / n_4. \qquad (9.15)$$

The best illustration of the above occurrence is given, again, by the Cherenkov light for which the speed of light in water is $c = c_o/n < c_o$, but electrons can propagate in the medium considered at the maximal causal speed $V_{Max} = c_o$, i.e., *electrons can propagate in water at a speed bigger than the local speed of light, exactly along Postulate II and Proposition IV.9.2.*

We now pass to the study of the physical origin of Postulates I and II and Propositions IV.9.1 and IV.9.2.

Within the context of Einstein's special relativity, we have motion of point-like particles in vacuum, and the only admissible forces are the conventional potential forces. Under these assumptions, it takes an infinite amount of energy to accelerate a massive particler to the speed c_o, as well known.

These conditions are fundamentally inapplicable to the isolorentz relativities. In fact, we now have motion of extended particles within a physical medium and, besides the conventional potential forces, we have the additional contact interactions between the particle considered and the medium itself. Now, for the latter forces the notion of potential energy has no physical meaning. As a result, massive particles can indeed attain speeds higher than c_o without any need of infinite energies.

In fact, there are astrophysical data indicating the possibility that jets of matter are emitted by exploding stars at speeds higher than c_o. These emissions could then be the result precisely of the contact interactions in the interior physical medium of the star.

As an incidental note, we *do not* believe that certain reported speeds of quasars exceeding the speed of light in vacuum are plausible, and we *do not* recommend these cases as possible illustrations of Postulate II. This is due to our basic assumption that Einsten's special relativity is exactly valid under the conditions conceived by its originators.

In fact, quasars move in the Universe in empty space, and, thus, under strictly Einstenian conditions. As a result, we expect that c_o is the maximal attainable speed for quasars (see later on for a possible alternative explanation of quasar redshift which does not imply a violation of Einstein's special relativity under Einstenian conditions). On the contrary, the physical conditions of the expulsion of jets of matter from exploding stars are of strictly interior dynamical character and, for this reason, they are recommended as a conceivable illustration of Postulate II. The understanding is that the jets are expected to be still in physical "contact" with the star matter, and not detacted from it.

The above two astrophysical cases illustrate the fact that *possible deviations from Einstein's special relativity, including causal speeds bigger than c_o are studied in this volume only under noneinsteinian conditions.*

We now pass to the classification of *isofourvectors* in our isominkowski space, which can be presented as follows (see Figure IV.9.1 for additional information)

$$\textit{Isotime-like,} \text{ when } x^{\hat{2}} < 0, \tag{9.16a}$$

$$\textit{Isonull,} \text{ when } x^{\hat{2}} = 0, \tag{9.16b}$$

$$\textit{Isospace-like,} \text{ when } x^{\hat{2}} > 0. \tag{9.16c}$$

Note that Postulate II is insensitive as to whether the isopoincare' transformations are linear or not, and centrally depends on the inhomogenuity between space and time, i.e., in the differences

$$\hat{b}_3 \neq \hat{b}_4, \quad \text{or} \quad n_3 \neq n_4. \tag{9.17}$$

The same situation holds for the mutation of the light cone of Fig. IV.9.1. In fact, irrespective of whether the b-quantities are linear or nonlinear (and local or nonlocal) in their variables, we have

$$V_{Max|\hat{b}_1=\hat{b}_2=\hat{b}_3=\hat{b}_4} = c_o \tag{9.18}$$

which is precisely the case of water (see next section for more details).

The fact that the *isoparticles of interior physical media traveling faster than c_0 are not tachyons*, is made clear by the following:

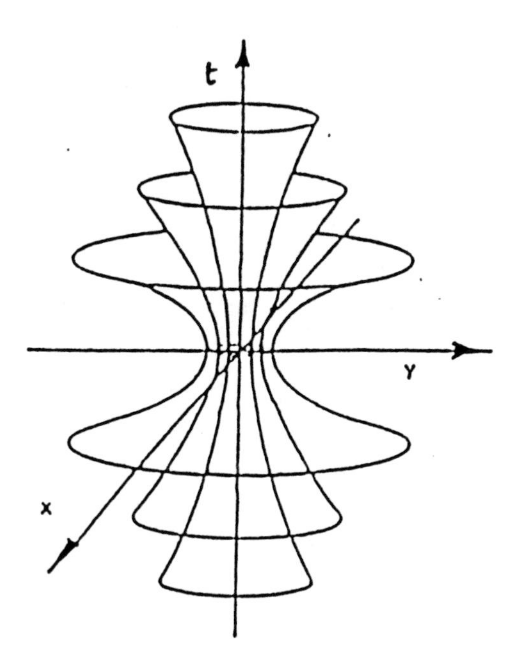

FIGURE IV.9.1: A geometrical view of the deformation (mutation) of the light cone caused by any alteration of the Minkowski metric, as submitted in Santilli (1983a), and interpreted as occurring for extended particles (or wavepackets) moving within inhomogeneous and anisotropic material media, or, equivalently, as due to nonlinear, nonlocal and nonhamiltonian interactions of the interior dynamical problem. The central cone represents the hypersurface of conventionally null four-vectors in Minkowski space $M(x,\eta,\Re)$. The internal cone represents the hypersurface of an isonull fourvector in isominkowski space $\hat{M}^{I}_{2}(x,\hat{\eta},\hat{\Re}) = \hat{M}(x,\hat{g},\hat{\Re})$ for maximal causal speeds smaller than c_0. The external hypersurface is that for a maximal causal speed bigger than c_0. Specific examples of physical media yielding the above two generalized cases are given in the next section. The reader should be reminded that the conventional mental attitude toward light cones is inapplicable under the physical conditions considered. In fact, traditionally, one imagines the light cones as holding in empty space, while the light cones of this figure must be strictly referredto, say, the interior of a nucleus or of a star. Also, the conventional light cones are the same everywhere in the Universe, while the light cones of this figure vary from point to point in space-time.

DEFINITION IV.9.3 (loc. cit.): Isotachyons are particles characterized by the isolorentz relativities which are expected to travel at speeds bigger than the maximal causal speed V_{Max},

$$V_{Isotachyons} > V_{Max} = \left|\frac{dr}{dt}\right|_{Max} = c_o \frac{\hat{b}4}{\hat{b}}, \tag{9.19}$$

In different terms, it is not sufficient for a particle to travel at speeds bigger than c_o to be a tachyon, because it could be a physical particle in interior dynamical conditions. To truly have a tachyon, one must have a local speed bigger than the maximal causal speed at the point considered, whether in the interior or in the exterior problem.

POSTULARE III *(loc. cit.): The dependence of the time intervals with speeds in the general and restricted isolorentz relativities follows the isotopic time-dilation*

$$\Delta t = \hat{\gamma}\, \Delta t_o = \Delta t_o \frac{1}{|1 - \hat{\beta}^2|^{\frac{1}{2}}}, \quad \hat{\beta}^2 = \frac{v^k \hat{b}_k{}^2 v^k}{c_o{}^2 \hat{b}_4{}^2} \tag{9.20}$$

while the dependence of space intervals with speed follows the isotopic space contraction

$$\Delta l = \hat{\gamma}^{-1} \Delta l_o = \Delta l_o |1 - \hat{\beta}^2|^{\frac{1}{2}}, \tag{9.21}$$

The above postulate appears to be useful for a better understanding of the stability of a nucleus. Again, we can provide here only the main idea and work out the operator details at some future time.

As well known, neutrons have a meanlife of about 15′, after which they decay in the familiar form

$$n \Rightarrow p^+ + e^- + \bar{\nu}_e. \tag{9.22}$$

But the neutrons are not at rest when members of a nuclear

structure. Thus, *the central issue of the problem of nuclear stability is the meanlife of the neutrons at the speeds when they are nuclear constituents.*

According to the current nuclear theories, the neutrons are assumed to be strictly Einstenian when member of a nuclear structure. Their meanlife τ then behaves with speeds according to the familiar law

$$\tau = \gamma\, \tau_o = \tau_o \frac{1}{(1 - v^2/c_o^2)^{\frac{1}{2}}}, \tag{9.23}$$

But the speeds of nuclear constituents are considerably lower than c_o (in fact, they are known to be much lower than the speeds of the atomic constituents). We can therefore conclude that, except for small relativistic corrections, the mean life of neutrons when members of a nuclear structure remains of the order of 15′.

This creates the problem of interpreting the stability of nuclei when their neutron constituents have a finite meanlife with decay (9.21). This problem has been studied via complex processes of neutron decays and their instantaneous regeneration, evidently to avoid local excesses of protons which would imply instability.

Now, these neutron regenerative processes are certainly plausible for neutrons in the *interior* of nuclei. However, a number of unresolved questions still persist, because at least a percentage of decays (9.21) from the neutrons of the nuclear *surface* should leave the nuclear structure, thus resulting in proton excesses. This is contrary to the experimental evidence of our macroscopic world whereby ordinary matter is stable without any production of particles. In turn, this leaves the problem of nuclear stability essentially unresolved to this day.

In the transition to the isospecial relativities the situation is fundamentally different. To begin, the value 15′ for the meanlife of the isoneutron at rest becomes questionable when the particle is a nuclear constituent because of field theoretical mutations (Santilli (1991d)). But even assuming at this rudimentary stage that such a meanife remains 15′, its behavior with speed is now mutated according to Postulate III in the isotopic form

$$\hat{\tau} = \tau_o \frac{1}{|\, 1 - v\hat{b}_3{}^2 v / t c_o{}^2 \hat{b}_4{}^2 t \,|^{\frac{1}{2}}}, \tag{9.24}$$

which may imply the limit value

$$\hat{\tau}\Big|_{v = V_{Max}} = \infty, \tag{9.25}$$

namely, *the isospecial relativities admit the possibility that neutrons become completely stable when members of a nuclear structure, via either a mutation of the meanlife at rest, or a mutation of the behavior of the meanlife with speed, or both.*

An interpretation of the stability of the nuclear structure and of the absence of decays (9.21) from ordinary matter would then follow.

We have reached in this way the following

PROPOSITION IV.9.3 (loc. cit.): Any (topology preserving) mutation of the Minkowski metric implies a necessary alteration of the behavior of the meanlife with speed which can be bigger, equal or smaller than the Einsteinian behavior,

$$\hat{\tau} = \hat{\gamma}\,\tau_o \;\substack{>\\=\\<}\; \tau = \gamma\tau_o, \tag{9.26}$$

depending on the local physical conditions of the interior medium considered.

It is appropriate here to recall that, as proved by Aringazin (1989), *isotopic time-dilation (9.24) is directly universal,* that is, capable of including all possible time dilation laws (universality), in the frame of the experimenter (direct universality).

This is essentially due to the arbitrariness of the functional dependence of the b-quantities. A virtually infinite number of possible, different, approximate laws can then be obtained from the exact isodilation law (9.24) via the use of different expansions, different coefficients and different truncations.

To illustrate this property, consider the case in (1+1)-dimension in which the $\hat{b}$-quantities are only dependence on the velocities (which is a most significant dependence). This allows the expansion (Aringazin (*loc. cit.*))

$$\hat{b} = \hat{b}(v) = 1 + \lambda_o + \lambda_1\gamma + \lambda_2\gamma^2 + \lambda_3\gamma^3 + \ldots\ldots\ldots \tag{9.27a}$$

$$\gamma = |1 - \beta^2|^{-\frac{1}{2}} < 1, \quad \beta = v^2/c_o^2, \quad \lambda_k \ll 1, \tag{9.27b}$$

By assuming $c_o = 1$ for convenience, isotopic law (9.25) can then be expanded, e.g., in the form

$$\tau = \tau_o \gamma\{1 + \lambda_o \gamma^2 + \lambda_1(1+\lambda_o)\gamma^3 + [\lambda_1{}^2 + \lambda_2(1 + \lambda_o)]\gamma^4 + ...\}, \qquad (9.28)$$

By using the above expansions, Aringazin (*loc. cit.*) then proved that the isotopic law (9.24) includes as particular cases:

1) the behavior by Blockhintsev (1962), Redei (1967), and others

$$\tau = \tau_o \gamma (1 + \lambda_o \gamma^2), \quad \lambda_o = 10^{25} a_o, \qquad (9.29)$$

where a_o is a fundamental length;

2) the meanlife by Nielsen and Picek (1983), i.e.,

$$\tau = \tau_o \gamma(1 + \lambda_o \gamma^2), \quad \lambda_o = 4\alpha/3, \qquad (9.30)$$

where α is the "Lorentz-asymmetry parameters" ;

3) the meanlife by Aronson *et al.* (1983), i.e.,

$$\tau = \tau_o \gamma(1 + b_\chi^{(N)} \alpha^{(N)}), \quad \alpha^{(N)} = E/m, \; N = 1,2, \qquad (9.31)$$

where E is the kinetic energy of the particle, m its mass, and the b's are certain slopes parameters;
as well as any other possible laws of the class considered.

It should also be mentioned that, in the same paper, Aringazin (1989) shows that all the remaining variations of the basic parameters of the $K^\circ-\bar{K}^\circ$ system (mass difference, CP parameters, etc.) are also particular cases of directly universal isospecial laws.

Postulate III may be relevant for the study of the behaviour of the meanlifes of unstable hadrons with speed (Sect. IV.3) and, as such, it is suitable for direct tests proposed in Chapter VII.

We pass now to the study of the notion of rest energy of an isoparticle. Consider the fundamental isoinvariant (IV.7.9), i.e.

$$p^{\hat{2}} = p^k \hat{b}_k{}^2 p^k - p_4{}^2 = - m_o{}^2 c_o{}^4 \hat{b}_4{}^4, \qquad (9.32)$$

where

$$p = (p^\mu) = mu = (m_o \hat{\gamma} c v, \quad m_o \hat{\gamma} c), \qquad (9.33)$$

DEFINITION IV.9.4 (loc. cit.): The energy $\hat{E}$ of an isoparticle on isominkowskispaces $\hat{M}^I(x,\hat{g},\hat{\Re})$ is characterized by the fourth component of the iso-four-momentum according to the rule

$$E = p_4, \tag{9.34}$$

and can be expressed in terms of the fundamental isoinvariant (9.32) in the form

$$E^2 = m_o{}^2 c^4 + p^k b_k{}^2 p^k, \tag{9.35}$$

We then have the following

***POSTULATE IV** (loc. cit.): The rest mass m_o of an isoparticle on an isominkowski space $\hat{M}^I(x,\hat{g},\hat{\Re})$ varies with speed according to the isotopic law*

$$\hat{m} = m_o \hat{\gamma} = \frac{m_o}{|1 - \hat{\beta}^2|^{\frac{1}{2}}}, \quad \hat{\beta}^2 = \frac{v^k \hat{b}_k{}^2 v^k}{c_o \hat{b}_4{}^2 c_o} \tag{9.36}$$

and the equivalent value of the energy $\hat{E}$ for at rest conditions is given by

$$\hat{E} = m_o c^2 = m_o c_o{}^2 \hat{b}_4{}^2(s,x,p,\mu,\tau,n,\ldots), \tag{9.37}$$

The most intriguing implications of the above postulate is the possibility of identifying the hadronic constituents with massive, physical, ordinary particles, of course, in an operator formulation of the isospecial relativities.

The best way to illustrate this possibility is via the structure model of the π° particle as a *"compressed positronium"* (Santilli (1978b), Sect. 5, and (1990), i.e., as a relativistic, two-body, closed nonhamiltonian system of one ordinary electron and one positron in conditions of total mutual immersion of their wavepackets down to the size of the charge distribution of the π° ($\approx$ 1F) The model is here symbolically indicated

$$\text{Positronium} = (e^+, e^-) \Rightarrow \pi^\circ = (\epsilon^+, \epsilon^-), \tag{9.38a}$$

$$M(x,\eta,\Re) \Rightarrow \hat{M}^I(x,\hat{g},\hat{\Re}), \tag{9.38b}$$

where $\epsilon^\pm$ represents the alterations of the electrons $e^\pm$ caused by short-range, nonlocal effects expected in the interior of the π°.

The use of conventional relativistic (or nonrelativistic) quantum mechanics does not permit the representation of the total energy of 135 MeV of the π° at rest, when the constituents have only a rest energy of 0.5 MeV, for technical reasons elaborated in the quoted literature.

Mathematically, this is due to the fact that, under the conditions considered, the indicial equation does not admit real solutions, and only mathematical models with complex total energies becomes possible.

Physically, the occurrence is clearly expressed by the fact that, in conventional relativistic quantum mechanics, *binding energies are necessarily negative,* as it is the case for the hydrogen atom, the deuteron, etc. The representation of model (9.38) with conventional quantum mechanics would then call for *positive binding energies,* owing to the excessive disparity between the total energy of the bound state and the total rest energy of the constituents).

Equivalently, we can say that the total energy needed for the electron to have a positive binding energy is so high, that it would prevent a quantitative interpretation of the relatively long meanlife of the π° ($\cong 10^{-16}$ sec).

As shown in Santilli (*loc. cit.*) the use of Postulate IV readily resolves the above problems. In fact, the isotopy $e^{\pm} \Rightarrow \epsilon^{\pm}$ essentially implies a form of "renormalization" of the rest energy of the electron when in interior dynamical conditions (only). In turn, this "renormalization" implies such an increase of the value of the rest energy of the constituents to render the indicial equations consistent, thus permitting the correct representation of the *real* total energy of 135 MeV, while mantaining a negative binding energy[23], and achieving a joint, quantitative interpretation of the meanlife as well as of *all* remaining characteristics of the particle.

What is rather intriguing is that such a renormalization of the mass is a direct consequence of the isotopy and appears already at the classical nonrelativistic level of Appendix III.A.

But, in the opinion of this author, the best test for the consistency and physical relevance of the isolorentz relativities is given by their capability to reach, in due time, a quantitative, operator, relativistic representation of Rutherford's (1920) historical conception of the neutron as a *"compressed hydrogen atom",* via the isotopy symbolically written

---

[23] In reality, the numerical value of the binding energy of model (9.38), even though negative, is close to zero, thus being considerably along current views on "asymptotic freedom".

$$\text{Hydr. Atom} = (p^+, e^-) \Rightarrow \text{neutron} = (p^+, \epsilon^-), \tag{9.39a}$$

$$M(x,\eta,\Re) \quad \hat{\Rightarrow} \quad \hat{M}^I(x,\hat{g},\hat{\Re}), \tag{9.39b}$$

studied at the operator level in Santilli (1990) and (1991f)), as well as in Animalu (1991a).

In addition to having the same problematic aspects of model (9.38) in the representation of the *real* total energy of the neutron, model (9.39) has a number of additional, rather serious difficulties, such as the achievement of a *Lie-isotopic total spin* $\frac{1}{2}$ from two constituents which, in Einstenian conditions have conventional spin $\frac{1}{2}$.

The latter problem appears to have been solved by Dirac (1971), (1972), who proposed a generalization of his celebrated equation which implies the mutation of the original spin $\frac{1}{2}$ of the electron precisely into the value zero, specifically, for the at rest conditions in the center of the system, as needed for Rutherford's electron. In particular, *"Dirac's generalization of Dirac's equation"* has an essential invariance under the isopoincaré symmetries, as studied in detail in santilli (1991e).

It is evident that, if model (9.38) and (9.39) are proven to be consistent in due time, they may well permit the identification of the hadronic constituents (isoquarks) with suitably altered forms of conventional massive particles freely produced in the spontaneous decays, as established in the preceding nuclear and atomic structures. In fact, model (9.39) would permit the identification of the isoquark $\tilde{d}$ with a mutated form ϵ^- of Rutherford's electron e^-.

Evidently, at the classical level of this paper we can only provide qualitative aspects. One way to illustrate Postulate IV in a preliminary way is via Nielsen-Picek metric (IV.3.21) for which

$$\hat{m} = m_o \left(1 - \frac{v^2(1 - \alpha/3)^2}{c_o^{\,2}(1 + \alpha)^2} \right)^{-\frac{1}{2}} \tag{9.40a}$$

$$\hat{E} = m_o\, c_o\, (1 + \alpha), \tag{9.40b}$$

We therefore have for a particle in the *interior* of a pion

$$\alpha \cong -3.8\times 10^{-3}, \quad \hat{m} < m, \quad \hat{E} < E, \tag{9.41}$$

and for for a particle in the interior of a kaon

$$\alpha \cong +0.6\times 10^{-3}, \quad \hat{m} > m, \quad \hat{E} > E. \tag{9.42}$$

The reader should be aware that metric (3.21) was computed as *a first approximation for low energies.* Nevertheless, results (9.41) and (9.42) above are sufficient to prove the following

PROPOSITION IV.9.4 (Santilli (1983a), (1989b)): Any (topology preserving) modification of the Minkowski metric implies a necessary alteration of the behavior $\hat{m}$ of the rest mass with energy and of the energy equivalence $\hat{E}$ of the rest mass, which can be bigger, equal or smaller than the corresponding Einstenian quantities

$$\hat{m} \gtreqless m, \qquad \hat{E} \gtreqless E, \tag{9.43}$$

depending on the local physical conditions of the medium considered.

We now study the redshift under noneinsteinian conditions.

DEFINITION IV.9.5 (loc. cit.): An "isoplanewave" is a conventional planewave in Minkowski space $M(x,\eta,\Re)$ under isotopic liftings to the isospaces $\hat{M}(x,\hat{g},\hat{\Re})$ with constant isometrics, i.e.,

$$\psi(x) = N\, e^{i K^{\mu} \eta_{\mu\nu} x^{\nu}} \Rightarrow \hat{\psi}(x) = \hat{N} * e^{i K^{\mu} \hat{\eta}_{\mu\nu} x^{\nu}}, \; N \in \Re, \;\; \hat{N} \in \hat{\Re}, \tag{9.44}$$

where K is an isonull vector, i.e.,

$$K^{\mu}\, \hat{\eta}_{\mu\nu}\, K^{\nu} = 0, \qquad K = (k, \omega/c), \quad \omega/c = 2\pi/\lambda. \tag{9.45}$$

Lifting (9.44) is essentially intended to represent the mutation of a plainwave in the transition from motion in vacuum to motion within an inhomogeneous and anisotropic medium, evidently assumed to be transparent to the wave considered.

Note that the lifting is here considered solely for the restricted case, because we are treating an apparently global effect of given media.

Suppose now that such an isoplanewave is detected by two observers S and S′, one at rest with respect to the medium, and the other in motion with respect to it, at a relative speed v along the x^3-axis.

As a specific case, the reader may think of ordinary light

propagating within our atmosphere which, being transparent, inhomogeneous and anisotropic, is an ideal interior medium for our isospecial relativities. Observer S can be an ordinary observer on our ground, and observer S′ can be either moving in the atmosphere or outside it.

Let α be the angle between k and x^3, and let k′, ω′, and α' be the corresponding quantities for S′.

From the manifest form-invariance of the isoplanewave under the Lorentz-isotopic transformations,

$$K^{\mu}\, \hat{\eta}_{\mu\nu}\, K^{\nu} = K'^{\mu}\, \hat{\eta}_{\mu\nu}\, K'^{\nu}, \tag{9.46}$$

it is then easy to see that

$$k'^1 = k^1 = k'^2 = k^2, \tag{9.47a}$$

$$k'^3 = \hat{\gamma}(k^3 - \beta k^4) = |k| \cos \alpha', \ \beta = v/c_o \tag{9.47b}$$

$$k'^4 = \hat{\gamma}(k^4 - \hat{\beta}\, k^3) = \omega'/c, \ \hat{\beta} = vb_3/c_o b_4. \tag{9.47c}$$

This leads to the following

POSTULATE V *(loc. cit.): The Doppler's frequency shift for electromagnetic waves propagating within an inhomogeneous and anisotropic physical medium transparent to it (isoplanewave) follows the isotopic laws*

$$\hat{\omega} = \omega \hat{\gamma}(1 - \hat{\beta} \cos\alpha), \tag{9.48a}$$

$$\hat{\gamma} = |1 - \hat{\beta}^2|^{-\frac{1}{2}}, \ \hat{\beta}^2 = \frac{vb_3{}^2 v}{c_o b_4{}^2 c_o}, \ \hat{\beta} = vb_3 / c_o b_4 \tag{9.48b}$$

with isotopic aberration law

$$\hat{\cos}\, \alpha' = \frac{\cos \alpha - \hat{\beta}}{1 - \hat{\beta} \cos \alpha}, \tag{9.49}$$

The above postulate offers genuine possibilities of resolving the now vexing problem of astrophysical bodies violating Einsteinian laws under Einsteinian conditions.

In fact, the redshift of certain far distant quasars has recently attained such high values to require the assumption, under the Einstenian redshit law

$$\omega' = \omega\, \gamma(1 - \beta \cos \alpha), \quad \beta = v / c_o, \tag{9.50}$$

that (portions of) quasars travel at speeds higher than c_o, up to speeds of the order of $10c_o$ or more. But the quasars travel in empty space. Thus their center-of-mass trajectory must be strictly Einsteinian. *The assumption of speeds bigger than c_o therefore constitutes a violation of Einstenian laws under Einsteinian conditions.*

In the original proposal of Postulate V (Santilli (1988c)) it was indicated that isoredshift (9.49) can indeed prevent such a manifestly unplausible violation of Einstein's special relativity, without eliminating the current expansion of the Universe.

In fact, light is emitted in the interior of the quasars and propagates first in the hyperdense, inhomogeneous and anisotropic atmopsheres surrounding them (estimated to be up to the order of hundred of thousands of miles and more). After leaving the quasar, light then propagates over very long distances in the Universe, to finally reach us.

Now, for specific values of the characteristics b-quantities for the medium considered, *Postulate V predicts that light can be redshifted by propagation within the quasars' inhomogeneous and anisotropic atmospheres.*

Moreover, space can be considered empty only in the neighborhood of our Solar systems. For large intergalactic distances, the space itself is no longer empty, but it is a physical medium characterized by particles, dark matter, radiations, etc. *Then, Postulate IV.9.V predicts that light can be additionally redshifted during its propagation over large intergalactic distances because of conceivable corrections caused by the lack of empty character of space.* The understanding is that, in this second case, we are referring to predictably small redshifts.

In summary, the redshift from far distant quasars, as measured on Earth, could be due to the superposition of:

1) a quantitatively nonignorable isotopic redshift (9.48) caused by propagation in the quasars' atmospheres;

2) a small isotopic redshift caused by propagation over intergalactic distances; and

3) a primary Einsteinian redshift (9.50) caused by the conventional expansion of the Universe.

The above features evidently permit the indicated possibility of preserving the expansion of the Universe, by avoiding violations of the special relativity under Einsteinian conditions.

Mignani (1991) conducted explicit calculations of the above possibilities, by computing preliminary explicit values for the characteristic b-quantities, which apparently confirm the above model. These astrophysical calculations are reviewed in Chapter VII in conjuction with proposals for the direct experimental test of Postulate V.

The preceding considerations imply the following

PROPOSITION IV.9.5 (loc. cit.): Any (topology-preserving) modification of the Minkowski metric implies a necessary mutation $\hat{\omega}'$ of the Doppler's redshift which can be bigger, equal or smaller than the Einstenian value ω'

$$\hat{\omega}' = \omega \, \hat{\gamma} \, (1 - \hat{\beta} \cos \alpha) \; \begin{matrix} > \\ = \\ < \end{matrix} \; \omega' = \omega \, \gamma \, (1 - \beta \cos \alpha), \qquad (9.51)$$

depending on the local conditions of the interior medium considered.

This completes our preliminary study of the isotopic liftings of the basic postulates of Einstein's special relativity for the most general possible nonlinear, nonlocal and nonhamiltonian isotopies.

Additional insights are provided in the next section where we shall review the results from a geometrical viewpoint. It is understood that some of the information on the isospecial relativities most crucial for experimental verifications are expected to result from the operator treatment of the studies.

It is also evident that the content of this section must be complemented with the isorotations $\hat{O}(3)$ of Sect. III.3, by therefore adding further possibilities of quantitative treatments of conditions that are unadmissible within the context of Einstein's special relativity, such as the purely classical representation of the actual, rotationally noninvariant shape of the particle considered, the admission of all its infinitely possible deformations, etc.

We can therefore say that

The isospecial relativities, if suitably developed and experimentally coinfirmed, can indeed provide an infinite family of coverings of Einstein's special relativity, in the sense that:

A) The isospecial relativities are constructed with mathematical methods (the Lie-isotopic theory) structurally more general then those of the conventional relativity (the conventional Lie's theory);

B) the isospecial relativities represent physical conditions (motion within inhomogeneous and anisotropic physical media, deformation of particles, etc.) which are structurally more general than those of the conventional relativity (point-like particles, motion in empty space, etc.); and, last but not least,

C) The isospecial relativities can approximate the conventional relativity as close as desired for $\hat{I}_2 \cong I$, and they all recover by construction the conventional relativity identically for $\hat{I}_2 \equiv I$.

A visual inspection of Postulates I-V proves the following important property.

THEOREM IV.9.1 (loc. cit.): All the infinitely possible general or restricted isospecial relativities on isospaces $\hat{M}^I(x,\hat{g},\hat{\Re})$ coincide with Einstein's special relativity at the abstract, realization-free level.

It is remarkable that, despite the general nonlinear and nonlocal dependence of the various physical quantities (invariant speed, maximal speed, meanlife, rest energy, etc.), Postulates I-V formally coincide with the corresponding Einstenian forms. In turn, this illustrate the rather unique function of the characteristic $\hat{b}$-quantities representing the interior media, as studied in more detail in the next section.

Above all, it is remarkable that *the "breaking of the barrier" of the speed of light in vacuum by causal signals is ultimately permitted by the Einsteinian axioms themselves*, only realized in a more general way.

This ultimate unity of isotopic and conventional postulates is the physical counterpart of the mathematical isomorphisms $\hat{P}(3.1) \approx P(3.1)$, and implies the following

COROLLARY IV.9.1.1 (loc. cit.): Locally, that is, at given values $\bar{x}$, $\bar{p}$, $\bar{\mu}$, $\bar{\tau}$, $\bar{n}$,.... of the local variables and quantities, the general isopoincare´ transformations coincide with the conventional Poincare´ transformations, up to a redefinition of the parameters,

$$\bar{x}' = \Lambda(u) * x|_{\bar{x},\bar{p},\ldots} \equiv \Lambda(u')\bar{x}, \qquad \hat{O}(3,1)|_{\bar{x},\bar{p},\ldots} \equiv O(3.1), \tag{9.52a}$$

$$\bar{x}' = \bar{x} + x^{\circ}\,\breve{b}|_{\bar{x},\bar{p},\ldots} \equiv \bar{x} + x'^{\circ}, \qquad \hat{T}(3.1)|_{\bar{x},\bar{p},\ldots} \equiv T(3.1). \tag{9.52b}$$

This is the relativistic generalization of the corresponding nonrelativistic property studied in the preceeding chapter, whereby the nonlinear, nonlocal and nonhamiltonian isogalilean laws for the motion of an extended particle within a hyperdense medium coincide with the historical Galilean law for the uniform motion in vacuum, with similar results for all other Galilean laws.

As concluding remarks, let us recall that Einsten's special relativity is based on the following:

<u>PRINCIPLE 1</u>: The homogenuity and isotropy of (empty) space;

<u>PRINCIPLE 2</u>: The general invariance of the speed of light in vacuum;

<u>PRINCIPLE 3</u>: The general invariance of the physical laws under the broadest possible linear and local group of isometries of the Minkowski space-time;

from which all other aspects of the relativity can be derived within inertial reference frames.

But, as stressed during the course of our analysis, inertial frames are a philosophical abstraction because they do not exist in our Earthly environment, nor they can be attained in our Solar or Galactic systems. Also, extended particles do not generally move in empty space, but within physical media. The covering principles submitted in (Santilli (1983a) and (1988c))) as an attempt to represent more general physical conditions, are:

<u>ISOPRINCIPLE $\hat{1}$</u>: The inhomogenuity and anisotropy of physical media, with the underlying space remaining homogeneous and

isotropic;

<u>*ISOPRINCIPLE $\hat{2}$:*</u> *The local invariance of the maximal, speed of causal signals within physical media, with the underlying invariant causal speed in vacuum remaining that of light; and*

<u>*ISOPRINCIPLE $\hat{3}$:*</u> *The local invariance of physical laws under the most general possible nonlinear and nonlocal groups of isometries of the isospace representing physical media, with the conventional linear and local isometries on the Minkowski space being always admitted as a particular case;*

from which all aspects of the general isospecial relativities can be derived, such as the selection, among the multiple infinity of noninertial frames of the Universe, of the subclass of equivalent frames characterized by the general, nonlinear and nonlocal isopoincaré symmetries.

Inertial frames are recovered as a particular case via the reduction of the general to the restricted isosymmetries, that is, (Sect. IV.3) via the averaging of the characteristic $\hat{b}$-functions of the medium considered to b-constants.

Finally, we should mention that the Lorentz-isotopic relativities considered in this section are particular cases of expected, still more general *Lorentz-admissible relativities* for the most general possible open-nonconservative conditions (Santilli (1981a)).

IV.10: ISORELATIVISTIC GEOMETRIZATION OF PHYSICAL MEDIA

In this section we shall study a central objective of our isominkowski spaces $\hat{M}^I(x,\hat{g},\hat{\Re})$: provide a geometrization of the inhomogenuity, anisotropy and nonlocality of interiort physical media. In turn, this task will assist in identifying the physical meaning of the *characteristics $\hat{b}$-quantities of interior physical media* . The geometrization here considered was first studied in Santilli (1988c). Additional studies can be found in Santilli (1991c).

The best way to conduct the study is by classifying all possible interior physical media permitted by isominkowski spaces of Class I, i.e., without gravitational effects, and with positive-definite isounits. This classification will then emerge as of fundamental character for the possible tests of the isospecial relativities proposed in Chapter VII.

Let us write the basic isoinvariant (IV.3.4b) on $M^I(x,\hat{g},\hat{\mathfrak{R}})$ in form (9.2a), i.e.,

$$x^{\mu}\,\hat{g}_{\mu\nu}\,x^{\nu} = \frac{1}{n_3^{\,2}}\,r^k\,r^k - \frac{1}{n_4^{\,2}}\,tc_o t, \tag{10.1}$$

where we have again assumed for simplicity that the space components of the $\hat{b}$-quantities are identical and

$$\hat{b}_1^{\,-1} = \hat{b}_2^{\,-1} = \hat{b}_3^{\,-1} = n_3 = n_3(s, x, p, \mu, \tau, n,..) > 0, \tag{10.2a}$$

$$\hat{b}_4^{\,-1} = n_4 = n_4(s, x, p, \mu, \tau, n,...) > 0. \tag{10.2b}$$

Under notation (10.1), the fundamental quantities of the isospecial relativities become

$$V_{MaX} = c_o\,\frac{b_4}{b_3} = c_o\,\frac{n_3}{n_4}, \tag{10.3a}$$

$$c = c_o\,b_4 = \frac{c_o}{n_4}, \tag{10..3b}$$

$$\hat{\beta} = \frac{n_4}{n_3}\,\beta, \tag{10.3c}$$

$$\hat{\gamma} = |1 - \frac{n_4}{n_3}\,\hat{\beta}^2|^{-\frac{1}{2}}, \tag{10.3d}$$

with similar expressions holding for other quantities.

The desired classification of admissible physical media can then be based on the values of the maximal causal speed and of c with respect to c_o

$$V_{Max} = c_o\frac{n_3}{n_4} \gtreqless c_o, \qquad c = \frac{c_o}{n_4} \gtreqless c_o. \tag{10.4}$$

In turn, such a classification can be reduced to the values n_3/n_4 and n_4 when compared to one, yielding the following nine different cases

ISORELATIVISTIC GEOMETRIZATION OF PHYSICAL MEDIA(Santilli (1988c))

TYPE 1: $n_3 = n_4, \quad n_4 = 1; \quad \hat{\beta} \equiv \beta, \quad \hat{\gamma} \equiv \gamma; \quad V_{Max} = c_o, \quad c = c_o.$

This is evidently the case of the special relativity, which is contained as a particular case of the Lorentz-isotopic relativity.

TYPE 2: $n_3 = n_4, \quad n_4 > 1; \quad \hat{\beta} \equiv \beta, \quad \hat{\gamma} \equiv \gamma; \quad V_{Max} = c_o, \quad c < c_o.$

This is the case of propagation of light within homogeneous and transparent fluids, such as water. A known illustration is that of the *Cherenkov light* mentioned in Sect. IV.9, for which light propagates at the speed

$$c = c_o/n_4 < c_o, \tag{10.5}$$

smaller than the speed of light in vacuum c_o, where n_4 is the familiar index of refraction. Nevertheless, ordinary particles such as the electrons have been measured to propagate at speeds higher than c.

This results in $V_{Max} > c$, by therefore providing a rather clear illustration of Postulate II. The validity of Postulate I is selfevident, because the assumption of $c = c_o/n_4$ as the invariant of the theory would lead to a series of inconsistencies within the context of the special relativity itself, let alone its isotopic generalizations. Postulates III, IV and V remain conventional because $\hat{\beta} \equiv \beta$, and therefore $\hat{\gamma} \equiv \gamma$ for the media considered.

Media of Type 1 imply the first and simplest possible generalization of the special relativity characterized by the *scalar isotopy*

$$x^\mu g_{\mu\nu} x^\nu = \frac{1}{n^2} x^\mu \eta_{\mu\nu} x^\nu. \tag{10.6}$$

We can therefore state that *the scalar isotopy of the Minkowski invariant can represent the transition from motion in vacuum to motion within a homogeneous and transparent medium.*

A form of anisotropy is however admitted by scalar isotopy (10.6).

In fact, Bogoslowski (1974) special relativity is precisely a particular case of isoinvariant (10.6).

TYPE 3: $n_3 = n_4, \quad n_4 < 1; \quad \hat{\beta} \equiv \beta, \quad \hat{\gamma} \equiv \gamma; \quad V_{Max} = c_o, \quad c > c_o.$

This is the first prediction of the isospecial relativities of a novel physical medium which is expected to be generally *opaque* to light (as well as to all electromagnetic waves), because $c > c_o$, in which case c acquires a purely geometric meaning, similar to that, say, of the element g_{44} of the general relativity. Nevertheless, ordinary massive particles can indeed propagate up to $V_{Max} = c_o$.

A conceivable canditate for the above media are given by *superconductors*. In fact, conductors are ordinary physical media for electrons. As a result, the value $V_{Max} = c_o$ becames plausible as a limit at superconductivity conditions, while these media are manifestly opaque to light.

An intriguing study of this case has been conducted by Animalu (1991b), who has shown the apparent isotopic structure of the Cooper pairs in superconductivity, thus opening the way to a possible relevance of our isotopic relativities in superconductivity.

Note that this is the last case of scalar isotopy (10.6) for which $\hat{\beta} \equiv \beta$ and $\hat{\gamma} \equiv \gamma$. Therefore, Postulates I and II are generalized, but Postulates III, IV, and V are not.

Needless to say, numerous additional cases of physical media corresponding to Type 3 are conceivable, e.g., fluids opaque to light, etc.

TYPE 4: $n_3 < n_4, \quad n_4 > 1; \quad \hat{\beta} > \beta, \quad \hat{\gamma} < \gamma; \quad V_{Max} < c_o, \quad c < c_o.$

This is the first case of a nontrivial isotopy other than the scalar form (10.6). We therefore expect in this case a nontrivial generalization for all Postulates I–V.

The first possible physical media of (Class I) Type 4 are given by *ordinary planetary or astrophysical atmospheres* which are transparent to light or to some electromagnetic waves. In fact, these atmospheres are manifestly inhomogeneous (e.g., because of the variation of the density with the distance from the center) and anisotropic (e.g., because of the intrinsic angular momentum which creates a preferred direction in the medium). These conditions ensure the lack of scalar isotopy (10.6) in favor of a nontrivial isotopy of the special relativity. The speed of the electromagnetic waves in the medium considered is evidently lower than that in vacuum, $c < c_o$, because of the local dependence on the density, i.e.,

$$c = c_o/n_4, \quad n_4 = n_4(x, \mu, \tau, n, ...), \tag{10.7}$$

although the function n_4 can be averaged to constants via methods of type III.3.56 for the description of global effects, e.g., for the characterization of the *average index of refraction* of our atmosphere.

Finally, an extended, massive particle cannot attain the speed c_o within the medium considered, even under the availability of infinite energy, V_{Max} being smaller than c_o because of drag effects caused by the medium itself.

The media under consideration therefore imply a first genuine isotopy of the special relativity with only locally definable quantities, which assume generally different values from space-time point to space-time point. In fact, beginning with the maximal possible deviations expected at the surface of the astrophysical object (maximal possible density of the atmosphere), we have continuously varying conditions (density), all the way up to a smooth recovery of Einsteinian conditions at the end of the atmosphere (null density). A nontrivial local realization of all generalized Postulates I-V follows.

An intriguing study of this case has been conducted by Mignani (1992) who has identified a conceivable application of the isotopic redshift law for light propagating in the hyperdense quasars' atmosphere. For details, we refer to experimental considerations of Chapter VII.

A second example of physical media of this type is given by ordinary conductors. In fact, in this case the medium is opaque to any electromagnetic wave, c has a mere geometrical meaning, while $V_{Max} < c_o$ because of drag effects caused by the metal medium on the current electrons. Specifically, by excluding here the limit case of superconductivity, *the experimental test of Postulate II for ordinary conductors would imply that electrons moving cannot attain the speed c_o in ordinary conductors even under the availability of infinite energies (infinite potential differences)*.

A third illustration is given in particle physics by Nielsen-Picek medium (III.3.21) for case (3.22), i.e., for the interior of pions at low energy, in which case

$$V_{Max} \cong c_o \frac{1 - 3.8 \times 10^{-3}}{} \cong 0.995\, c_o < c_o, \tag{10.8a}$$

$$1 + 1.3\times10^{-3}$$

$$c = c_o / n_4 = c_o(1 - 3.8\times10^{-3}) \cong 0.996\, c_o < c, \tag{10.8b}$$

In this case c has a mere geometrical meaning because, at our classical approximation, photons and neutrinos are not admitted and the medium is opaque to light. Also note that, according to Postulate IV, ordinary massive particles cannot be accelerate inside a pion up to c_o even under infinite energies.

TYPE 5: $n_3 < n_4, \quad n_4 = 1; \quad \hat{\beta} > \beta, \quad \hat{\gamma} < \gamma; \quad V_{Max} < c_o, \quad c = c_o.$

This case is geometrically equivalent to Type 4, yet it is refers to possible media in which electromagnetic waves can locally propagate at speeds c_o, or just $c = c_o$ acquires a pure geometric value.

TYPE 6: $n_3 < n_4, \quad n_4 < 1; \quad \hat{\beta} > \beta, \quad \hat{\gamma} < \gamma; \quad V_{Max} < c_o, \quad c > c_o.$

Conceivably, this is a *classical geometrization of the medium in the interior of nuclei* as originally proposed in Santilli (1982a), (1983a), and elaborated in more details in Santilli (1989). In fact, nuclei are manifestly inhomogenous and anisotropic, thus justifying the activation of our isotopies. Also, according to this interpretation, nuclei are (classically) opaque to all electromagnetic waves, the understanding is that photons, neutrinos and other particles are indeed admitted and they can indeed propagate inside nuclei, but only after suitable quantum field theoretical extensions. The quantity c is therefore a purely geometrical quantity, and its value bigger than c_o merely expresses the fact that space is filled up with hyperdense nuclear matter.

The value $V_{Max} < c_o$ is inferred from the very good approximation of nonrelativistic quantum mechanics in nuclear physics recalled in Sect. IV.9. It is then argued that, according to Postulate II, nucleons cannot attain the speed of light in vacuum when members of a nuclear structure, even under the hypothetical availability of infinite energies.

Needless to say, the full technical evaluation of the above interpretation and its confrontation with experimental evidence requires an operator formulation of the isotopic lifting of the special relativity.

Other classical examples of this type of physical media are also conceivable, e.g., astrophysical atmospheres so dense to be completely opaque to all electromagnetic waves (to justify the value $c > c_o$).

Note that the physical media of (Class I), Types 4, 5 and 6 share the same characteristics $\hat{\beta} > \beta$ and $\hat{\gamma} > \gamma$ because they both have the values $n_3 < n_4$. Their difference is characterized by $n_4 > 1$ for Type 4, $n_4 = 1$ for Type 5 and $n_4 < 1$ for Type 6. Therefore, they have different Postulates I and II, and the same Postulates III, IV and V.

TYPE 7: $n_3 > n_4, \quad n_4 > 1; \quad \hat{\beta} < \beta, \quad \hat{\gamma} > \gamma; \quad V_{Max} > c_o, \quad c < c_o.$

Certain *interior astrophysical conditions* can provide an illustration of this type of geometrization, e.g., the jets of matter reported to have been emitted from astrophysical bodies at speeds bigger than c_o. For these conditions we have atmospheres which can still be trasperent to light, and thus c can represent its local speed; nevertheless, the medium is subjected to such extreme turbulences to create contact interactions capable of propagating physical matter at speeds higher than c_o.

A number of other examples of physical media of this type are also possible, e.g., ion interior gravitational problems, but they will be investigated at some future time.

This is the first "breaking of the barrier" of speed of light in vacuum by a causal event we find in this isorelativistic geometrization.

TYPE 8: $n_3 > n_4, \quad n_4 = 1; \quad \hat{\beta} < \beta, \quad \hat{\gamma} > \gamma; \quad V_{Max} > c_o, \quad c = c_o.$

This case is geometrically equivalent to media of Type 7, with the sole difference of permitting the value $c = c_o$.

TYPE 9: $n_3 > n_4, \quad n_4 < 1; \quad \hat{\beta} < \beta, \quad \hat{\gamma} > \gamma; \quad V_{Max} > c_o, \quad c > c_o.$

An illustration of the above media is provided in particle physics by Nielsen-Picek metric (IV.3.21) for case (IV.3.23), i.e., for the interior of kaons, for which we have

$$V_{Max} = c_o \frac{n_1}{n_2} \cong c_o \frac{1 + 0.6\times10^{-3}}{1 - 0.1\times10^{-3}} \cong 1.0008\, c_o > c_o \tag{10.9a}$$

$$c \cong 1.0006\, c_o > c_o. \tag{10.9b}$$

where c evidently is a pure geometric quantity.

Note that the transition from the case of pions, Eq.s (10.8), to that of kaons, Eq.s (10.9), parallels the corresponding increase of density. We can therefore expect that *all remaining hadrons characterize media of (Class I) Type 9.* This creates the possibility that *hadronic constituents are massive physical particles moving at speeds higher than that of light in vacuum.* In turn, if such a conjecture could be verified, it would offer a realistic possibility of achieving a true confinement of quarks, as discussed in Sect. IV.9.

In the transition from hadrons to the core of stars, we have an evident increase of density because, in addition to the mutual wave overlapping of the constituents existing in the interior of hadrons, we have their compression. *The core of stars is therefore expected to costitute physical media of (Class I) Type 9 .*

This is the case predicted by our isospecial relativities in which a physical massive particle, such as a proton, could be locally accelerated by internal, short range, nonlocal forces to speeds higher than c_0 without any need of infinite energies, evidently because the notion of energy has no meaning for contact interactions. The understanding is that we are referring to local speeds, i.e., instantaneous speeds at a given point in the interior of the star considered, and not to globally defined constants such as c_0.

It is then evident that higher values of $V_{Max} > c_0$ are attained under higher short range nonlocal interactions, such as those in the interior of a star undergoing gravitational collapse.

At the limit of our internal conditions, i.e., for a star at the theoretical limit of gravitational collapse into a singuilarity,our isopecial relativities predict the possibility of infinite causal speeds, under which all distinctions between relativistic and Newtionian mechanics cease to exist

Note that media of Types 7, 8 and 9 have diffrent values of n_4 bigger, equal or smaller than one, but are otherwise geometrically equivalent. As a result, thay have different Postulates I and II, but the same Postulates III, IV, and V.

The above isorelativistic classification of physical media into nine types of Class I will soon result to be of particular physical value in proposing experimental tests of the isospecial relativities, inamsuch as it permits the identification of media which are readily available in our environment (such as our atmospheres which are of Type4), from other media which are not yet within our direct experimental capabilities (such as the gravitational ollapse which are of Type 9).

IV.11: CONCLUDING REMARKS

As pointed out in Sect. III.9, the isogalilean relativities need no experimental verification in our classical interior environment, because they are constructed from given equations of motion of our physical reality, by therefore verifying by construction the phenomena for which they have been built, e.g., deformations of extended charge distributions, etc.

By comparison, the isospecial relativities studied in this chapter do need indeed direct experimental verifications, because they predict a variety of novel phenomena within the physical conditions of their applicability: relativistic dynamics of extended particles and electromagnetic waves propagating within inhomogeneous and anisotropic physical media.

These experimental verifications will be studied in Chapter VII. In this section we shall conclude our presentation with a few remarks regarding the abstract identity of the isospecial relativities with the conventional one. In turn, this property, and the compatibility of the isospecial with the isogalilean relativities studied in Chapter VI, are at the foundations of the plausibility of said novel predictions.

Consider a conventional relativistic free particle in Minkowski space $M(x,\eta,\Re)$ represented via the canonical action with subsiadiary constraint

$$\delta A = \delta\int_{t_1}^{t_2}(p\,dx - Hds) =$$

$$= \delta\int_{t_1}^{t_2}(p_\mu\,\eta^{\mu\nu}\,dx_\nu - H\,ds) = 0, \tag{11.1a}$$

$$H = p^2 / 2\lambda - \tfrac{1}{2}c_o{}^4\lambda = p_\mu\,\eta^{\mu\nu}\,p_\nu / 2\lambda - \tfrac{1}{2}c_o{}^4\lambda \tag{11.1.b}$$

where λ represents the subsidiary constraint

$$x^\mu\,\eta_{\mu\nu}\,x^\nu - 1 = 0. \tag{11.2}$$

The isorelativistic representations of nonlinear, nonlocal and nonhamiltonian systems have been constructed in such a way to coincide with systems (11.1) whenever there is no action-at-a-distance force.

In fact, systems (11.1) admit the infinite number of geometrically equivalent, but physically different isotopes on isominkowski spaces $\hat{M}(x,\hat{\eta},\hat{\Re})$, $\hat{\eta} = T\eta$, $\hat{\Re} = \Re\hat{1}$, $\hat{1} = T^{-1} > 0$

$$\delta \hat{A} = \delta \int_{t_1}^{t_2} (p * dx - \hat{H}\, d\hat{s}) = \delta \int_{t_1}^{t_2} (p\, T\, dx - \hat{H}\, d\hat{s})$$

$$= \delta \int_{t_1}^{t_2} (p_\mu\, \hat{\eta}^{\mu\nu}(\hat{s}, x, \dot{x}, ...)\, dx_\nu - \hat{H}\, d\hat{s}) = 0, \qquad (11.3a)$$

$$\hat{H} = p^{\hat{2}} / 2\lambda - \tfrac{1}{2} c^4 \lambda = p_\mu\, \hat{\eta}^{\mu\nu}(\hat{s}, x, \dot{x}, ...)\, p_\nu / 2\lambda - \tfrac{1}{2} c^4 \lambda, \qquad (11.3b)$$

with isoconstraint[24]

$$x^\mu\, \hat{\eta}_{\mu\nu}(\hat{s}, x, \dot{x}, ...)\, dx^\nu - 1 = 0. \qquad (11.4)$$

The abstract identity of systems (11.1) and (11.3), as well as of the corresponding constraints (11.2) and (11.4), are evident. The same identity persists under the addition of external electromagnetic fields, as in Eq.s (IV.7.36), as well as for a system of particles, as in Eq.s (IV.8.3).

Nevertheless, Eq.s (11.1) represents a *free particle*. On the contrary, systems (11.3) represent *an extended and deformable particle moving within an inhomogeneous and anisotropic medium*. Aa an example, one may think at a free proton for system (11.1), and at the same proton when inside a star undegoling gravitational collapse for systems (11.3).

The important aspect for these concluding remarks is that system (11.1) verifies the conventional Postulates of the special relatvity, while systems (11.3) verify the generalized Postulates I–V of the isospecial relatujvities (Sect. IV.9).

As an example, the proton (11.1) travels at a speed smaller or equal to c_o. On the contrary, the same proton according to system (11.3) can travel at speed bigger, equal or smaller than c_o, depending on the local physical conditions of the star's core (density, temperature, etc.).

Exactly the same occurrence holds for the electromagnetic waves which, when traveling in vacuum obey the Einsteinian laws, including that of the Doppler's redshift, but when traveling in physical media obey the laws of the isospecial relativities, including the isodoppler's redshift. At the abstract level, however, their equations of motion coincide and so do the related laws.

This illustrates a central aspect of the analysis of these volumes, the fact that ***all the predictions of the isospecial relativities, including the expected capability of physical massive particles to attain local, interior speeds higher***

[24] Unlike the isogalilean case (see footnote[7] p. 91), there is no need to use a full isocontact structure in isorelativistic mechanics, $p * dx - \hat{H} \odot ds = p\, T_x\, dx - \hat{H}\, T_s\, ds$, because it is equivalent to the isosymplectic form $p\, T'\, dx - \hat{H}\, ds$ with scaled isotopic element $T' = T_x/T_s$.

than c_o, the expectation that quasars' light is redshifted by propagation in their atmosphere, etc., are all admitted by Einstein's axioms themselves, only realized in more general way.

APPENDICES

APPENDIX A: ISODUAL ISOSPECIAL RELATIVITIES

Consider an isominkowski spaces of Class I

$$M^I(x,\hat{g},\hat{\mathfrak{R}}), \quad \hat{g} = T\eta, \quad \hat{\mathfrak{R}} = \mathfrak{R}\,\hat{1}, \quad \hat{1} = T^{-1} > 0, \quad \eta \in M(x,y,\mathfrak{R}), \quad (A.1)$$

which, as now familiar, is the carrier space for our reporesentation of extended particles and electromagnetic waves propagating within physical media.

The isotopic techniques permit the identification of a rather intriguing image of isospaces (A.1), called *isodual isominkowski spaces* (Santilli (1988c))

$$\hat{M}^d(x,\hat{g}^d\hat{\mathfrak{R}}^d), \quad \hat{g}^d = T^d\eta, \quad \hat{\mathfrak{R}}^d = \mathfrak{R}\,\hat{1}^d, \quad \hat{1}^d = T^{d-1} < 0, \quad (A.2)$$

characterized by the *isorelativistic isoduality law*

$$\hat{1}^d = -\hat{1}, \quad \text{or} \quad T^d = -T. \quad (A.3)$$

Isoduals $\hat{M}^d(x,\hat{g}^d,\hat{\mathfrak{R}}^d)$ evidently admit the separation

$$x^{\hat{2}d} = x^\mu\, \hat{g}^d_{\mu\nu}\, x^\nu = -\, x^\mu\, \hat{g}_{\mu\nu}\, x^\nu \quad (A.4)$$

As such, they geometrize and generalize the conventional alternatives of selecting the Minkowski metric η = diag. (1, 1, 1, − 1) or η' = diag. (−1, −1, −1, 1).

The above isoduality is nevertheless significant because it is a new operation independent from conventional or isotopic inversions

$$x \Rightarrow x' = \hat{P}*\hat{T}*x \equiv \hat{P}^d *^d \hat{T}^d *^d x = -x. \qquad (A.5)$$

Also, isoduality requires the necessary use of the isotopic techniques for its identification, trivially, because it needs the generalized notion of unity even for the conventional case in which $\hat{I}^d = - I = \text{diag.}\ (-1, -1, -1, -1)$.

Finally, isoduality is far from merely mapping the conventional or generalized metric into their opposite, because it implies a change of the underlying field. In fact, consider the conventional Minkowski space which can be isomorphically written

$$M(x,\eta,\hat{\Re}), \quad \eta = \text{diag.}\ (1, 1, 1, -1), \qquad (A.6a)$$

$$\hat{\Re} = \Re\, I, \quad I = \text{diag.}\ (1, 1, 1, 1) \qquad (A.6b)$$

Then its *isodual Minkowski space* is given by

$$\hat{M}^d(x,\hat{\eta}^d,\hat{\Re}^d), \quad \eta^d = T^d\eta = \text{diag.}\ (-1, -1, -1, +1), \ (A.7a)$$

$$\hat{\Re}^d = \Re\, \hat{I}^d, \ \hat{I}^d = \text{diag.}\ (-1, -1, -1, -1). \qquad (A.7b)$$

The above structure is evidently based on the *isodual isoreals* $\hat{\Re}^d$ which are essentially *the image of the conventional field $\Re$ of real numbers under the assumption of −1 as "unit"*. In turn, this implies the reversal of the sign of all quantities defined in the original space, including the absolute values, e.e.,

$$|x - y| \in \mathfrak{M}(x,\eta,\hat{\Re}) \quad \Rightarrow \quad |x - y|^d = -|x - y| \in \hat{M}^d(x,\hat{\eta}^d\hat{\Re}^d) \quad (A.9)$$

A similar although generalized situation occurs for the full isospaces $M(x,\hat{g},\hat{\Re})$ and their isoduals $\hat{M}^d x,\hat{g}^d\hat{\Re}^d)$.

The isoponcaré symmetries and related isotransformations (Theorem IV.6.1) on $\hat{M}^I(x,\hat{g},\hat{\Re})$

$$\hat{P}(3.1) = \hat{O}(3.1)\otimes\hat{T}(3.1): \qquad x \Rightarrow x' = \hat{\Lambda}*x + x^\circ\tilde{b}^{-2}, \quad x^\circ > 0, \ (A.10)$$

admits the *isodual isopoincaré symmetries* and related *isodual isotransformations*

$$\hat{P}^d(3.1) = \hat{O}^d(3.1) \otimes \hat{T}^d(3.1): \quad x \Rightarrow x' = \hat{\Lambda} * x - x^\circ \hat{b}^{-2}, \quad x^\circ > 0, \quad (A.10)$$

respectively, which are precisely characterized by tlaw (A.3).

By comparing Eq.s (A.9) and (A.10) we see that a rudimentary image of isoduality for conventional Poincaré transformations is givebn by the change of sign of the parameters $x^\circ \Rightarrow x^{\circ\prime} = - x^\circ$. The understanding is that this is insufficient for the full isosymmetries, and the use of the isodual isounits is necessary for consistency.

Isosymmetries $\hat{P}^d(3.1)$ and related isodual isotransformations constitute fully acceptable symmetries of relativistic systems within physical media, and characterize a new class of relativities, tentatively called *isodual isospecial relativities*.

In fact, exactly as it occurred at the isogalilean level (Sect. III.8), *the law of isorelativistic isoduality is a universal symmetry of nature, in the sense that, whenever the isopoincaré symmetry is verified, so is its isodual*. As a particular case for conventional relativistic formulations, we learn that, *whenver the conventional Poincaré symmetry P(3.1) holds, so is its isodual $P^d(3.1)$*.

Along similar lines, *whenever the isospecial relativities hold, their isodual also holds too, thus offering a duality of relativities for the description of relativistic systems in both the interior and the exterior problems*

The above occurreneces are not trivial as the conventional replacement $\eta \Rightarrow \eta^d = - \eta$, because of the underlying duality of the field,. These comments therefore indicate ***the apparent existence of two, novel, separate Universes characterized by exactly the same physical laws which are not interconnected by space-time inversions or any other conventional mapping, but instead by our isoduality.***

APPENDIX IV.B: ISOTOPIC LIFTINGS OF DIRAC'S CONSTRAINTS

The isotopic liftings from the conventional to the *general* isopoincaré symmetries $\hat{P}(3.1)$ (Sect. IV.6) have fundamental implications for the Lagrangian formalism.

In fact, the Lagrangians L of conventional relativistic theories are *first-order*, in the sense that they depend on derivatives of the local variables up to the first-order, e.g., for for one point-particle without spin on $M(x,\eta,\Re)$

$$L = L(x, \dot{x}) \; = \; - m_o(-\dot{x}^{\mu}\eta_{\mu\nu}\dot{x}^{\nu})^{\frac{1}{2}}, \tag{B.1}$$

Then, on account of the fundamental invariant

$$ds^2 \; = \; - \; dx^{\mu}\eta_{\mu\nu}dx^{\nu} \; = \; -1, \tag{B.2}$$

these Lagrangians are *degenerate (singular)*, in the sense that

$$\text{Det}(\frac{\partial^2 L}{\partial\dot{x}^{\mu}\partial\dot{x}^{\nu}}) = 0, \tag{B.3}$$

In the transition to the general Poincaré-isotopic symmetries, the Lagrangians become of *second-order*, in the sense that they now generally depend on the derivatives of the local variables up to and including the second-order, e.g, for the isotopes of particle (B.1)

$$\hat{L} = \hat{L}(x,\dot{x},\ddot{x},..) = - m_o[\dot{x}^{\mu}\hat{g}_{\mu\nu}(x,\dot{x},\ddot{x},...)\dot{x}^{\nu}]^{\frac{1}{2}}, \tag{B.4}$$

Then, despite the fundamental isoinvariant (IV.7.4), i.e.

$$d\hat{s^2} \; = \; - \, dx^{\mu}\hat{g}_{\mu\nu}dx^{\nu} \; = \; -1, \tag{B.5}$$

the latter Lagrangians are *nondegenerate (regular)*, in the sense that, in general,

$$\text{Det}(\frac{\partial^2\hat{L}}{\partial\dot{x}^{\mu}\partial\dot{x}^{\nu}}) \neq 0. \tag{B.6}$$

The above occurrence can be seen in a number of ways, such as: the arbitrariness of the functional dependence of the isometric $\hat{g}$; the appearance of acceleration-dependent forces already in the case of nonrelativistic, two-body, closed, nonhamiltonian systems (Appendix III.A); or, more rigorously, the fact that the Lagrangian counterpart of Birkhoffian mechanics is precisely of second-order nondegenerate type (see Santilli (1982a), p.38).

By keeping in mind the complexity of second-order Lagrangian formalism, this illustrates the reason why the Lagrangian approach is rarely used in Birkhoffian mechanics, the first-order Pfaffian actions (IV.7.13) being much simpler, as well as analytically, algebraically and geometrically preferable.

Despite that, *Dirac's ((1950), (1958), (1964)) method for subsidiary constraints* remains useful for the study of the consistency of the central systems of our analysis, the closed nonhamiltonian systems with subsidiary constraints, as well as for the identification of the hypersurface of definition of the systems and of their symmetries.

The isotopic liftings of Dirac's method on isospaces $\hat{M}^{I}(x,\hat{g},\hat{\mathfrak{R}})$ was studied in Santilli (1989a), and we shall only outline it here for brevity.

Consider a given closed nonhamiltonian system (IV.2.5) represented in terms of the Birkhoffian variational principle (IV.7.13) with subsidiary constraints

$$\delta\int_1^2 ds[p^{a\mu}\hat{g}_{\mu\nu}\dot{x}^{a\nu} + \dot{\Phi}_\chi\lambda_\chi - H(s,x,p,\lambda,..)] = 0, \qquad \text{(B.7)}$$

where

$$H = H_o + H_C \approx H_o, \qquad \text{(B.8a)}$$

$$H_o = \{\sum_a[p^{a\mu}\hat{g}_{\mu\nu}p^{a\nu}/2\lambda_a - \tfrac{1}{2}\lambda_a + V(\hat{x_{ab}^2})\}, \qquad \text{(B.8b)}$$

$$H_C = C_k(s, x, p, ..)\,\lambda_k, \qquad \text{(B.8c)}$$

and the C's are subsidiary constraints of type (IV..2.5c)–(IV.2.5e) under study in this appendix. The above quantity H is called the *effective Hamiltonian* in Dirac's terminology, and the last equality in Eq. (B.8a) is Dirac's *weak equality*.

For the system to be consistent, the C-constraints must be conserved along an actual path, i.e., they must verify the conditions

$$\dot{C}_k = [C_k \,\hat{,}\, H] = [C_k \,\hat{,}\, H] \approx = [C_k \,\hat{,}\, H_o] + \lambda_i\,[C_k \,\hat{,}\, C_i], \qquad \text{(B.9)}$$

where the isocommutations rules are given by Eq.s (IV.4.23).

The Dirac's method now applies under a trivial isotopy. In fact, it is possible that conditions (B.9) are:

A) strongly verified along an actual path; or

B) verified via conditions on the multipliers; or

C) verified via a set of *secondary constraints*

$$\chi_\nu(x,p,..) = 0, \quad \nu = 1,2,..,m. \qquad \text{(B.10)}$$

By conducting the Dirac's iterative procedure, one remains at the end with a number of (independent) *first class constraints*

$$\psi_i(x,p,...) \approx 0, \;\; i = 1,2,...,r \tag{B.11}$$

and a number of (independent) *second class constraints*

$$\phi_\mu(x,p,...) \approx 0, \;\; \mu = 1,2,...,s. \tag{B.12}$$

Introduce the s×s matrix $\hat{C}$ with elements

$$\hat{C}_{\mu\nu} = [\phi_\mu \hat{,} \phi_\nu]. \tag{B.13}$$

It is easy to see that it is nonsingular (under the regularity of the isotopic brackets, as assumed throughout our analysis). Then the *isotopic liftings of Dirac's brackets* on $M^I(x,\hat{g},\hat{\mathfrak{R}})$ are given by

$$[A \hat{,} B]^* = [A \hat{,} B] + [A \hat{,} \phi_\mu] \hat{C}^{\mu\nu} [\phi_\nu \hat{,} B], \tag{B.14a}$$

$$(\hat{C}^{\mu\nu}) = (\hat{C}_{\alpha\ni})^{-1}. \tag{B.14b}$$

and verifies the Lie algebra axioms as in the conventional case.

The replacement of the isotopic brackets (IV.4.23) with the isotopic Dirac's brackets (B.14) then implies that we are dealing only with first class constraints with the new effective Hamiltonian

$$H' = H_o + \phi_\mu \lambda_\mu \tag{B.15}$$

where the multipliers λ_μ have the explicit structure

$$\lambda_\mu = \hat{C}^{\mu\nu} [\phi_\nu \hat{,} H_o], \tag{B.16}$$

thus showing their explixit functional dependence. But the first class constraints can be added anywhere without altering the equations of motion.

Thus, the total effective Hamiltonian is given by

$$\hat{H} = H_o + \phi_\mu \lambda_\mu + \psi_i \lambda_i, \tag{B.17}$$

where the multipliers λ_i are arbitrary functions, and the new terms express certain gauge degrees of freedom of the formulation. The rest of Dirac's theory then follows.

It is intriguing to note that *the commutators actually used in Dirac's treatment of conventional relativistic systems are already of isotopic type.* In fact, the *conventional* Dirac's brackets are Lie–

isotopic brackets. We therefore have the following important property.

PROPOSITION IV.B.1 (Santilli (1988c)): Dirac's relativistic formulations with subsidiary constraints possesses an essential Birkhoffian structure and, in particular, are characterized by the contravariant Birkhoff's equations (4.9), i.e.,

$$\frac{da^i}{ds} = \hat{\Omega}^{ij}(a) \frac{\partial \hat{H}(s,a,..)}{da^j}, \quad a = (x, p), \quad i,j = 1,2,..,8, \tag{B.18}$$

where: $\hat{H}$ is given by Eq.s (B.17); Birkhoff's tensor is given for Hamiltonian relativistic systems by

$$\hat{\Omega}^{ij} = \omega^{ij} + \omega^{ir} \frac{\partial \phi_\mu}{\partial a^r} \hat{C}^{\mu\nu} \frac{\partial \phi_\nu}{\partial a^s} \omega^{sj}; \tag{B.19}$$

ω^{ij} is the canonical Lie tensor (IV.4.14); while for nonhamiltonian systems we have the form

$$\hat{\Omega}^{ij} = \Omega^{ij} + \Omega^{ir} \frac{\partial \phi_\mu}{\partial a^r} \hat{C}^{\mu\nu} \frac{\partial \phi_\nu}{\partial a^s} \Omega^{sj}. \tag{B.20}$$

where Ω^{ij} is Birkhoff's tensor (IV.4.10).

In different terms, with his originality of thought, Dirac showed that the covering Birkhoffian mechanics is *necessary* for the study of the *conventional* relativistic systems, Eq.s (B.19), thus setting the foundations for a rather natural inclusion of the nonhamiltonian generalizations, Eq.s (B.20).

As a further comment we note that the Lagrangian for the *restricted* isopoincaré symmetries (Sect. IV.6) is of conventional type, i.e., it is of *first-order and, therefore, degenerate* as the conventional Lagrangian (B.1). Conventional methods then apply, although the first-order Pfaffian approach remains preferable on a number of counts, e.g., analytically, algebraically and geometrically, as well as from the viewpoint of the mapping into operator formulations.

The study of the isotopic, first-order, degenerate Lagrangians is here left to the interested reader for brevity.

APPENDIX IV.C: THE "NO NO-INTERACTION THEOREM"

One of the most significant implications of the generalization of relativistic Hamiltonian mechanics into the Birkhoffian form is the elimination of the so-called "No Interaction Theorem" (see, e.g., Mann (1974), or Kracklauer (1976), and quoted papers).

In turn, this implies a property of conventional treatments that does not appears to have been identified in the existing literature to our best knowledge, namely, the fact that *Dirac's formulation of conventional relativistic systems with subsidiary constraints also bypasses the No Intractions Theorem,* owing to its essential Birkhoffian structure (Appendix IV.B).

These aspects were also considered in Santilli (1988c) and are here briefly reviewed via the geometric approach.

Consider a vector field Υ on the space $\Re_s \times T^*M(x,\eta,\Re)$, where s is the proper time, M is the conventional Minkowski space, and T^*M is the conventional cotangent bundle. Suppose that Υ is a *Hamiltonian vector-field,* i.e., it verifies the property

$$\omega_2 \rfloor \Upsilon = -dH, \tag{C.1}$$

where ω_2 is the exact, canonical, symplectic two-form (Sect. IV.4). Then, the "No intreraction theorem" (Kraklauer (*loc. cit.*)) states that the condition that Υ is invariant under the conventional Poincaré group P(3.1) implies that all acting forces are identically null.

It is important to review the mechanism underlying the above occurrence. In essence, the first condition on the Hamiltonian structure of the vector field implies the following form

$$\Upsilon = \sum_a \left(u^{a\mu}\frac{\partial}{\partial p^{a\mu}} + K^{a\mu}\frac{\partial}{\partial x^{a\mu}}\right) + \frac{\partial}{\partial s}, \tag{C.2a}$$

$$u^{a\mu} = \partial H/dp^{a\mu}, \quad K^{a\mu} = -\partial H/\partial x^{a\mu} = K^{\alpha\mu}{}_{SA}, \tag{C.2b}$$

The second condition on the P(3.1)-invariance implies the existence of local parametrization

$$s = c^1 x^{a1} = c^2 x^{a2} = c^3 x^{a3}, \tag{C.3}$$

for given c-constants, and the identities

$$\frac{\partial}{\partial s} = c^1 \frac{\partial}{\partial x^{a1}} = c^2 \frac{\partial}{\partial x^{a2}} = c^3 \frac{\partial}{\partial x^{a3}}, \tag{C.4}$$

which, when plotted in Eq.s (C.1), imply

$$K^{a1} = K^{a2} = K^{a3} \equiv 0. \tag{C.5}$$

namely, all acting forces are null.

Consider now the *nonhamiltonian* vector field Γ of Eq.s (IV.8.1) on isospaces $\hat{\mathfrak{R}}_s \times T^*\hat{M}^I_2(x,\hat{g},\hat{\mathfrak{R}})$ of Sect. IV.4, and suppose that Γ is a *Birkhoffian vector-field*, i.e., it verifies the generalized property (IV.4.5)

$$\hat{\Omega}_2 \rfloor \Gamma = - dB, \tag{C.6}$$

where $\hat{\Omega}_2$ is the exact, symplectic-isotopic two-form (IV.4.18) on $T{*}\hat{M}^I(x,\hat{g},\hat{\mathfrak{R}})$.

Then, the imposition of the invariance of Γ under a Poincaré-isotopic symmetry $\hat{P}(3.1)$ does indeed imply the local parametrization (C.3) and identities (C.4), because of the preservation of the parameters and generators under isotopy.

However, when identities (C.4) are inserted in the Birkhoffian condition (C.6), they no longer imply the null value of all forces, trivially, because of the generalized Birkhoffian structure of the vector field. The above simple generalization of the proof of the "No Interaction Theorem" of Kraklauer (*loc. cit.*) then implies the following:

THEOREM IV.C.1 ("No-No-Interaction Theorem", Santilli (1988c)): Let Γ be a Birkhoffian vector field on isospace $\hat{\mathfrak{R}}_s \times T^*\hat{M}^I(x,\hat{g},\hat{\mathfrak{R}})$. Then, the invariance of Γ under a Poincaré-isotopic symmetry $\hat{P}(3.1)$ allows nontrivial, linear and nonlinear, local and nonlocal, as well as selfadjoint and nonselfadjoint interactions, unless the exact, symplectic-isotopic two-form on $T^*\hat{M}(x,\hat{g},\hat{\mathfrak{R}})$ reduces to the canonical form, $\hat{\Omega}_2 \equiv \omega_2$

Note that conventional *potential* interactions which are prohibited in the Hamiltonian relativistic setting, now become permitted in the Birkhoffian formulation, of course, jointly with much more general interactions.

The reader should be warned that the true understanding of Theorem IV.C.1 requires a knowledge of the Birkhoffian mechanics, e.g., its transformation theory and the techniques of representing forces via the generalized Lie tensor, rather than the Hamiltonian (see also below).

Theorem IV.C.1 can also be proved in a variety of other ways. Another simple geometric proof is that based on the study of *geodesic trajectories* (Sect. II.12). In fact, *relativistic Hamiltonian trajectories are equivalent to geodesic trajectories in our physical space-time* , thus implying the possible elimination of all acting forces.

When we pass to the study of relativistic trajectories within physical media, the geometrization characterized by the isometric $\hat{g}$ implies that *the trajectories of relativistic Birkhoffian systems can be made geodesic in their isospace, but their projections in our physical space-time are intrinsically nongeodecic.*

As a matter of fact, it is easy to prove the following

COROLLARY IV.C.1.1: The projection of the trajectory of a $\hat{P}(3.1)$-invariant Birkhoffian vector-field Γ on our physical space-time is nongeodesic even when all potential interactions are null.

At a still deeper level of study, we should recall that conventional relativistic Hamiltonian mechanics considers only one class of interactions, the local and potential ones, while at the covering Birkhoffian level we have the same interactions plus nonlinear, nonlocal and nonselfadjoint ones. A technical understanding of our "No No Interactions Thorem" cannot be achieved without a knowledge of the role and interplay of the above two classes of interactions.

This latter aspect can be best studied via the transformation theory. Recall that the transformations establishing the equivalence of a generic relativistic trajectory with the geodesic one are *canonical transformations* . In fact, the conventional "No Interaction Theorem" can be equivalently formulated via the following known property

THEOREM IV.C.2 (see, e.g., Mann (1974)): Within the context of relativistic Hamiltonian mechanics for Hamiltonians $H = T(p) + V(x)$, and under sufficient smoothness and regularity conditions, there always exist a canonical transformation, i.e. a transformation preserving the canonical structure

$$a = (x,p) \Rightarrow a' = (x', p'), \tag{C.7}$$

$$\omega^{ij} \Rightarrow \Omega^{ij} = \frac{\partial a'^i}{\partial a^r} \omega^{rs} \frac{\partial a'^j}{\partial a^s} \equiv \omega^{ij}, \tag{C.8}$$

under which the potential V of all forces is identically null,

$$V'(a') \equiv 0. \tag{C.9}$$

An equivalent formulation does indeed exist under isotopy. In fact, by using the *transformation theory of Birkhoffian mechanics* (Santilli (1982a), Chap. 5,), one can easily prove the following property.

THEOREM C.3 (Santilli (1988c): Within the context of relativistic Birkhoffian mechanics with Hamiltonian $H = T(p) + V(x)$, and under sufficient smoothness and regularity conditions, there always exists a "canonical" transformation, i.e., a transformation preserving Birkhoff's tensor

$$a = (x,p) \Rightarrow a' = (x', p'), \tag{C.10a}$$

$$\hat{\Omega}^{ij}(a) \Rightarrow \hat{\Omega}'^{ij}(a') = \frac{\partial a'^i}{\partial a^r} \Omega^{rs}[a(a')] \frac{\partial a'^s}{\partial a^s} \equiv \hat{\Omega}^{ij}(a'), \tag{C.10b}$$

under which all potentials are identically null,

$$V'(a') \equiv 0. \tag{C.11}$$

but the interactions represented by the Birkhoffian tensor $\hat{\Omega}^{ij}$ remains unaffected.

Thus, the physically important point is the impossibility of performing the transition

$$\hat{\Omega}^{IJ} \Rightarrow \omega^{ij}. \tag{C.12}$$

in a way compatible with the observer. By recally that the central role of the generalized Birkhoff's tensor is that of representing nontrivial interactions, the above occurrence implies the following

LEMMA IV.C.1 (Santilli (1982a)): No "canonical" transformation of Birkhoff's equations can render identically null all forces represented by Birkhoff's tensor.

At this point too the understanding of the above property and of its implications for *conventional* relativistic theories requires a technical knowlege of Birkhoffian mechanics.

In fact, *Birkhoffian mechanics can represent all forces, whether selfadjoint or not, via Birkhoff's tensor*; in which case the Birkhoffian is merely the generator of time evolution. As a result, our "No No Interaction Theorem" can be used to admit *conventional* selfadjoint interactions, provided that they are all represented via Birkhoff's tensor.

The best way to illustrate this occurrence is via the conventional electromagnetic interactions in vacuum without any nonselfadjoint force. In this case Birkhoffian mechanics allows the replacing of the familiar minimal coupling rule $p_\mu \Rightarrow P_\mu = p_\mu + eA_\mu$ with the reprersentation of electromagnetioc interactions via the Birkhoff's tensor (Santilli (*loc. cit.*), p. 98 and ff.)

$$\hat{\Omega}^{ij} = \begin{pmatrix} 0 & I \\ -I & e(\partial_\mu A_\nu - \partial_\nu A_\mu) \end{pmatrix}, \tag{C.13a}$$

$$H = p^2/2m , \tag{C.13b}$$

in which case the Birkhoffian is reduced to the "free" term $H = p^2/2m$.

Lemma IV.C.1 then illustrates the impossibility of eliminating the electromagentic interactions. Intriguingly, even though the equations of motion for systems (C.13) coincide with the conventional ones by construction, *their symmetry is no longer the conventional Poincaré symmertry P(3.1), but its isotope $\hat{P}$(3.1) characterized by Birkhoff's tensor (C.13a).* Its study is here left to the interested reader for brevity.

To avoid conceivable misrepresentations, let us note that the conventional "No Interaction Theorem" remains fully correct under the analysis of this Appendix, within the arena of its formulation and proof. It becomes merely inapplicable to the covering relativistic Birkhoffian mechanics and its Poincaré-isotopic symmetries.

Lemma IV.C.1 also allows Dirac's relativistic formulation with subsidiary constraints to admit nontrivial interactions. In fact, *Dirac's fundamental tensor is not canonical*, but given by form (B.19) with an essential, generalized, Birkhoffian structure. Then, (contrary to a rather popular erroneous belief) the applicable transformation theory is no longer Hamiltonian but Birkhoffian, and the applicable symmetries are no longer conventional, but of necessary Lie-isotopic type.

The generalized structure of Dirac's tensor and its preservation under the transformation theory ensure the lack of geodesic motion of Dirac's trajectories in our space-time, with the consequential inapplicability of the "No Interaction Theorem".

In closing this appendix we would like to mention that the most effective way of studying the effect of nonselfadjoint forces to the "No Interaction Theorem" is via the Riemannian-isotopic geometry of Sect. II.11, and the notion of isogeodesics of Sect. II.12. We shall therefore have the opportunity of returning to the subject in the next chapter on gravitation, by confirming the results at a higher geometric level.

CHAPTER V: ISOTOPIC GENERALIZATIONS OF EINSTEIN'S GRAVITATION

V.1: STATEMENT OF THE PROBLEM

Einstein's (1916) intuition of the fundamental role of the *Riemannian geometry* (Riemann (1868)) for the representation of the gravitational field, is one of the most brilliant advances in physical knowledge of this century.

Despite that, *Einstein's general theory of relativity*, or *Einstein's gravitation* for short (see, e.g., Pauli (1921) in the English translation of 1958) remains afflicted by a considerable number of rather fundamental problematic aspects which are unresolved at this writing.

In this chapter we are primarily interested in the study of

THE INTERIOR GRAVITATIONAL PROBLEM, i.e., the gravitational theory that is applicable in the interior of the minimal surface S^o encompassing all matter of the body considered, including its *possible atmosphere (Sect. I.1).*

The complementary problem is then evidently

THE EXTERIOR GRAVITATIONAL PROBLEM, i.e., the gravitational theory that is applicable in the exterior of said minimal surface S°.

In Chapter I we established the inequivalence of the interior and exterior problems, and the irredicibility of the former to the latter.

In Chapter II we outlined the novel mathematical tools for the quantitative treatement of the interior problem at the nonrelativistic, relativistic and gravitational levels.

The generalized relativities for the classical nonrelativistic interior problem were then studied in Chapter III, and their relativistic extension in Chapter IV, with particular reference to the identification of the space-time symmetries that are applicable to *extended-deformable bodies moving within inhomogeneous and anisotropic physical media.*

These generalized interior relativities are now assumed as *locally* valid for the interior gravitational treatment of this chapter, and their knowledge is assumed hereon.

The lack of exact character of Einstein's gravitation for the interior problem appears to be beyond reasonable scientific doubts for numerous independent reasons each one warranting a suitable generalization of the theory.

To begin, *the Riemannian geometry is local-differential,* as well known. Such a geometry is evidently ideal for the representation of the geodesic motion of a point-like test particle in the homogeneous and isotropic vacuum of the exterior problem. The same characteristics, however, are fundamentally at variance with the ultimate nonlocal nature of the interior problem.

A typical case is that for the core of a star undergoing gravitational collapse, where the wavepackets of a large number of constituents are, not only in a state of mutual penetration as expected in the hadronic structure (Sect. I.1), but also under *extremely dense compression.* The need for a suitable nonlocal-integral generalization of the Riemannian geometry for a more adequate representation of the physical reality of the interior gravitation is beyond reasonable doubts.

Similarly, the geodesic behavior of a test particle in the Riemannian space of the exterior gravitational problem is equally established on solid grounds. However, visual experimental observation, e.g., for a body in free fall toward Earth, equally establishes the *deviations* from the Riemannian geodesic behavior at the instant of penetration in our atmosphere. The need to generalize the notions of parallel transport and geodesic in a form directly applicable to the interior problem, is therefore also established beyond reasonable doubts.

At a deeper technical inspection, one can see that these geodesic deviations are due to *contact forces* (Sect. I.2) which, for sufficiently high speeds, are of the type of a power expansion in the velocity up to the tenth power and more (as routinely done in contemporary rocketry). Moreover, such a power series expansion are well known in engineering to be an approximation of nonlocal-integral forces. Still at a deeper inspection, the nonrelativistic forces of our interior physical reality, say, for a satellite during re-entry in Earth's atmosphere, are not only nonlinear and nonlocal, but also nonlagrangian-nonhamiltonian, as well as nonnewtonian[25].

The need to generalize the notion of geodesic in such a way to incorporate arbitrary forces existing in the classical reality, is also beyond reasonable doubts.

In fact, it is well known that the analytic structure of Einstein's gravitation permits the recovering, under the PPN and other limits, of only a *subclass* of Newtionian forces, those of local, potential, variationally selfadjoint type (Helmholtz (1887), Santilli (1978e)). Assuming the validity of certain modifications of the PPN limit to estract velocity dependent forces, the emerging nonselfadjoint forces have a necessarily limited dependence in the velocities, trivially, as required by the Christoffel's symbols (Sect. II.11).

This establishes also beyond reasonable doubts the need to generalize the geometric structure of Einstein's gravitation into a form admitting all conceivable forces of the interior trajectories of the physical reality.

A further reason for the lack of exact character of the theory in the interior problem is its *local Lorentz character*. In fact, such a character is evidently essential in the exterior problem for the representation of the stability of the planetary orbits, trivially, from the exact local rotational symmetry. However, the insistence of the exact, local, Lorentz character also for the interior problem implies excessive approximations of the physical reality of the type of the perpetual motion within a physical environment (Sect. IV.1).

This establishes the need to generalize Einstein's gravitation into a form locally admitting rotational, Galilei and Lorentz noninvariant

[25] We assume the reader is familiar with these terms, as defined in Chapter I.

unstable trajectories for the interior problem (only).

After all, visual observations of the interior of astrophysical objects, such as Jupiter (see Fig. V.1.1), establish the existence of interior vortices with monotonically varying angular momenta which are simply beyond the representational capability of Einstein's gravitation.

Numerous additional problematic aspects were studied in the preceding chapters, such as the homogeneous and isotropic character of Einstein's gravitation which is at variance with the evident, inhomogeneous and anisotropic structure of interior physical medium; the need to achieve a relativity representing the deformability of the extended test particles of our interior physical reality; and others. All these problematic aspects are also applicable to the interior gravitational problem of this chapter, and their knowledge is herein assumed.

In summary, Einstein's general theory of telativity cannot be considered the final theory for the interior gravitational problem because of numerous insufficiencies of the ultimate structure of the theory, including:

a) Insufficiencies of the fundamental Riemannian geometry, owing to its strictly local character when compared to the evidence of the ultimate nonlocality of the structure of gravitation;

b) Insufficiencies of the conventional analytic structure, owing to its maximal first-order character in the velocities with consequential established inability to recover the arbitrary forces of the interior physical reality;

c) The local Galilei's or Poincaré symmetries, owing to their consequential inability to represent nonconservative interior trajectories resulting in excessive approximations of physical reality;

d) The strictly homogeneous and isotropic character of the theory, which is manifestly unable to represent the established inhomogenuity and anisotropy of interior physical media;

e) The additional insufficiencies reviewed in the preceding chapters, which are now locally valid for the analysis of this paper.

The hopes of bypassing these problematic aspects by reducing a macroscopic interior trajectory to its elementary constituents, have been proved to have *no physical value* , because not realizable on technical grounds δue to the "No Reduction Theorems" I.3.1 and I.3.2 (Sect. I.3).

These clear occurrences force the physics community to face its duty: initiate the scientific process of trial and error in the construction of a generalization of Einstein's gravitation which represent more closely the interior physical reality.

For this purpose, we have introduced the isotopic generalizations of the affine geometry (Sect. II.10) and of the Riemannian geometry (Sect. II.11) which are capable of admitting arbitrary, linear and nonlinear, local and nonlocal, Lagrangian and nonlagrangian, Newtonian and nonnewtonian forces, as well as of containing the conventional affine and Riemannian geometries, respectively, as particular cases. The consequential generalized notions of parallel transport and geodesic under these arbitrary forces are presented in Sect. II.12.

These covering mathematical tools allowed us to submit in Santilli (1988d) the isotopic generalizations of Einstein's general theory of relativity for the interior problem, called *isogravitations* [26] under the conditions of:

A) Recovering identically Einstein's gravitation for the exterior problem;=

B) Being locally Lorentz-isotopic or Poincarré-isotopixc in character; and

C) Admitting at the nonrelativistic limit all conceivable forces of the interior problem of our physical reality.

By specific intent, no *exterior* experiment to test the proposed new theory is submitted in this volume because the generalized interior theory recovers, by construction, conventional gravitational theories whenever the exterior conditions are recovered.

In Chapter VII we shall indeed submit a number of experimental tests, but they are solely referred to interior physical conditions.

A number of independent contributions in the interior gravitational problem must be acknowledged from these introductory words. First, in three pioneering papers, Gasperini (1984a, b, c) submitted a generalization of Einstein's gravitation that is locally Lorentz-isotopic (see Appendix V.A) in the sense of Santilli's (1978a) proposal of the Lie-

[26] The plural stands to denote an infinite number of possible, different, isogravitations for each given total gravitational mass, evidently because of the infinitely possible interior conditions.

isotopic theory, and the foundations of the Lorentz-isotopic relativities (Santilli (1983a)).

Gasperini's notion of a generalized gravitational theory of local Lorentz-isotopic character, with the consequential, generalized principle of equivalence, will remain central notions of our study.

However, Gasperini formulated his gravitation in a conventional Riemannian space, thus being unable to treat nonlocal-integral forces. In the next section we shall present the generalization of Gasperini's studies in the Riemannian-isotopic spaces (Sect. II.11) submitted by Santilli ((1988d), (1991b)).

Such a generalization is needed on a number of grounds, such as the need to achieve geometrically unique notions of geodesic and parallel transport for the interior and exterior problems, the identity of interior and exterior gravitational theories at the abstract, coordinate-free level, and others.

Moreover, Gasperini (*loc. cit.*) formulated his gravitation everywhere in space-time, thus resulting in predictable, severe limitations on the magnitude of the allowed isotopy from available gravitational experiments.

In Santilli (1988d) we therefore restricted Gasperini's studies to the interior gravitational problem, and the same position will be adopted in this chapter. This second modification of Gasperini's original proposal is also needed to permit the recovering of Einstein's exterior gravitation in its entirety, as well as to eliminate any restriction on the magnitude of the possible interior isotopies.

In the main text of this chapter we shall solely study the *Lie-isotopic general treatment of the interior gravitation as a closed-isolated system.* The reader should however be aware of *the broader Lie-admissible approach to gravitational, interior, open-nonconservative trajectories*, as anticipated in the appendices of Chapter II.

In aother pioneering paper, Gasperini (1983) submitted a more general gravitations, this time, of locally *Lorentz-admissible* character in the sense of Santilli (1981a), and stressed his view that

> *The ultimate formulation of gravitation encompassing all others as particular cases, is expected to be precisely of Lie-admissible character.*

Numerous additional contributions with a direct relevance to the studies of this chapter have been identified in Sect. I.5, including those by Jannussis (1985), Gonzalez-Diaz (1986), Nishioka ((1983), (1987)), and others.

Regrettably, we are unable to study the more general Lie-admissible formulation of gravitation to prevent a prohibitive length of this monograph.

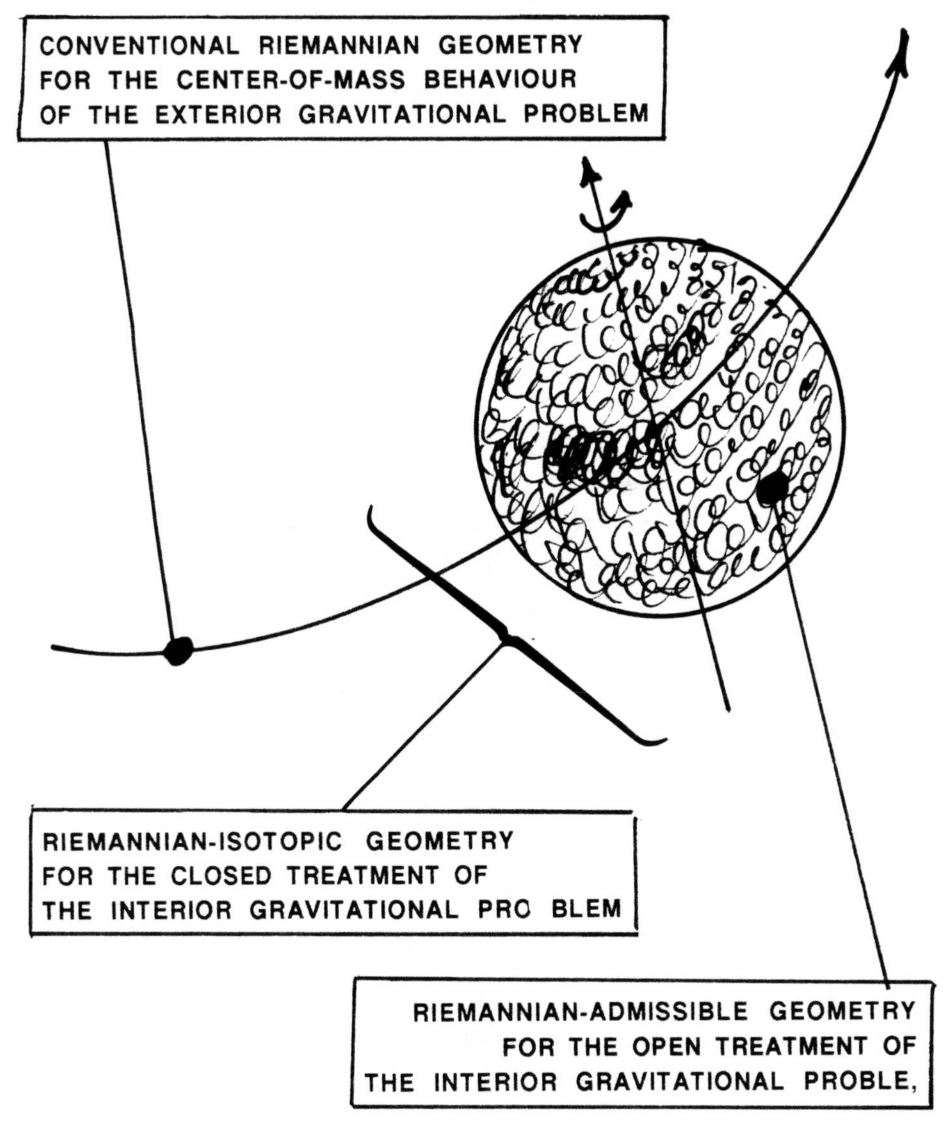

FIGURE V.1.1: We initiated these studies with the remainder (Fig. I.1.1) that the birth of contemporary relativities for exterior problems can be identified with the historical visual inspection of the Jovian system by

Galileo Galilei back in 1609. The birth of the generalized relativities for interior problem can be identified with the visual observation, this time, of Jupiter's structure. A main objective of Chapters III and IV have been the outline of nonrelativistic and relativistic structure models of Jupiters which recover conventional relativities for the center-of-mass behavior, while permitting a structurally more general, nonlinear, nonlocal and nonlagrangian interior dynamics. In this chapter we shall complete these classical studies by presenting a gravitational model of Jupiter's structure which, on one side, is capable of recovering identically the Einsteinian behavior of Jupiter's center-of-mass trajectory in the solar system, while, on the other side, it admits locally nonconservative interior trajectories (such as Jupiter's vortices with continuously varying angular momenta). The study of this dichotomy via the Riemannian geometry for the exterior problem and its Riemannian-isotopic generalization for the interior problem is grossly insufficient from a physical viewpoint on a number of grounds, such as: both approaches imply total, conventional, conservation laws; both approaches imply an intrinsically reversible dynamics; etc. These deficiencies can be resolved via the additional study of the local interior gravitational problem as an open-nonconservative system in term of the broader Riemannian-admissible geometry of Appendix II.C. In fact, the latter geometry is ideally suited to represent an extended testparticle in unstable interior conditions while considering the rest of the system as external, because of its capability of directly representing open-nonconservative conditions, of being intrinsically irreversible, and structurally compatible with the Riemannian-isotopic geometry (which is characterized by the symmetric part attached to the Riemannian-admissible geometry). According to a rather general view, current knowledge in gravitation (based on the Riemannian geometry only) is believed to be exactly valid under whatever conditions exist in the Universe. On the contrary, we believe that our current knowledge in gravitation is at its first infancy, owing to a large number of fundamental unresolved problems, with the understanding that current views provide a first approximation of undeniable validity. These comments are introduced here to stess that, by no means, the isotopic interior generalizations of Einstein's exterior gravitation revieweduin this chapter should be considered as final, for physics is a discipline that will never admit final theories.

We now pass to a few comments on Einstein's *exterior* gravitation. In the introductory words of our relativistic studies (Sect. IV.1), we stressed the *exact* validity of the special relativity in the arena of its original conception, and its *approximate* validity even when inapplicable because of noneinsteinian conditions.

The situation is somewhate different for the general relativity owing to a considerable number of problematic aspects accumutaled thoughtout this century, which are fundamentally unresolved at this writing owing to their lack of proof or disproof by experts in the field in the technical literature.

It is evident that due scientific caution requires the resolution of at least the most important of these problematic aspects prior to any final jugment on the exact or approximate validity of Einstein's exterior gravitation[27].

The literature in the field is so vast, to discourage even a partial outline. An overview, including societal aspects caused by the lack of resolution of the problematic aspects over a protracted period of time, has been recently provided by Weiss (1991).

In this section we shall consider only the following three problematic aspects because they have a direct connection with our interior studies.

The first is due to Einstein's historical conception of the exterior gravitational field in vacuum as reducible to pure geometry without source. Such a conception is in direct and irreconcilable disagreement with the charged structure of matter, and Maxwell's electrodynamics.

In fact, it is well established in elementary particle physics that mass has a primary electromagnetic origin. This implies the existence of an electromagnetic source tensor of the gravitational field which is of first-order in magnitude even for celestial bodies with null total charge and null total electromagnetic moments, as studied in detail by Santilli (1974) (see Sect. V.3 for a review).

The second problematic aspect is that raised by Yilmaz's studies ((1958), (1971), (1977), (1979), (1980), (1982a, b), (1984), (1989), and (1990a, b)), on the apparent need for a stress-energy tensor in the right-hand-side of the field equations in vacuum. These studies have remained largely ignored in the technical literature and, therefore, unresolved.

The third aspect considered in this chapter is that related to the forgotted *Freud identity* . As the reader will recall from Sect. II.11, Freud (1939) discovered an identity which is independent of the others

[27] In the opinion of this author, gravitation is the least rigorous of all branches of physics, because of numerous nonscientific aspects that have existed over considerable periods of time (Weiss (1991)). The most unreassuring of them is the widespread claim on the *exact validity of Einstein's gravitation under whatever conditions exist in the Universe, and for both the exterior and the interior problems, without a disproof of the problematic aspects published in the technical literature.* But Science is a collegial effort which, as such, never achieves final resolutions via the judgment of only one segment of the community. The ignorance of the problematic aspects of Einstein's gravitation by such segment does not constitute their disproof, and only leaves the field in a state of "suspended animation", without any possibility of truly scientific resolution one way or the other.

(such as the *contracted Bianchi identity* reviewed in Sect. II.11), and which must be verified by any gravitational theory for consistency, exactly as it is the case for the other identities.

The Freud identity did not escape Pauli who reviewed and discussed it in considerable details (see Pauli (1958), Sect.s 23, 57 and 61, pp. 70 and 216 in particular). The identity was thereafter ignored by subsequent books in the field, with only one exception given by Carmeli *et al.* (1990), where the Freud identity is mentioned, although without the identification of its consequences. The identity finally received a mathematically regorous treatment by Rund (1991), whose analysis was adopted in our presentation of Seect. II.11.

Despite these advances, the Freud identity raises a number of physical problems that are still unresolved at this writing. In fact, Yilmaz (1990b) "rediscovered" the Freud identity by stating that, when imposed to the field equations, it implies the necessary presence in the right-hand-side of the equations of two tensors, the stress-energy tensor proposed by Yilmaz for decades ((1958), (1971), (1977), (1979), (1980), (1982a, b), (1984), (1989), and (1990a, b)), and a second tensor with all the characteristics of the electromagnetic source of the gravitational field, as pointed out by Santilli (1974) and (1988d).

These aspects are of such evident relevance, to require their critical examination by experts at large.

It should be stressed that our objectives are limited to the interior gravitational problem, and the latter issues pertaining to the exterior gravitation are only reviewed in this chapter for the reader's convenience without any resolution one way or the other.

A true understanding of this chapter requires the prior knowledge of *all* the preceding parts of these monographs.

V.2: ISOTOPIC LIFTINGS OF EINSTEIN'S GRAVITATION

In this section we shall assume Einstein's general theory of gravitation[28] as being correct for the exterior gravitational problem

[28] Among a literature in the field is so large to prevent any comprehensive outline here, this author still prefers Pauli ((1921) in its 1981 English edition) because of its correct scientific language, as well as its content still broader than sebsequent reviews (e.g., because of Pauli's treatment of the Freud identity ignored by subsequent books).

gravitational problem only, and study its infinite class of possible isotopic generalizations for the interior problem under the name of *isogravitations*, as proposed by Santilli (1988d), (1991b,c)).

Let us begin by recalling the three basic representation spaces used in the conventional approach to gravity:

1) The carrier space of Newtonian exterior mechanics, the Kronecker product of *Euclidean spaces* (Chapter III)

$$\Re_t \times E(r,\delta,\Re) \ : \ \delta = \text{diag.}\,(1,1,1), \tag{2.1a}$$

$$r = (r^i),\ r^2 = r^t \delta r = r^i \delta_{ij} r^j = r_1{}^2 + r_2{}^2 + r_3{}^2\,, \tag{2.1b}$$

$$i = i,2,3\ \ (=x,y,z)$$

over the reals $\Re$;

2) The carrier space of relativistic exterior mechanics, the familiar *Minkowski space* (Chapter IV)

$$M(x,\eta,\Re) \ : \ \eta = \text{diag.}\,(1,1,1,-1), \tag{2.2a}$$

$$x = (x^\mu) = (r, x^4), \quad x^4 = c_o t, \quad r \in E(r,\delta,\Re)$$

$$x^2 = x^t \eta x = x^\mu\, \eta_{\mu\nu}\, x^\nu = x_1{}^2 + x_2{}^2 + x_3{}^2 - x_4{}^2 = R^2\,; \tag{2.2b}$$

$$\mu, \nu = 1,2,3,4, \quad R \in \Re,$$

where c_o represents hereon the *speed of light in vacuum;* and

3) The carrier space of Einstein's exterior gravitation, the *Riemannian space* in the conventional (3.1)-space-time with a symmetric connection and null torsion (Sect. II.11)

$$R(x,g,\Re) \ : \qquad g = g(x) = (g_{\mu\nu}) = (g_{\nu\mu}), \tag{2.3a}$$

$$x = (x^\mu) = (r, x^4); \qquad x^2 = x^t g(x)\, x = x^\mu g_{\mu\nu} x^\nu = R^2, \tag{2.3b}$$

$$\Gamma^1{}_{\mu\rho\nu} = \tfrac{1}{2}\left(\frac{\partial g_{\mu\rho}}{\partial x^\mu} + \frac{\partial g_{\rho\nu}}{\partial x^\nu} - \frac{\partial g_{\mu\nu}}{\partial x^\rho} \right) \neq 0, \tag{2.3c}$$

$$\tau_\mu{}^\rho{}_\nu = \Gamma^2{}_\mu{}^\rho{}_\nu - \Gamma^2{}_\nu{}^\rho{}_\mu \equiv 0. \tag{2.3d}$$

As well known, the largest groups of linear and local-differential

Rund (1975).

isometries of space (2.1) is the *Galilei symmetry* $G_\delta(3.1)$. The largest group of linear and local isometries of the Minkowski space (2.2) is the *Poincaré symmetry* $P_\eta(3.1)$. The largest group of isometries of the Riemannian space (2.3) is known only locally, i.e., in the neighborhood of a point and it is given also by $P_\eta(3.1)$.

The first contribution by the Lie-isotopic techniques is therefore for *conventional* gravitational theories, and can be formulated as follows.

THEOREM V.2.1 (Santilli (1988d)): The largest possible nonlinear but local-differential group of isometries of the conventional Riemannian spaces (2,3) with metric g(x) are the isopoincaré symmetries $\hat{P}_g(3.1)$ with isounits $\hat{I} = T^{-1}$, $g = T\eta$. For the case of Einsteinian theories of gravity, $T > 0$ and all isosymmetries $\hat{P}_g$ (3.1) are locally isomorphic to the conventional Poincaré symmetry P_η (3.1).

PROOF: All possible metrics of Riemannian spaces can be written

$$g(x) \quad = \quad T(x)\,\eta, \tag{2.4}$$

where T is nowhere singular, symmetric and real valued, in which case Theorem IV.6.1 holds, characterizing the isopoincaré symmetries $\hat{P}_g(3.1)$ with respect to the isounit

$$\hat{I} = T^{-1}. \tag{2.5}$$

The symmetry transformations are then nonlinear but local, from the corresponding functional dependence of the isounit on the coordinates. For the case of Einstenian theories of gravitation, T is positive-definite, $T > 0$, in which case Theorem IV.6.1 implies, and $\hat{P}_g(3.1) \approx P_\eta(3.1)$. Q.E.D.

We assume the reader is familiar with the fact that the above Theorem (and Theorem IV.6.1) are not purely mathematical results, but permit the computation of the symmetry transformations in their explicit form from the sole knowledge of each given metric g(x).

Also, we should recall the conventional nonrelativistic limit

$$P_\eta(3.1)\big|_{c_o/R \Rightarrow \infty} \Rightarrow G_\delta(3.1). \tag{2.6}$$

Via the use of Appendix IV.A, we can then say that the Poincaré-isotopic symmetries $\hat{P}_g(3.1)$ admit a nonrelativistic limit into a symmetry

$\hat{G}_{\tilde{\delta}}(3.1)$ which are also locally isomorphic to the Galilei symmetry $G_{\delta}(3.1)$ whenever $\tilde{\delta} > 0$ (as it is the case for Einstenian gravitational theories),

$$\hat{P}_g(3.1)_{|c_o/R \Rightarrow \infty} \Rightarrow \hat{G}_{\tilde{\delta}}(3.1) \approx G_{\delta}(.3.1). \tag{2.7}$$

Finally, we should recall that $\hat{P}_g(3.1)$ admits the conventional symmetry $P_{\eta}(3.1)$ as a local relativistic symmetry, and $G_{\delta}(3.1)$ as a local nonrelativistic symmetry.

This essentially summarizes the geometrical structure of the local and global symmetries of Einstein's exterior gravitation.

Our objective is to construct an infinite class of *symmetry-preserving isotopies* of Einstein's gravitation for the interior dynamical problem, and interpret the isotopies, as now familiar, as representing the transition from motion in vacuum to motion within a physical medium.

Note that the *axiom-preserving isotopies* cannot be introduced for Einstein's gravitation because of the technical difficulties caused by Einstein's tensor discussed in Sect. II.11, but they will be submitted in the next section for the "axiomatically completed" Einstein's theory.

To identify the structure of the isotopies of Einstein's gravitation, let us review the three basic isospaces of our analysis:

1') The carrier space of the nonrelativistic[29] interior mechanics, the *isoeuclidean spaces* for nonlinear, nonlocal, nonlagrangian and nonnewtonian trajectories (Chapter III)

$$\hat{\mathfrak{R}}_t \times \hat{E}(r,\hat{\delta},\hat{\mathfrak{R}}): \quad \hat{\delta} = \hat{\delta}(r, \dot{r}, \ddot{r}, ...) = T_{\hat{\delta}}(r, \dot{r}, \ddot{r},..) \,\delta = (\hat{\delta}_{ij}) = (\hat{\delta}_{ji}), \tag{2.8a}$$

$$r^{\hat{2}} = r^t \hat{\delta}\, r = r^i \,\hat{\delta}_{ij}(r,\dot{r},\ddot{r}, ..)\, r^j, \tag{2.8b}$$

$$\hat{\mathfrak{R}} = \mathfrak{R}\,\hat{1}_{\hat{\delta}}, \quad \hat{1}_{\hat{\delta}} = T_{\hat{\delta}}^{-1}, \quad \hat{\mathfrak{R}}_t = \mathfrak{R}\hat{1}_t; \quad \hat{1}_{\hat{\delta}} > 0, \quad \hat{1}_t > 0; \tag{2.8c}$$

2) The carrier space of relativistic interior mechanics, the *isominkowski spaces* of Class I for the relativistic description of nonlinear, nonlocal, nonlagrangian and nonlorentzian trajectories (Chapter IV)

$$\hat{M}^I(x,\hat{\eta},\hat{\mathfrak{R}}): \; \hat{\eta} = \hat{\eta}(x, \dot{x}, ...) = T_{\eta}(x, \dot{x},..)\, \eta = (\hat{\eta}_{\mu\nu}) = (\hat{\eta}_{\nu\mu}), \tag{2.9a}$$

$$x^{\hat{2}} = x^t \,\hat{\eta}(x, \dot{x}, \ddot{x}, ..)\, x = x^{\mu}\, \hat{\eta}_{\mu\nu}(x, \dot{x}, \ddot{x}, ..)\, x^{\nu} = R^2, \tag{2.9b}$$

[29] We continue to avoid the term "Newtonian" for the interior dynamical problem, and use instead the term "nonrelativistic", because of the presence of nonnewtionian forces discussed in Chapter I.

$$\hat{\Re} = \Re\,\hat{1}_{\hat{\eta}}, \quad \hat{1}_{\hat{\eta}} = T_{\hat{\eta}}^{-1} > 0, \quad x, \eta \in M(x,\eta,\Re), \quad R \in \Re. \tag{2.9c}$$

Finally, we introduce

3′) The carrier space of our gravitational interior theory, the *isoriemannian spaces* in (3.1)-dimension with a symmetric isoconnection and a null isotorsion in the geometrical space, but a non-null torsion in the physical space of the observer (Sect. II.11)

$$\hat{R}(c,\hat{g},\hat{\Re}): \hat{g} = \hat{g}(x, \dot{x}, ...) = T_g(x, \dot{x}, ...)\, g(x)$$

$$= (\hat{g}_{\mu\nu}) = (\hat{g}_{\nu\mu}) = (T_\mu{}^\sigma g_{\sigma\nu}), \tag{2.10a}$$

$$x^{\hat{2}} = x^t\, \hat{g}(x, \dot{x}, ...)\, x = x^\mu\, \hat{g}_{\mu\nu}(x, \dot{x}, ...)\, x^\nu = R^2, \tag{2.10b}$$

$$\hat{\Gamma}^1{}_{\mu\rho\nu} = \tfrac{1}{2}\left(\frac{\partial\hat{g}_{\mu\rho}}{\partial x^\mu} + \frac{\partial\hat{g}_{\rho\nu}}{\partial x^\nu} - \frac{\partial\hat{g}_{\mu\nu}}{\partial x^\rho}\right) \tag{2.10c}$$

$$\hat{\tau}_\mu{}^\rho{}_\nu = \hat{\Gamma}^2{}_\mu{}^\rho{}_\nu - \hat{\Gamma}^2{}_\nu{}^\rho{}_\mu \equiv 0, \tag{2.10d}$$

$$\tau_\mu{}^\rho{}_\nu = T_\mu{}^\sigma\, \Gamma^2{}_\sigma{}^\rho{}_\nu - T_\nu{}^\sigma\, \Gamma^2{}_\sigma{}^\rho{}_\mu \neq 0. \tag{2.10e}$$

The largest possible nonlinear and nonlocal groups of isometries of the isoeuclidean space (2.8) are given by the *isogalilean symmetries* $\hat{G}_{\hat{\delta}}(3.1)$ of Sect. III.5. The largest possible nonlinear and nonlocal groups of isometries of the isominkowski spaces (2.9) are given by the isopoincaré symmetries $\hat{P}_{\hat{h}}(3.1)$ of Sect. IV.6. The same methods then leads to the following

THEOREM V.2.2 (loc. cit.): The largest possible nonlinear and nonlocal groups of isometries of the isoriemannian spaces $\hat{R}(x,\hat{g},\hat{\Re})$ are the Poincaré-isotopic symmetries $\hat{P}_{\hat{g}}$ (3.1) for isometrics $\hat{g} = Tg$ with (nowhere singular and Hermitean) isounits $\hat{1} = T^{-1}$, which result to be locally isomorphic to the conventional Poincaré symmetry P_η (3.1) for all infinitely possible, positive-definite isotopic elements T_g.

Therefore, the fundamental space-time symmetry of Einstein's gravitation is not lost in our transition to interior gravitational problems, but merely realized in the most general known, nonlinear and nonlocal, Lie-isotopic form (see Fig. V.2.1 for more details).

In particular, as hown in Sect. VI.3, we have the property in full analogy with property (2.7),

$$\hat{P}_{\hat{\eta}}(3.1)\Big|_{c_o/R \Rightarrow \infty} = \hat{G}'_{\delta'}(3.1), \qquad (2.11)$$

Similarly, the general symmetry $\hat{P}_{\hat{g}}(3.1)$ admits, locally, the relativistic isosymmetry $\hat{P}_{\hat{\eta}}(3.1)$ and the nonrelativistic isosymmetry $\hat{G}_{\hat{\delta}}(3.1)$. In this way, every major aspect of the conventional theory has been shown to admit an infinite number of corresponding isotopic liftings.

More particularly, isotopic spaces (2.8), (2.9) and (2.10) provide an infinite number of *coverings* of the corresponding conventional spaces (2.1), (2.2) and (2.3), respectively; similarly, isosymmetries $\hat{P}_{\hat{g}}(3.1)$, $\hat{P}_{\hat{\eta}}(3.1)$ and $\hat{G}'_{\hat{\delta}'}(3.1)$ provide an infinite number of coverings of the corresponding conventional symmetries $P_g(3.1)$, $P_\eta(3.1)$ and $G_\delta(3.1)$; where the term "coverings" are used in a sense indicated earlier.

Let us finally keep in mind the notion of *interior and exterior problems,* as reviewed at the beginning of Sect. V.1.

We are now sufficiently equipped to present the *infinite number of isotopic liftings of Einstein's gravitation for the interior gravitational problem,* called *Einstein-isotopic gravitations* , or *isogravitations* for short, which can be introduced via the following generalized equations on isoriemannian space $\hat{R}(x,\hat{g},\hat{\mathfrak{R}})$ in standard units (Santilli (*loc. cit.*))

$$\delta\hat{A} = \delta\int d^4x\, \hat{\Delta}^{\frac{1}{2}}(\hat{R} - 8\pi\hat{M}) =$$

$$= \delta\int d^4x\, \Delta^{\frac{1}{2}}(\hat{g}^{\mu\nu}\hat{g}^{\rho\sigma}\hat{R}_{\mu\nu\rho\sigma} - 8\pi\hat{g}^{\mu\nu}\hat{M}_{\mu\nu}) = 0, \qquad (2.12a)$$

$$\hat{g} = T_g g, \qquad T_g > 0, \qquad (2.12b)$$

$$T_{g\,|\,r > R^{\circ}} = I = \text{diag.}\,(1,1,1,1), \qquad (2.12c)$$

where:

1) Eq. (2.12a) represents the isotopic action on $\hat{R}(x,\hat{g},\hat{\mathfrak{R}})$ (Theorem II.11.5), with $\hat{R}$ being the isocurvature isoscalar and $\hat{M}$ the isoscalar of the conventional matter tensor computed and contracted on $\hat{R}(x,\hat{g},\hat{\mathfrak{R}})$. This first condition ensures the achievement of a generalized theory

of the interior gravitational problem, with particular reference to the admission of nonlinear, nonlocal, nonlagrangian and nonnewtonian internal trajectories, as well as a direct representation of the inhomogenuity and anisotropy of the interior physical media, as evidently permitted by the isotopic element T_g.

II) Condition (2.12b), is imposed to preserve the topological structure of the exterior treatment $g = T_\eta \eta$, $T_\eta > 0$, also in the interior problem with $\hat{g} = T_g g$, $T_g > 0$, thus allowing the fundamental preservation of the Poincaré symmetry as the universal symmetry for local and global, interior and exterior conditions. And

III) Condition (2.12c) implies that the isotopic element T_g acquires the trivial unit value $I = \text{diag.}\ (1,1,1,1)$ everywhere in the exterior problem, by therefore guaranteeing that the isogravitations recover the conventional Einstein's gravitation identically everywhere in the exterior problem.

An inspection of the various metrics then implies the following physical consequence.

LEMMA V.2.1 (loc. cit.): In the transition from the exterior to t interior isogravitations (2.12), there is the transition from a local-differential dynamics with (variationally) selfadjoint interactions describing motion in the homogeneous and isotropic vacuum, to a nonlocal-integral dynamics with selfadjoint and nonselfadjoint interactions describing motion within generally inhomogeneous and anisotropic interior physical media.

In different terms, Einstein's exterior gravitation represents the trajectories of dimensionless test particles in vacuum which, as such, can only have a local-differential geometry with action-at-a-distance dynamics, as well known.

In the transition to our interior problem, we have instead the representation of extended (and therefore deformable) test bodies moving within resistive media which, as such, demand a nonlocal-integral geometry and a contact dynamics.

The direct representation of interior physical media is evidently ensured by the isometrics $\hat{g}(x, \dot{x}, \ddot{x}, \mu, \tau, n, \ldots)$ which can represent physical notions and events essentially beyond the representational capabilities of Einstein's gravitations, such as:

a) the variation of the density μ of the interior medium with the distance from the center (inhomogenuity);

b) a preferred direction in the interior medium caused, e.g., by the intrinsic angular momentum of the body (anisotropy);

c) the local variation of the index of refraction n,

d) the local dependence of the speed of light $c = c_o b_4(x, \dot{x}, \mu, \tau, ...)$ on the physical conditions of the medium considered (when transparent);

e) the maximal local causal speed as the true local invariant of the theory (recall from Chapter IV that the speed of light cannot be invariant because it varies from transparent medium to transparent medium, and can be *smaller* than the maximal causal speed of a massive particle, as it happens for the Cerenkov light in water which travels at speed smaller than that of the electrons;

and numerous additional features typical of interior dynamics studied in the preceding chapters.

The representation of all the above local properties is ensured by the fact that the isometric $\hat{g}$ can also be interpreted as *the isotopy of an isotopy*, i.e.,

$$\eta \Rightarrow \hat{\eta} = T_\eta \eta \Rightarrow \hat{g} = T_{\hat{\eta}} \hat{\eta} \equiv T_g g. \tag{2.13}$$

By recalling that the Lie-isotopic liftings of Lie's symmetries preserve, by construction, the original generators (and parameters), the physical implications of the above results is expressed by the following property directly originating from the Lie-isotopic theory (II.6).

THEOREM V.2.3 (loc. cit.): All global and local, conventional and isotopic Poincaré symmetries of isogravitations (2.12) admit as generators the same, conventional, total conserved quantities

$$P^\mu = \int M^{\mu 0} d^3x, \tag{2.14a}$$

$$J^{\mu\nu} = \int (x^\mu M^{\nu 0} - x^\nu M^{\mu 0}) d^3x. \tag{2.14b}$$

Note that in conventional presentations of Einstein's gravitation, the above total conservation laws are deduced via rather complex methods, while in our approach the same conservation laws are directly derived from the global Poincare symmetry.

Moreover, the *conventional* total conservation laws (2.14) occur under a *generalized* interior dynamics. By recalling the central notions of closed nonselfadjoint systems of Chapters III and IV and their isosymmetries, we have the following

COROLLARY V.2.3.1 (loc. cit.): Isogravitation (2.12) characterizes the gravitational extension of closed-isolated systems with nonselfadjoint internal forces, in which the total, conventional conservation laws are ensured by the general isopoincaré symmetry.

Again, as it had been the case at the nonrelativistic and relativistic levels, we are dealing with a subclass of *gravitational closed nonselfadjoint systems* , those without subsidiary constraints. The understanding is that the most general class is that defined on the hypersurface of subsidiary constraints, this time defined in an isoriemannian manifold.

The extension of model (2.12) with *bona-fide* subsidiary constraints, e.g., to define the total quantities on a conventional Riemannian manifold (see the corresponding relativistic case (IV.2.5)), is here left for brevity to the interested reader.

Next, we consider the isotopic form of the principle of equivalence. For this purpose, recall the isonormal coordinates of Sect. II.11, under which we have the reduction of the isometric $\hat{g}(x, \dot{x}, \ddot{x},...)$ to the tangent isometric $\hat{\eta}(\dot{x}, \ddot{x},...)$ of the isominkowski spaces $\hat{M}^{I}(x,\hat{\eta},\hat{\mathfrak{R}})$.

The following formulation is then rather natural (see Appendix V.A for Gasperini's formulation in a conventional Riemannian space).

ISOTOPIC PRINCIPLE OF EQUIVALENCE FOR ISOGRAVITATIONS (2.12) (loc. cit.): Gravitational effects on an isoriemannian space $\hat{R}(x,\hat{g},\hat{\mathfrak{R}})$ can be locally made to disappear by transforming the isometric $\hat{g}$ into that $\hat{\eta}$ of the tangent isominkowski space oof Class I, or in the neighborhood of an isonormal point y° at which

$$\Gamma^{2}{}_{\rho}{}^{\mu}{}_{\sigma}(y^{\circ}) = 0. \qquad (2.15)$$

The primary novelty is that, in the conventional case the test particle becomes locally free, while in our case the test particle remains under the action of the contact nonpotential interactions in the neighborhood of the point considered.

The best way to appraise and understand isogravitations (2.12) is by considering a "test particle" in the gravitational field of a given astrophysical object, say, Jupiter.

As shown in Sect. II.12, the trajectory must be geodesic for both the exterior and the interior motion. In different terms, in the transition from the exterior to the interior motion the carrier space is generalized, but the geodesic character of the trajectory persists. We evidently have the transition from one geodesic expression in one space to another geodesic expression in structurally more general space.

Consider first the local, exterior, Newtonian approximation. Then, in the neighorhood of a point of the orbit, Galilei's historical transformations hold, say, along the third axis

$$r^3 \Rightarrow r'^3 = r^3 + v^{\circ 3}\, t. \tag{2.16}$$

and the trajectory is, locally, the geodesic straight line in Euclidean space $E(r,\delta,\Re)$.

In the transition to the relativistic setting, we have the local validity of the equally historical Lorentz boosts

$$x'^3 = \gamma(x^3 - \beta x^4),\ x'^4 = \gamma(x^4 - \beta x^3), \tag{2.17a}$$

$$\beta = v/c_o, \quad \gamma = (1 - \beta^2)^{-\frac{1}{2}}. \tag{2.17b}$$

and the trajectory remains, locally, a geodesic straight line, this time, in Minkowski space.

Finally, in the transition to the full exterior gravitational setting in Riemannian space $R(x,g,\Re)$, the trajectory is no longer a straight line, but it is characterized by the familiar geodesic equations (Sect. II.12)

$$\frac{d^2x^\mu}{ds^2} + \Gamma^{2\ \mu}_{\ \rho\ \sigma} \frac{dx^\rho}{ds} \frac{dx^\sigma}{ds} = 0. \tag{2.18}$$

as experimentally verified.

However, when performing the transition to the interior gravitational problem, that is, when the extended test particle penetrates within Jupiter's atmosphere, it is also experimentally established that the the trajectory is no longer of geodesic type (2.18), because of the emergence of the new forces studied in these volumes.

The order of magnitude of the violation of laws (2.18) should also be kept in mind to prevent attempts at approximations of dubious value.

In fact, for sufficiently high speeds, the resistive forces may dependent up to the 10-th power of the speed and more, as routinely done in rocketry (Sect. I.1.3).

The inability of Einstein's gravitation to represent variationally nonselfadjoint forces of this type is established beyond any reasonable doubt, and so is the inability of the conventional theory to provide a quantitative representation of the *deviations* from geodesic motion (2.18) in the interior problem.

Our isotopic liftings of Galilei's relativity of Chapter III, Einstein's special relativity of Chapter IV, and that of Einstein's gravitation of this chapter appear to be particularly suited for the characterization of interior dynamics. In fact, in nonrelativistic approximation, the trajectory remains an isogeodesic line[30] in isoeuclidean space $\hat{\mathfrak{R}}_t \times \hat{E}(r,\hat{\delta},\hat{\mathfrak{R}})$, but the linear and local Galilei's boosts (2.16) are replaced by the nonlinear and nonlocal isogalilean boosts (III.6.6)

$$r'^3 = r^3 + t\, v^{o3} \tilde{B}_3^{\,2}(r, \dot{r}, \ddot{r}, ...). \qquad (2.19)$$

Even though the conventional Galilei's *transformation* (2.16) is no longer a local symmmetry, the isotopic formulation is such to preserve the exact Galilei's *symmetry* at the higher isotopic level.

The preservation of the local symmetry at the abstract level in the transition from transformation (2.16) to its coverings (2.19) evidently depends on the preservation of the geodesic character of the motion, of course, when formulated in the appropriate isospace.

In the transition to a relativistic setting of the interior problem, the trajectory remains an isogeodesic line, this time, on isominkowski space $\hat{M}(x,\hat{\eta},\hat{\mathfrak{R}})$, and we have the exact validity of the covering nonlinear and nonlocal isolorentz boosts (IV.5.29), i.e.

$$x'^3 = \hat{\gamma}(x^3 - \beta x^4), \qquad x'^4 = \hat{\gamma}(x^4 - \hat{\beta} x^3), \qquad (2.20a)$$

$$\beta = v/c_o, \quad \hat{\beta} = v b_3(x, \dot{x}, \ddot{x}, ..) \,/\, c_o b_4(x, \dot{x}, \ddot{x}, ..)\,, \qquad (2.20b)$$

$$\hat{\beta}^2 = v b_3^{\,2} v / c_o b_4^{\,2} c_o, \quad \hat{\gamma} = |1 - \hat{\beta}^2|^{-\frac{1}{2}}, \qquad (2.20c)$$

which, being a covering of isotransformations (2.19), enjoy the same

[30] As indicated in Sect. II.12, *the isogeodesic image of a straight line is not necessarily a straight line even in the absence of external potential fields and on a flat space*, because it dependends on local conditions (shape of body, density of exterior medium, etc.), as established by experimental evidence and as represented by variational principle (II.12.19).

properties of direct universality.

Again, the conventional Lorentz *transformations* are manifestly violated for the interior conditions considered for the numerous reasons studied in Chapter IV.

Nevertheless, our isotopic theory restores the exact character of the Lorentz *symmetry*, although in lesser trivial realizations. In turn, the preservation of the underlying symmetry ensures that of the geodesic motion in the new space, and viceversa.

Finally, in our full gravitational treatment of the interior problem on an isoriemannian spaces $\hat{R}(x,\hat{g},\hat{\mathfrak{R}})$, the isogeodesics are given by the solutions of the isodifferential equations (Sect. II.12)

$$\frac{\hat{d}^2 x^\mu}{\hat{d}s^2} + \hat{\Gamma}^{2}{}_{\rho}{}^{\mu}{}_{\sigma}(x,\dot{x},\ddot{x},..)\, T^{\rho}{}_{\alpha}(x,\dot{x},\ddot{x},..)\, \frac{\hat{d}x^\alpha}{\hat{d}s}\, T^{\sigma}{}_{\beta}(x,\dot{x},\ddot{x},..)\, \frac{\hat{d}x^\beta}{\hat{d}s} = 0. \qquad (2.21)$$

The direct universality of our gravitational theory for representing all possible interior dynamical conditions considered, is then reduced to a mere selection of the isotopic element T, as we shall illustrate later on in specific cases.

The local Poincaré-isotopic character has been discussed earlier, thus activating Gasperini's (1983) isotopies in their entirety, when again restricted to the interior problem only and formulated in a conventional Riemannian space (Appendix V.A).

The generally non-null value of torsion in the frame of the experimenter, Eq.s (II.11.91), is evident, thus activating all studies by Rapoport-Campodonico (1990) and others, also in the interior problem.

The main features of isogravitations (2.12) will be outlined in the oncluding remarks.

V.3: ISOTOPIC ORIGIN OF GRAVITATION

As indicated in Sect. V.1, Einstein's gravitation is afflicted by truly serious problematic aspects of rather numerous and diversified nature, including:

A) Problems of geometric consistency caused by the forgotted

Freud identity (II.11.34);

B) Geometric incompleteness of Einstein's tensor (Lemma II.11.2) caused by the lack of invariance of the contracted Bianchi identity under isotopies;

C) Apparent incompatibility with Maxwell's electrodynamics predicting a first-order electromagnetic source of the gravitational field in vacuum (Santilli (1974);

D) Numerous theoretical and experimental problems caused by the lack of stress-energy tensor (Yilmaz (1958), (1971), (1977), (1979), (1980), (1982), (1989), (1990a, b));

E) Inherent structural difficulties preventing basic advancements, such as the formulation of an unambiguous quantization, or the achievement of a grand unification of all interactions;

as well as others.

Our analysis would be grossly incomplete without submitting the most general possible theory of isogravitation which is conceivable with the knowledge gained in these volumes, in the hope of resolving, evidently in due time, at least some of the above problematic aspects.

As we shall see, the critical examination of Einstein's gravitation on rigorous geometric terms creates new possibilities for further advances that would be otherwise precluded. As an example, it shifts the attention from the *description* of the gravitational field, to the *problem of the ultimate physical origin of the gravitational field itself.*

In turn, the latter profile apparently permits the elimination of the now vexing problem of "unification" of the gravitational and electromagnetic fields, via their "identification".

Consider an astrophysical body with null total electromagnetic phenomenology (null total charge, null total electric and magnetic moments). Then, Einsteins field equations for the exterior problem in a conventional Riemannian space R(x,g,$\Re$) on (3.1)-space-time dimensions have the familiar form

$$G^{\mu\nu} = R^{\mu\nu} - \tfrac{1}{2}g^{\mu\nu} R = 0. \tag{3.1}$$

representing Einstein's central geometric conception of the gravitational field reduced to pure geometry without source.

According to Santilli (1974), this conception of gravitation is incompatible with Maxwell's electrodynamics in the following sense. Even though the total electromagnetic *quantities* are null, the total electromagnetic *energy-momentum tensor* $T^{\mu\nu}{}_{elm}$ resulting from the contributions of each individual charged constituent of matter is, not null, and actually so large that it can account for the entire gravitational mass m of the astrophysical body considered (this is called the *strong assumption, loc. cit.*).

The following generalization of Eq.s (3.1) was then submitted (*loc. cit.*)

$$G^{\mu\nu} = R^{\mu\nu} - \tfrac{1}{2} g^{\mu\nu} R = 8\pi\, T^{\mu\nu}{}_{elm}, \qquad (3.2a)$$

$$m = \int d^3x\, T^{oo}{}_{elm}, \qquad (3.2b)$$

$$\mathrm{tr}\, T^{\mu\nu}{}_{elm} = 0, \qquad (3.2c)$$

where Eq.s (3.2a) represent the nowhere null electromagnetic tensor originating in the structure of matter (see below), Eq.s (3.2b) represent the strong assumption, and Eq.s (3.2c) represent a known property of the electromagnetic energy-momentum tensor.

One can directly arrive at Eq.s (3.2) by recalling that the physical origin of the masses of all elementary particles is mostly of electromagnetic nature. This technically means that most of the mass tensor $M^{\mu\nu}$ must be replaced by a suitable electromagnetic tensor $T^{\mu\nu}{}_{elm}$. When summing up the contributions from a large number of massive constituents in the atomic, nuclear and subnuclear structures, the existence of a first-order electromagnetic tensor in the exterior of bodies with null total charge is than an incontrovertible consequence, in evident disagreement with Eqs. (3.1) (see Fig. V.3.1 for more details).

It should be indicated here that the incompatibility of Eq.s (3.1) with Maxwell electrodynamics is irreconcilable, in the sense that the tensor $T^{\mu\nu}{}_{elm}$ can be rendered ignorable only by modifying Maxwell's theory or assuming that all charges constituting the body are at rest, as well as at very close mutual distances (see below). Thus, either one accepts Einstein's equations (3.1), in which case Maxwell's theory must be modified, or one accepts Maxwell's theory, in which case Einstein's gravitation must be modified.

For the purpose of attempting the identification of the origin of the gravitational field, it is important to briefly review the main lines of Santilli (*loc. cit.*).

The ideal particle image of the gravitational body with null total charge and dipole moments is the π° particle. The author therefore computed the total electromagnetic field of the π° via the use of conventional relativistic techniques, beginning with the Lienard-Wieckert potential in conventional Minkowski space $M(x,\eta,\Re)$

$$A_m{}^\alpha(x) = -q\,\frac{v_m{}^\alpha}{d_m}\,. \tag{3.3}$$

where m stands for retarded or advanced, the v's are the velocities of the charges and d is a suitable distance (see below).

Under the assumption that the π° is a bound state of two constituents of point-like charges $\pm q$ (which is in agreement with current quark theories on the π° structure), the total potential outside the π° due to the interior elementary charges is given by

$$_qA^\mu_{\pi^0}(x) = -q\sum_{nm}\epsilon_n\epsilon_m C_{nm}\frac{v^\mu_{nm}}{d_{nm}} = -q\left\{\left[C_{+\mathrm{Ret}}\frac{v^\mu_{+\mathrm{Ret}}}{d_{+\mathrm{Ret}}} - C_{+\mathrm{Adv}}\frac{v^\mu_{+\mathrm{Adv}}}{d_{+\mathrm{Adv}}}\right]\right.$$

$$\left. - \left[C_{-\mathrm{Ret}}\frac{v^\mu_{-\mathrm{Ret}}}{d_{-\mathrm{Ret}}} - C_{-\mathrm{Adv}}\frac{v^\mu_{-\mathrm{Adv}}}{d_{-\mathrm{Adv}}}\right]\right\} = \sum_{nm} C_{nm}A^\mu_{nm}(x), \tag{3.4}$$

where: the observer is located at a (space-time) point x in the exterior of the π°; the charges are located at the points y_{mn}, with n indicating the positive or negative charge; the distances of the charges from the observers are therefore four in total and denoted with $D_{mn} = x - y_{mn}$;

$$A^\mu_{nm}(x) = -q\epsilon_n\epsilon_m\frac{v^\mu_{nm}}{d_{nm}}\,. \tag{3.5a}$$

$$d_{nm} = D_{nm}\cdot v_{nm} = -\epsilon_m\gamma c D_n\left(1 - \epsilon_m\frac{\mathbf{D}_{nm}\cdot\mathbf{v}_{nm}}{D_n c}\right). \tag{3.5b}$$

$$D^\alpha_{nm} \equiv (\epsilon_m\,|\,\mathbf{D}_{nm}\,|;\ \mathbf{D}_{nm}) = (\epsilon_m D_n\ ;\ \mathbf{D}_{nm}); \tag{3.5c}$$

$$D^2_{nm} = 0; \qquad |\,\mathbf{D}_{n\,\mathrm{adv}}\,| = |\,\mathbf{D}_{n\,\mathrm{ret}}\,| = D_n\ ;$$

$$\epsilon_n = \begin{cases} -1 & \text{for positive charge,} \\ +1 & \text{for negative charge;} \end{cases} \tag{3.5d}$$

the C's are constants verifying the conditions

$$\begin{aligned} C_{+\mathrm{Ret}} + C_{+\mathrm{Adv}} &= 1, \\ C_{-\mathrm{Ret}} + C_{-\mathrm{Adv}} &= 1, \end{aligned} \tag{3.6}$$

and are usually assumed to take the value ½ to give equal weight to the advanced and retarded contributions.

The exterior electromagnetic field of the π° due to its structural charges is then given by

$${}_qF^{\alpha\beta}(x) = \sum_{nm} C_{nm}\, {}_qF^{\alpha\beta}_{nm}(x), \tag{3.7}$$

$${}_qF^{\alpha\beta}_{nm}(x) = q\epsilon_n\epsilon_m \left\{ \frac{c^2[D^\alpha, v^\beta]_{nm}}{d^3_{nm}} - \frac{[D^\alpha, a^\beta]_{nm}}{d^2_{nm}} + \frac{D_{nm}\cdot a_{nm}}{d^3_{nm}}\, [D^\alpha, v^\beta]_{nm} \right\},$$

where

$$[D^\alpha, v^\beta]_{nm} = D^\alpha_{nm} v^\beta_{nm} - D^\beta_{nm} v^\alpha_{nm}\,. \tag{3.8}$$

with first expressions

$${}_qF^{\alpha\beta}_{nm,1/D^2}(x) = q\, \frac{c^2\epsilon_n\epsilon_m}{d^3_{nm}}\, [D^\alpha, v^\beta]_{nm}\,, \tag{3.9a}$$

$${}_qF^{\alpha\beta}_{nm,1/D}(x) = q\epsilon_n\epsilon_m \left\{ \frac{D_{nm}\cdot a_{nm}}{d^3_{nm}}\, [D^\alpha, v^\beta]_{nm} - \frac{1}{d^2_{nm}}\, [D^\alpha, a^\beta]_{nm} \right\}. \tag{3.9b}$$

The total electromagnetic energy-momentum tensor of the π° due to the charged structure of the constituents is then given by expressions of the type

$${}_{1q}T^{\alpha\beta}_{\pi^0} = \frac{q^2c^4}{4\pi} \sum_{nn'} \left\{ \frac{1}{d_n^6}\, [c^2 D_n^\alpha D_n^\beta + (D\cdot v)_n \{D^\alpha, v^\beta\}_n \right.$$

$$- \frac{(1-\delta_{nn'})}{d_n^{\,3} d_{n'}^{3}} [(D_n \cdot v_{n'})\{D_{n'}^{\alpha}, v_n^{\beta}\} - \tfrac{1}{2}(v_n \cdot v_{n'})\{D_n^{\ \alpha}, D_{n'}^{\beta}\}$$

$$- \tfrac{1}{2}(D_n \cdot D_{n'})\{v_n^{\ \alpha}, v_{n'}^{\beta}\}] - \tfrac{1}{2} g^{\alpha\beta} \frac{(D_n \cdot v_n)^2}{d_n^{\ 6}}$$

$$- g^{\alpha\beta} \frac{(1-\delta_{nn'})}{d_n^{\,3} d_{n'}^{3}} [(D_n \cdot D_{n'})(v_n \cdot v_{n'}) - (D_n \cdot v_{n'})(D_{n'} \cdot v_n)]\Big\},$$

$$\{A^{\alpha}, B^{\beta}\} = A^{\alpha}B^{\beta} + A^{\beta}B^{\alpha}. \tag{3.10}$$

For the case of the magnetic moments of the constituents, the potential can be written

$$A_m^{\ \alpha}(x) = -q \frac{v_m^{\ \alpha}}{d_m} - \frac{c}{d_m} \left[\frac{d}{d\tau} \frac{\mu^{\alpha\beta} D_\beta}{d}\right]_{\tau=\tau_m} \tag{3.11}$$

$$= {}_q A_m^{\ \alpha}(x) + {}_\mu A_m^{\ \alpha}(x),$$

where

$${}_q A_m^{\ \alpha}(x) = -q \frac{v_m^{\ \alpha}}{d_m} \tag{3.12a}$$

$${}_\mu A_m^{\ \alpha}(x) = \left[c^3 \frac{\mu^{\alpha\rho} D_\rho}{d^3} + c(D \cdot a) \frac{\mu^{\alpha\rho} D_\rho}{d^3} - c \frac{\dot{\mu}^{\alpha\rho} D_\rho}{d^2}\right]_{\tau=\tau_m}. \tag{3.12b}$$

The total electromagnetic field of the π° due to the intrinsic magnetic moments of its constituents can then be written

$${}_\mu F^{\alpha\beta} = {}_\mu F^{\alpha\beta}_{1/D^3} + {}_\mu F^{\alpha\beta}_{1/D^2} + {}_\mu F^{\alpha\beta}_{1/D} \tag{3.13}$$

where

$${}_\mu F^{\alpha\beta}_{1/D^3} = \frac{2c^3}{d^3} \mu^{\alpha\beta} - \frac{3c^5}{d^5} (\mu^{\alpha\rho} D^{\beta} - \mu^{\beta\rho} D^{\alpha})\, D_\rho\,; \tag{3.14a}$$

$${}_\mu F^{\alpha\beta}_{1/D^2} = -\frac{2c}{d^2} \dot{\mu}^{\alpha\beta} + \frac{2c}{d^3} (D \cdot a)\, \mu^{\alpha\beta} - \left(\frac{6c^3}{d^4} + \frac{6c^3}{d^5} (D \cdot a)\right) (\mu^{\alpha\rho} D^{\beta} - \mu^{\beta\rho} D^{\alpha})\, D_\rho$$

$$+ \frac{3c^3}{d^4} (\dot{\mu}^{\alpha\rho} D^{\beta} - \dot{\mu}^{\beta\rho} D^{\alpha})\, D_\rho + \frac{c}{d^3} (\mu^{\alpha\rho} a^{\beta} - \mu^{\beta\rho} a^{\alpha})\, D_\rho$$

$$+ \frac{c}{d^3} (\dot{\mu}^{\alpha\rho} D^{\beta} - \dot{\mu}^{\beta\rho} D^{\alpha})\, v_\rho + \frac{2c}{d^3} (\dot{\mu}^{\alpha\rho} v^{\beta} - \dot{\mu}^{\beta\rho} v^{\alpha})\, D_\rho\,,$$

$$ {}_{u}F^{\alpha\beta}_{1/D} = \left[\frac{c(a\cdot D)}{d^4} - \frac{3c(D\cdot a)^2}{d^5}\right](\mu^{\alpha\rho}D^\beta - \mu^{\beta\rho}D^\alpha)\,D_\rho $$

$$ + \frac{3c(D\cdot a)}{d^4}(\dot\mu^{\alpha\rho}D^\beta - \dot\mu^{\beta\rho}D^\alpha)\,D_\rho - \frac{c}{d^3}(\ddot\mu^{\alpha\rho}D^\beta - \ddot\mu^{\beta\rho}D^\alpha)\,D_\rho\,. $$

$$ {}_{k}F^{\alpha\beta}_{\pi^0}(x) = {}_{k\eta}F^{\alpha\beta}_{\pi^0}(x) + {}_{ku}F^{\alpha\beta}_{\pi^0}(x), \qquad k = 1, 2, 3. $$

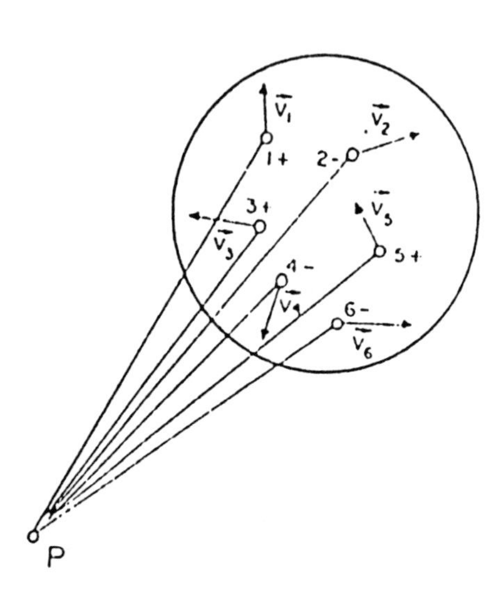

FIGURE V.3.1: The model of a star with null total charge as a gas of charged constituents, submitted by Santilli (1974) to illustrate the inconsistency of Einstein's reduction of the exterior gravitation to pure geometry without source. The model also suggests the ultimate origin of the gravitational field as being primarily of electromagnetic nature (plus contributions from short range interactions discussed later on in this section). In few words, even though the total charge is null, Maxwell's electromagnetism establishes beyond any reasonable doubt that the charged structure of thebody generates an electromagnetic tensor which is of first-order in magnitude, and nowhere null in the exterior gravitational problem, in direct conflict with Einstein's Eq.s (3.1). The admission of exterior gravitational sources has a number of predictable, ratherfundamental implications. In fact, it first shifts the emphasis from the "description" of the gravitational field, to the study of the its origin. Secondly, it permits the elimination of the now vexing problem of "unification" of the gravitational and electromagnetic fields via their "identification". Moreover, it permits the achievement of an axiomatically correct theory which is inclusive of the forgotten Freud identity (Sect. II.11), as shown later on in this section.

The total electromagnetic field of the π° is then given by

$$T^{\alpha\beta}_{\pi^0} = \frac{1}{4\pi}\left(F^{\alpha\mu}_{\pi^0} F^{\beta}_{\pi^0\mu} + \tfrac{1}{4} g^{\alpha\beta} F^{\rho\sigma}_{\pi^0} F_{\pi^0\rho\sigma}\right). \tag{3.15}$$

where the sum over all possible forms is understood, with total energy-momentum tensor

$$T^{\alpha\beta}_{\pi^0} = \frac{1}{4\pi}\left(F^{\alpha\mu}_{\pi^0} F^{\beta}_{\pi^0\mu} + \tfrac{1}{4} g^{\alpha\beta} F^{\rho\sigma}_{\pi^0} F_{\pi^0\rho\sigma}\right). \tag{3.16}$$

Explicit calculations conducted in Santilli (*loc. cit.*) then confirm that the mass of the π° has a primary origin of of electromagnetic nature, as expected, with an exterior, first-order, nowhere null source tensor $T^{\mu\nu}{}_{elm}$.

The extension of the model to an astrophysical body is straighforward and merely done by summing up the total electromagnetic field (3.15) due to a large number of charged constituents, thus resulting in a first-order electromagnetic source which requires modifications (3.2).

Note that Eq.s (6.2a) equally hold without any modification when the total charge of the body is null.

Needless to say, the electromagnetic field is indeed the primary, but not the total source of the π° mass, owing to the presence of the additional short range interactions, the weak and strong interactions in the interior of the nuclei and of their constituents.

This leads to the formulation in (*loc. cit.*) of the so-called ***weak assumption***, for which the gravitational mass of a celestial body is not entirely due to the electromagnetic fields of the particles constituents,

$$m \approx \int d^3x \, T^{\mu\nu}{}_{elm}, \tag{3.17}$$

but rather to all interactions in the structure of matter

$$m = \int d^3x \, T^{\mu\nu}{}_{elm} + \int d^3x \, t^{\mu\nu}{}_{sr}. \tag{3.18}$$

In this case Einstein's equations (3.1) are replaced by the more general form

$$G^{\mu\nu} = R^{\mu\nu} - \tfrac{1}{2} g^{\mu\nu} R = 8\pi(T^{\mu\nu}{}_{elm} + t^{\mu\nu}{}_{sr}) \tag{3.19}$$

A number of experiments were then submitted in (*loc. cit.*) to test the fundamental assumption of all gravitational theories according to which the electromagnetic field is a source of the gravitational field (e.g., by using interferometric measures connected to the large magnets currently available at various laboratories, and other means).

In a way independent from the above studies, Yilmaz ((1958), (1971), (1979), (1982), (1984), (1987) (1990)) has submitted numerous arguments over several decades according to which Einstein's equations (3.1) are incomplete (also for the case of the exterior field of a body with null total electromagnetic quantities), because of the lack of the *stress-energy tensor* $t^{\mu\nu}{}_{stress}$ Yilmaz has therefore advocated the following generalization of Eq.s (3.1)

$$G^{\mu\nu} = R^{\mu\nu} - \tfrac{1}{2} g^{\mu\nu} R = 8\pi \, t^{\mu\nu}{}_{stress} \tag{3.20}$$

and worked out a new theory of gravitation, here called *Yilmaz's exterior gravitation*, in numerous details.

Some of the most relevant problematic aspects of Eq.s (3.1) pointed out by Yilmaz (*loc. cit.*) are the following:

1) The apparent inability by Einstein's gravitation to recover the Newtonian gravity (because it apparently recovers the so-called *Hookes mechanics* in which the sun has infinite inertia (Yilmaz (1984));

2) Problematic aspects of Einstein's gravity in recovering the Newtonian 532″ advancement of the perihelion of Mercury (note that clear proofs apparently exist only for the relativistic advancement of 43″) (Yilmaz (1984), (1989), (1990a));

3) Problematic aspects for Einstein's gravitation to achieve true compatibility with the special relativity, particularly in the

reconstruction of the reletivistic conservation laws (Yilmaz (1980), (1982a, b), (1989)); and others.

Regrettably, we are not in a position to review these studies for brevity. We shall limit ourselves to indicate how the stress-energy tensor arises rather naturally in the interior gravitation via the isoriemannian geometry.

With Theorem II.11.5 we have identified the most general possible interior field equations with source. We can now reinterpret these results via the following

THEOREM V.3.1: (ISOTOPIC ORIGIN OF GRAVITATION, Santilli (1988d)): The most general possible isogravitations on isoriemannian spaces $R(x,\hat{g},\hat{\Re})$ for interior gravitational problems verifying the contracted isobianchi identity and the isofreud identiy can be expressed via the following variational principle

$$\delta\hat{A} = \delta\int d^4x[\hat{R} - 8\pi(\hat{T}_{elm} + \hat{t}_{sr})] =$$

$$= \delta\int d^4x[\hat{g}^{\mu\nu}\hat{g}^{\rho\sigma}\hat{R}_{\mu\nu\rho\sigma} + 8\pi\hat{g}_{\mu\nu}(\hat{T}^{\mu\nu}{}_{elm} + \hat{t}^{\mu\nu}{}_{sr})] = 0 \qquad (3.21a)$$

$$\hat{g} = T_g g, \qquad T_g > 0, \quad g \in R(x,g,\Re), \qquad (3.21b)$$

$$T_g{}_{|r>R^\circ} = I = \text{diag.}\ (1,1,1,1). \qquad (3.21c)$$

with Euler-Lagrange equations

$$\hat{E}^{\mu\nu} = \hat{R}^{\mu\nu} - \tfrac{1}{2}\hat{g}^{\mu\nu}\hat{R} - \tfrac{1}{2}\hat{g}^{\mu\nu}\hat{\theta} - 8\pi(\hat{\tau}^{\mu\nu} + \hat{t}^{\mu\nu}{}_{s.r.int} =$$

$$= \hat{R}^{\mu\nu} - \tfrac{1}{2}\hat{g}^{\mu\nu}\hat{R} - 8\pi(\hat{T}^{\mu\nu}{}_{elm} + \hat{t}^{\mu\nu}{}_{s.r.int}) = 0, \qquad (3.22a)$$

$$\hat{T}^{\mu\nu}{}_{elm} = \hat{\tau}^{\mu\nu} - \hat{g}^{\mu\nu}\hat{\theta} / 16\pi, \qquad tr\ T^{\mu\nu}{}_{elm} = 0, \qquad (3.22b)$$

$$tr\ t^{\mu\nu}{}_{s.r.int} \neq 0, \qquad (3.22c)$$

thus identifying the gravitational field with the electromagnetic and short range interactions at the origin of matter,

$$M^{\mu\nu}_{matter} \equiv \hat{T}^{\mu\nu}_{elm} + \hat{t}^{\mu\nu}_{s.r.int}. \qquad (3.23)$$

Isogravitations (3.22) evidently preserve all basic properties of forms (2.12), including the preservation of the general and local, interior and exterior, exact isopoincaré symmetries, the geodesic character of the trajectories, etc.

The most important mathematical advancement of the formers over the latters is the achievement of a geometrically consistent theory, which avoids the incompleteness of Einstein's tensor under isotopies identified in Sect. II.11.. The most important physical advancement of isogravitations (3.22) with respect to forms (2.12) is the possibility of eliminating the vexing problem of *"unification"* of the gravitational and electromagnetic fields, and replacing it with their "identification" , thus permitting a study of the origin of gravitation.

In particulart, isogravitations (3.22) permits the identification of the possible origin of Yilmaz's stress-energy tensor with the weak and strong interactions in the strutture of matter (Santilli (1988d)).

Note that *there is no quantitative difference between isogravitations (3.22) and (2.12) as far as the interior problem is concerned.* However, the two isogravitations are structurally different in the *exterior* problem.

In fact, the formers reduce, by construction, to Einstein's source-free conception of exterior gravitation in vacuum

$$G^{\mu\nu} = R^{\mu\nu} - \tfrac{1}{2} g^{\mu\nu} R \equiv 0, \qquad (3..24a)$$

$$\hat{M}^{\mu\nu}_{matter \,|\, r > S^{\circ}} = \hat{T}^{\mu\nu}_{elm} + \hat{t}^{\mu\nu}_{s.r.int.} \equiv 0, \qquad (3.24b)$$

while the latter models reduce to the following generalized equations

$$G^{\mu\nu} = R^{\mu\nu} - \tfrac{1}{2} g^{\mu\nu} R = 8\pi\,(T^{\mu\nu}_{elm} + t^{\mu\nu}_{stress}), \qquad (3.25a)$$

$$\hat{M}^{\mu\nu}_{matter \,|\, r > S^{\circ}} = T^{\mu\nu}_{elm} + t^{\mu\nu}_{stress}, \qquad (3.25b)$$

or to the simpler, yet nontrivially generalzed form

$$G^{\mu\nu} = R^{\mu\nu} - \tfrac{1}{2} g^{\mu\nu} = 8\pi\, T^{\mu\nu}_{\ \ elm}, \tag{3.26a}$$

$$\hat{M}^{\mu\nu}_{\ \ matter\,|\,r > S^\circ} = T^{\mu\nu}_{\ \ elm}\ , \quad \hat{t}^{\mu\nu}_{\ \ stress\,|\,r > S^\circ} = 0, \tag{3.26b}$$

in case the interior source due to the short range, weak and strong interactions has no first-order image in the exterior problem or, equivalently, in case the necessarily nonnull stress-energy tensor of the interior problem has no nonnull image in the exterior.

As stressed earlier, these studies are solely devoted to the *interior* gravitational problem. As such, we have merely identified above the most salient alternatives and pointed out the possible contributions to the exterior problem from our interior isogeometries. However, the issue of which of the three *exterior* models (3.24), (3.25), (3.26) is correct will not be addressed. It is hoped that experts in the field will finally confront these now vexing problematic aspects of the exterior gravitation and eventually resolve them.

We can now present a few preliminary comments on some expected implications of isogravitations for black holes and singularities at large. As well known, these topics have been studied until now via a conventional, local-differential Riemannian geometry. The central issue to be addressed is therefore the identification of the implications for black holes and singularities at large which are expected from the nonlocal and non potential interactions of the interior gravitation at large, and of the collapsing gravitational problem in particular (see the comments in Sect. V.1).

An effective way to study the problem is to identify, first, the new perspective offered by our isotopic methods for conventional gravitational singularities, and then pass to their nonlocal and nonpotential, interior generalization.

The central notion recommended for the analysis is the representation of a conventional Riemannian space $R(x,g,\mathfrak{R})$ as an isotope of the Minkowski space $M(x,\eta,\mathfrak{R})$ (Sect. II.3)

$$R(x,g,\mathfrak{R}) \approx \hat{M}(x,\hat{\eta},\hat{\mathfrak{R}}), \tag{6.27a}$$

$$g(x) = T(x)\,\eta, \qquad \hat{\eta}(x) = T\,\eta \equiv g, \qquad \hat{\mathfrak{R}} = \mathfrak{R}\,\hat{1}, \qquad \hat{1}(x) = T^{-1}. \tag{6.27b}$$

In fact, by recalling that the Minkowski metric $\eta = \text{diag.}\ (1,1,1,-1)$ has no singularities, the above representation focuses the attention on the fact that *the singularities of the Riemannian metric $g = T\eta$ are in actuality the singularities of the isotopic element T and, thus, the* *zeros of the isounit $\hat{1}$* (Santilli (1988d))

$$g(x) \Rightarrow \infty, \qquad T(x) \Rightarrow \infty, \qquad \hat{1}(x) \Rightarrow 0 \tag{3.28}$$

We reach in this way the new result that *the isounits can directly represent gravitational collapse via its possibler zeros* . Stated in different terms, all conventional singularities of Einstein's gravitation, such as that of the Schwartzchild metric, can be equivalently represented via the singularities of the isotopic element T or, still equivalently, via the zeros of the isounit $\hat{1} = T^{-1}$.

But structures (3.27) and (3.28) represent exterior gravitation in vacuum. To represent more closely the physical reality of interior gravitation, we therefore perform a further isotopy. We recover again the result of Sect. II.3, namely, the most general possible isoriemannian space $\hat{R}(x,\hat{g},\Re)$ is an isotope of the Minkowski space $M(x,\eta,\Re)$, this time expressed as an isotopy of isotoppy (3.27)

$$\hat{\Re}(x,\hat{g},\hat{\Re}) \approx \hat{M}(x,\hat{\eta},\hat{\Re}), \tag{3.29a}$$

$$\hat{g}(x,\dot{x},\ddot{x},...) = \hat{T}(x,\dot{x},\ddot{x},...)\,\eta = T'\,(x,\dot{x},\ddot{x},...)\,g(x) = T'(x,\dot{x},\ddot{x},...)\,T(x)\,\eta, \tag{3.29b}$$

$$\hat{\eta}(x,\dot{x},\ddot{x}) = T\eta \equiv \hat{g}, \quad \hat{1}(x,\dot{x},\ddot{x}) = T^{-1}. \tag{3.29c}$$

We can therefore say that *the possible singularities of isogravitations with isometrics $\hat{g} = T\eta$ are those of the isotopic elements T, that is, the possible zeros of the isounits $\hat{1} = T^{-1}$* (Santilli (*loc. cit.*)).

It is easy to see that *the conventional singularities of the exterior gravitation, e.g.,. that of the Schwartzchild metric g, are not necessarily preserved under the additional of internal nonlocal and nonpotential forces* Stated differently, the singularity $g(x) \Rightarrow \infty$ is not necessarily preserved under the isotopy $g(x) \Rightarrow \hat{g}(x,\dot{x},\ddot{x},...) = T(x,\dot{x},\ddot{x},...)\, g(x)$.

We can therefore conclude these introductory comments to the problem of singularities in isogravitations by saying that *the conventional black holes and singularities of type (3.28) characterized by local-differential-potential methods and geometries may well result to be only a limit geometrical approximation of "brown holes" or "near singilarities" under the addition of internal nonlocal and nonpotentuial interactions because, in the final analysis, infinities are not expected to exist in our physical Universe.*

The resolution of the problem whether or not singularities truly persist for interior isotopies (3.29) requires a deeper knowledge of the representation of interior physical media via isotopic elements T. As such, the study will be conducted at a future time.

Needless to say, the interior gravitational models presented in this chapter are at their very initiation and so much remains to be done. Among a considerable number of open aspects the attentive reader can readily identify, we mention in particular:

A) The need to achieve a dual representation, not only of the stability of the center-of-mass orbits vz the instability of the interior trajectories as permitted by isogravitations (2.12) and (3.22),[31] but also of the reversibility of the center-of-mass trajectory vz the intrinsic irreversibility of the interior dynamics;

B) The need to re-examine with care and scientific objectivity all currents interior theories derived from contemporary local-differential models based on the conventional Riemannian geometry, whose results may be directly affected by the ultimate nonlocality of the structure of gravitation, such as "black holes", the "big bang", etc.;

C) Study the new class of singularities offered by all isotopic theories: the zeros in their isounits identified earlier; and others.

The mature understanding of these novel perspective is that ***isotopies can at best provide relatively small corrections to the existing theories, and not radical revisions,*** as illustrated in Chapter VII, e.g., in regard to the current view on the size of the Universe, and the quasars' redshift.

V.4: EXAMPLES OF ISOGRAVITATIONS

In this section we shall present a few examples of the interior gravitational theory of the preceding sections.

[31] The representation of lacally decaying trajectories via isogravitations is so direct and simple to appear trivial. Consider a local, tangent, interior plane with related isopoincaré symmetries and isorotational subsymmetries $\hat{O}(3)$. Its local isocasimir invariant can be written (Sect. III.3) $\hat{J}^2 = J \hat{\delta} J = \text{const.}$ Local isometrics of the type $\hat{\delta} = \exp \gamma t$ then imply the monotonic decrease of the angular momentum of the type $J^2 = JJ = \text{const.} \times \exp(-\gamma t)$, as desired.

It is sufficient for this objective to consider the *isoriemannian spaces* $\hat{R}(x,\hat{g},\hat{\Re})$ of Sect. II.11 for the case of a *diagonal isotopic element* T_g, or *isounit* $\hat{I}_g$ of the type

$$\hat{g}(x,\dot{x},\ddot{x},\mu(x),\tau(x),n(x),...) = T_g(x,\dot{x},\ddot{x},\mu(x),\tau(x),n(x),...)\, g(x), \qquad (4.1a)$$

$$\hat{I}_g = T_g^{-1} = \text{diag.}\,(n_1^2, n_2^2, n_3^2, n_4^2)\,, \qquad (4.1b)$$

$$n_\mu = n_\mu(x,\dot{x},\ddot{x},\mu(x),\tau(x),n(x),...) > 0, \qquad (4.1c)$$

$$g \in R(x,g,\Re), \qquad x = (x^\mu) = (x^k, x^4)\,,\ x^4 = c_o t, \qquad (4.1d)$$

$$\mu = 1, 2, 3, 4, \qquad k = 1, 2, 3$$

where c_o represents the speed of light *in vacuum* , and $R(x,g,\Re)$ is a conventional Riemannian space.

As the reader will recall, the isounits $\hat{I}_g$ depend on the derivative of the coordinates x with respect to an independent (invariant) parameter s of arbitrary order, as well as on the local density $\mu(x)$ of the interior medium considered, its temperature $\tau(x)$, its possible index of refraction $n(x)$, and any needed additional quantity..

Parametrization (4.1) is essentially that for our *relativistic geometrization of interior physical media* of Sect. IV.10. In fact, in the conventional normal coordinates in which the Riemannian metric $g(x)$ recovers the conventional Minkowski metric

$$g \Rightarrow \eta = \text{diag.}\,(1, 1, 1, -1), \qquad (4.2)$$

parametrization (4.1] implies the invariant separation on the local isominkowski spaces of Class I

$$\hat{M}^I(x,\hat{\eta},\hat{\Re}):\ \hat{\Re} = \Re\,\hat{I}_g, \quad \hat{I}_g = T_g^{-1}, \qquad (4.3)$$

with invariant isoseparation

$$x^{\hat{2}} = x^\mu \hat{\eta}_{\mu\nu} x^\nu$$
$$= x^1 n_1^{-2} x^1 + x^2 n_2^{-2} x^2 + x^3 n_3^{-2} x^3 - x^4 n_4^{-2} x^4, \qquad (4.4)$$

in which case the index of refraction n can be identified with the characteristic function n_4, and the speed of light within the medium considered (when transparent) is given by the familiar expression

$$c = c_o/n_4. \qquad (4.5)$$

This illustrates the reason for assuming, not only the isounit $\hat{1}_g$, but also the individual functions n_μ to be positive definite.

In the above parametrization, the *maximal causal speed* , say, along the third axis, is given by Postulate IV.9.II

$$V_{Max} = c_o \frac{n_3}{n_4}, \qquad (4.6)$$

while for the Lorentz-isotopic transformations (IV.5.20) we have the expressions, again, for speeds v along the third axis

$$\hat{\beta}^2 = \frac{n_4^2}{n_3^2}\,\beta, \qquad \hat{\gamma} = |1 - \hat{\beta}^2|^{-\frac{1}{2}}. \qquad (4.7)$$

Moreover, in order to separate the relativistic from the Newtonian contributions, it is recommendable as a first step to assume $n_1 = n_2 = n_3$, in which case isoseparation (4.4) can be written

$$x^{\hat{2}} = \frac{1}{n_3^2} x^k x^k - \frac{1}{n_4^2} t\, c_o^2 t, \qquad (4.8)$$

With the above notation, the content of this section can be interpreted as providing a *gravitational geometrization of interior physical media*, with particular reference to the classification of physical media into the following nine classes (Sect. IV.10)

$$n_3 \lesseqgtr n_4, \qquad n_4 \lesseqgtr 1. \qquad (4.9a)$$

$$V_{Max} \lesseqgtr c_o, \qquad c \lesseqgtr c_o. \qquad (4.9b)$$

The reader should recall that, unless the medium is transparent, the quantity $c = c_o/n_4$ has a sole geometric meaning without representing any physical speed, and that media of Type 7 are precisely those expected in the core of a star undergoing gravitational collapse.

Before passing to specific examples of isogravitation, the following comments appear recommendable to prevent possible misrepresentations. First, one should observe the physical reality as it appears, say, in Jupiter's structure, whereby we have local, internal variations of energy, linear momentum, angular momentum and other physical quantities *without affecting the total conservation law.* These phenomena are mere internal exchanges which are such to balance each other and result in the global stability of the system.

Moreover, *the total mass of Jupiter is the conventional expression* from (V.2.14), i.e.

$$m = \int d^3x \ T^{oo}_{total}, \qquad (4.10)$$

without any contribution from the internal nonpotential exchanges of energy. In actuality, the nonpotential nature of the internal forces can be best illustrated precisely with their lack of contribution to the total energy of the system.

This occurrence has been geometrically represented via the Poincaré-isotopic symmetry in which the generators remain the conventional ones (Theorem V.2.3), and it implies that the internal characteristic quantities n_μ have no impact or measurable effect whatever in the exterior geometry.

The above comments are important to prevent possible attempts of measuring the *interior* behavior of gravitation with *exterior* experiments, say the precession of the perihelion of planets moving at large distances in vacuum, or the bending of light rays in the exterior gravitational field of an astrophysical body with isostructure (4.1).

On the contrary, in order to test the *interior* gravitation, one must necessarily conduct *interior* experiments, such as the measure of the redshift of light propagating in the *interior* of an inhomogeneous and anisotropic medium.

Further notions which are useful to prevent misrepresentations are those related to the nature of the internal forces and their reference frame. To begin, the reader should be aware that *test particles at rest in the interior physical medium experience no contact nonpotential forces.*

This is the reason why the primary functional dependence of the drag forces (read, isounit or isotopic element) is in the *velocities*, because for null velocities, the drag forces are notoriously null.

Needless to say, the coordinate dependence must also be considered, but it is generally related to other aspects of the theory, such as the local behavior of the density $\mu(x)$ (e.g., its decrease with the distance from the center), etc.

In fact, drag forces do depend on the local density and, therefore, indirectly on the coordinates.

A further reason of misrepresentations is the use of improper or physically nonrealizable reference frames. *The fundamental reference frame in the study of the interior gravitation is the frame at rest with the medium considered,* generally assumed at the center of the astrophysical body. Other frames must be selected with care.

In fact, as stressed in the preceding chapters, *inertial frames are a philosophical abstraction* because they are not realizable in actual experiments on our Earthly environment, nor do they exist in our planetary or Galactic systems. While conventional relativities are conceived strictly for intertial systems, the isotopic relativities are conceived to identify the equivalence class of real reference frames, that is, frames which are in *noninertial* conditions, such as our conventional laboratory frames on Earth.

The important point is to acknowledge the physical evidence that *the reference frames at rest with the interior of astrophysical bodies are noninertial,* with no experimentally established exception known at this writing.

Major misrepresentations are then conceivable, particularly from an experimental viewpoint, if one attempts to elaborate experimental data on an abstract inertial reference frame to test a theory that is intrinsically noninertial in conception and realization.

Also, *the symmetries of inertial frames are necessarily linear and local, as well known. On the contrary, the effects to be measured are fundamentally nonlinear and nonlocal in coordinates, velocities and other quantities.* This is precisely the reason for our efforts in reaching nonlinear and nonlocal generalizations of the conventional, linear and local Lorentz transformations.

Thus, if one attempts the setting up of experiments on noninertial and nonlinear interior gravitational effects, via the conventional inertial and linear settings of contemporary physics, major expertimental misrepresentations can evidently result.

In conclusion, an infinite number of reference frames can evidently be considered for experiments on interior gravitations. The point is that, prior to claiming the outcome of the tests as "experimental results", these frames must be equivalent NOT to inertial frames, but to the primary frame at rest with the atrophysical body considered, i.e., must belong to the infinite class of equivalence of the nonlinear Lorentz-isotopic transformations characteristics of the medium at hand.

Lacking these precautions, the transition from an intrinsically noninertial, nonlinear and nonlocal setting to the familiar inertial, linear and local contest of contemporary experiments evidently implies the possibility of alteration of the experimental results.

The above context is rendered more complex by the fact that *our isotopic spaces are geometrical structures for a more rigorous treatment of interior conditions, but they are not the physical space-time of the experimenter.*

In fact, the notions of isominkowski spaces $\hat{M}(x,\hat{\eta},\hat{\Re})$ and isoriemannian spaces $\hat{\Re}(x,\hat{g},\hat{\Re})$ have been conceived to reach the geometric spaces in which the conventional Poincaré symmetry can be reconstructed as exact. But the physical spaces for actual measures remain the conventional spaces $M(x,\eta,\Re)$ or $R(x,g,\Re)$ where the Poincaré symmetry is violated by the effects considered.

As a result, the calculations obtained on isotopic spaces must be projected on actual spaces to separate theoretical from actual effects.

This is exactly the case of torsion which is null in the geometric space $\hat{\Re}(x,\hat{g},\hat{\Re})$, but non-null when projected in the physical space $\Re(x,g,\Re)$ (Lemma II.11.7).

To state this occurrence differently, *the interior effects we are studying are deviations from conventional Lorentzian predictions. If the experimenter uses our geometric isospaces as physical spaces, no deviation whatever occurs because, in the latter spaces, the Lorentz symmetry has been reconstructed as exact.*

It is only by reprojecting the isolinear and isolocal predictions of our isospaces in conventional spaces that experiments can be properly formulated and conducted.

Finally, in order to have a genuine generalization of Einstein's special and general relativity with internally verifiable effects, one must have

$$n_3 \neq n_4. \tag{4.11}$$

In fact, for $n_3 = n_4$, $\hat{\beta} \equiv \beta$, $\hat{\gamma} \equiv \gamma$, the Lorentz-isotopic transformations coincide with the conventional ones, and no meaningful departure has occurred from conventional settings.

The reader should be aware that the value $n_3 = n_4$ implies the so-called *scalar isotopies*

$$\hat{g} = T(x,\dot{x},...)\, g(x), \tag{4.12}$$

where $T = n_3^{-2} = n_4^{-2}$ is a multiplicative scalar function to the Riemannian metric g. Thus, even though isotopies (4.12) are certainly intriguing and worth studying, *our primary interest in this section is for isotopies (4.1) which are not reducible to the scalar isotopy (4.12)*.

At a deeper inspection, it is possible to show that the non-null value of torsion in the physical space of the experimenter is precisely related to the non-unit value of the characteristic B ratio, in which case from Lemma II.11.7 we have

$$T_{\mu}{}^{\rho}{}_{\nu} = \hat{\Gamma}^{2}{}_{\mu}{}^{\rho}{}_{\nu} - \hat{\Gamma}^{2}{}_{\nu}{}^{\rho}{}_{\mu} = \Gamma^{2}{}_{\sigma}{}^{\rho}{}_{\nu}\, T^{\alpha}{}_{\mu} - \Gamma^{2}{}_{\alpha}{}^{\rho}{}_{\mu}\, T^{\alpha}{}_{\nu} \neq 0. \tag{4.13}$$

In this way one can link departures from Einsteinian settings caused by non-null torsion to our isotopic departure $n_3 \neq n_4$.

As a matter of fact, and as stressed during our analysis, an empirical check for the nontriviality of the considered interior gravitational experiment is to verify that it is indeed characterized by a non-null torsion, again, in the physical space of the experimenter.

As a result of the above preliminary clarifications, we are now in a position to study a few illustrative examples.

CONSTANT ISOTOPIES. As indicated earlier, the characteristic n-functions can be subjected to a suitable average into constants

$$n_{\mu} = \text{const}, \ \mu = 1,2,3,4 \tag{4.14}$$

and provide a first, significant approximation of the interior problem.

In particular, even though the boundary values of the n-functions are one, their average is not one, thus leading to significant deviations from Einsteinian conditions.

A first example of constant isotopies is provided by the Nielsen and Picek metric (IV.3.21) computed for the interior of pions and hadrons at low energies via the use of the Higgs sector of conventional gauge theories. For such a metric we have

$$n_3^{-2} \approx 1 + 1.2\text{x}10^{-3}, \ n_4^{-2} \approx 1 - 3.8\text{x}10^{-3} \ \text{for pions,} \tag{4.15a}$$

$$n_3^{-2} \approx 1 - 0.2\text{x}10^{-3}, \ n_4^{-2} \approx 1 + 0.6\text{x}10^{-3} \ \text{for kaons.} \tag{4.15b}$$

The metric does represent the noneinstenian conditions $n_3 \neq n_4$ and it is therefore acceptable for the analysis of this section. In particular, the characteristic ratio between the space and time parts are given by

$$B^2 = \frac{n_3^2}{\qquad} \approx 0.995 \ \text{for pions,} \tag{4.16}$$

$$n_4^2$$

and

$$B^2 = \frac{n_3^2}{n_4^2} \approx 1.008 \quad \text{for kaons,} \tag{4.17}$$

with value expected to increase with the increase of the density for heavier hadrons (Sect. IV.9).

Another set of values was obtained by Mignani (1991) via the use of the isotopic redshift law applied to a number of quasars, under the preliminary condition that they are at rest with respect to the associated galaxy.

By averaging Mignani's values reproduced in Eq.s (IV.9.52), we obtain the constant average ratio

$$B = \frac{n_3}{n_4} \approx 84.64, \tag{4.18}$$

where one should recall that all Mignani's B-values are positive and bigger than one.

With the use of the above numerical values, one can first inspect Gasperini's (1984b) explicit examples of constant local isotopies, such as the isogeodesic trajectory

$$(1 - \frac{4}{3}\alpha)\frac{d^2x^\mu}{ds^2} + \frac{4}{3}\alpha\, \delta_4^\mu \frac{dx^4}{ds} + \Gamma^2{}_\nu{}^\rho{}_\mu \frac{dx^\nu}{ds}\frac{dx^\mu}{ds} = 0, \tag{4.19}$$

where α is given by Eq.s (A.25) and the Γ's are the conventional connection coefficients; or the approximate (to first order terms) interior orbit equation

$$u = \frac{m}{h^2}[1 + \cos(\phi - \phi_o - \Delta\phi_o)], \tag{4.20}$$

where h and ϕ_o are integration constants, and the *isoprecession* of the orbit after a full revolution ($\phi = 2\pi$) within the medium is given by [loc. cit.]

$$6\,\pi\, m \qquad\qquad 10\, L$$

$$\Delta\phi_O \big|_{\phi = 2\pi} = \frac{}{L}\left(1 - \alpha \frac{}{9M}\right); \qquad (4.21)$$

or the approximate modification of the Schwartzschild metric (*loc. cit.*)

$$ds^2 = -\left(1 - 2\frac{m}{r}\right)^{1+2\alpha/3} dt^2 + \frac{dr^2}{1 - 2m/r}$$

$$+ r^2(d\theta^2 + \sin^1\theta \, d\phi). \qquad (4.22)$$

The first physical meaning of the above models is for motion of extended test particles within an inhomogeneous and anisotropic ($n_3 \neq n_4$) atmosphere *of very light density* ($n^\mu \approx 1$), whereby one can see the deviations from the conventional trajectories, stable orbits and Schrartzschild line element caused by a small, constant drag.

One can then increase the density of the medium considered, that is, increase the average value of the n-constants, and see the proportionately higher deviations caused by drag effects.

In the limit case of extremely high density, as in the core of a collapsing star, the test particle is not expected to be able to complete one full orbital revolution within the medium considered in a finite period of time.

This point serves to stress that the isotopies under consideration are purely internal and, as such, they cannot be restricted from conceivable upper bounds originating from external experiments.

As indicated in Appendix A, Gasperini's isotopies (4.20), (4.21) and (4.22) are still partial inasmuch as they are *local* on a conventional Riemannian space.

A number of examples of the full nonlocal and nonpotential examples of isogravitations will be presented at some later time.

V.5: CONCLUDING REMARKS

It appears that isotopic techniques do indeed permit the achievement of the desired objectives: the formulation of a generalized theory for the *interior* gravitational problem which is capable of recovering identically the correct gravitational theory for the exterior problem, while being directly compatible with the preceding isospecial and isogalilean relativities (see bext chapter).

Despite that, we have to lament the inability to reach a definite model because of the unavailability at this writing of the definite theory for the *exterior* gravitational problem accepted by the scientific community at large. In turn, this has forced us to submit different isotopies capable of recovering the corect exterior theory, whatever that theory will finally be.

Despite these open issues, we can conclude by saying that isogravitations (2.12) and/or (3.22) are capable of:

1) Identifying, apparently for the first time, the general isopoincaré symmetry $\hat{P}_g$(3.1) for the characterization of the conventional Einstein's gravitation, via the embedding of the curvature in the isounit of the theory;

2) Directly representing the conventional total conservation laws (2.14) via the generators of the isopoincaré symmetry $\hat{P}_g$(3.1);

3) Not being detectable from the outside, because of the conventional exterior character of the total conservation laws, as inherent in all closed nonselfadjoint systems;

4) Recovering Einstein's gravitation identically in the exterior problem;

5) Verifying all exterior experiments verified by Einstein's gravitation;

6) Implying nonlocal-integral generalizations of conventional geometries in the transition from the exterior to the interior problem, the affine geometry and the Riemannian geometry;

7) Representing the most general possible linear or nonlinear, local or nonlocal, Lagrangian or nonlagrangians, Newtonian or nonnewtonian interior trajectories;

8) Preserving the symmetries and geodesic characte in the transition from the exterior to the interior problem.

9) Predicting a new series of interior phenomena which can be subjected to direct experimental verification, such as the apparent isotopic deviations from the Einsteinian Doppler's redshift for light propagating within inhomogeneous and anisotropic transparent media (Sect. VII.3), and other interior effects.

10) Offering a fundamentally new approach to singularities and black holes via the zeros of the isounits with due consideration for nonlocal internal effects;

11) Offering truly novel possibilities for unambiguous operator formulation of gravitation[32] *and grand-unification*[33]*;*

---

[32] As well known, the historical difficulty to achieve an unambiguous quantization of gravity is that, on one side, Einstein's gravitation has an identically null Hamiltonian, while, on the other side, quantum mechanics is fundamentally dipendent on the existence of the Hamiltonian. The isotopies of quantum mechanics (hadronic mechanics) offer truly novel possibilities for achieving an unambiguous operator formulation of gravity precisely because they bypasses the need for a Hamiltonian, and embed the geometry of gravitation in the isounit of the operator theory. In Sect. II.6, Eq.s (II.6.23-26) of Volume I we presented the basic equations of hadronic mechanics. In footnote[10] of p. 94 of this volume we recalled the main idea of the (naive) mapping of Hamilton-isotopic mechanics into hadronic mechanics, i.e., the mapping of the isoaction $\hat{A}$ into $- i \hat{1} \log \psi(t, r)$, where $\hat{1} = T^{-1}$ is the isounit, with consequential map of the isotopic Hamilton-Jacobi equations into the isoschrödinger's equation $i\partial_t\psi(t,r) = H * \psi(t, r) = \hat{E}*\psi(t,r) = E \psi(t, r)$. Recall now the fundamental structure of conventional and isotopic, gravitational metrics $\hat{g} = T \eta$, where η is the conventional Minkowski metric. Then, the *isooperator formulation of gravitation* here considered is given by the embedding of gravitation in the isounit of hadronic mechanics (Santilli (1989d). An intriguing aspect of this approach is that it has escaped identification until now owing to the simplicity of the isotopic theory. In fact, the formulation suggests that the representation of gravitation is "hidden" in the modular-associative structure of the conventional eigenvalue equations "$H \psi$" via its modular isotopy "$H*\psi$" = "$H T \psi$", where T is precisely the gravitational element of the decomposition $\hat{g} = T \eta$, of course, in its proper operator form. In different terms, the operator formulation of gravitation proposed in Santilli (1989d) is based on the condition that it coincides with conventional quantunm mechanics at the abstract, realization-free level, where all distinctions between "$H \psi$" and "$H*\psi$" cease to exist.

[33] As well known, the current unified gauge theories have not been completely successful in incorporating gravitational and strong interactions. The approach submitted in Santilli (1989d) for an *isograndunification* encompassing all known interactions is via the isotopy of the conventional unified theories of weak and electromagnetic interactions, that is, via the generalization of their trivial unit into our isounits. The approach leaves the main results of contemporary theories essentially unaffected and, in addition, can directly represent operator formulations of gravitational interactions (see the preceding footnote), as well as the nolocal short range component of the strong interactions due to wave overlappings. One of the intriguing possibilities of this "isograndunification" is that of turning conventionally

12) Permitting, apparently for the first time, a study on the "origin" of gravitation, by eliminating the now vexing problem of "unification" of the gravitational and electromagnetic fields, and replacing it with their "identification".

We would like to close this chapter by praising Albert Einstein, not only for his historical discoveries, but also for his scientific caution and honesty. In fact, Einstein's expressed quite clearly his doubts and reservations on his gravitational theory, by comparing the left-hand-side of his gravitational equations to the left wing of a house made of *fine marble*, and the right-hand-side of his equations to the right wing of a house made of *bare wood*. A reinspection of the theory permitted by the more general isoriemannian geometry has confirmed the correctedness of the left-hand-side of the equations for the exterior problem and the unsettled character of their right-hand-side.

It is important to note that our isogravitations essentially allow the reconciliation of the historical legacy of the Founding Fathers of contemporary physics on the ultimate nonlocal structure of matter (and the consequential need for a nonlocal interior geometry), with the established local-differential character of a yet unresolved exterior gravitation. The important point is that, whatever the final exterior theory is, it will indeed be the exterior limit of our interior isogravitations owing to the "direct universality" of the isoriemannian geometry.

The understanding is that the identification of all the implications of our nonlocal interior gravitation for the contemporary conception of the Universe, such as for black holes or gravitational collapse, will predictably require some time.

---

divergent perturbative series into convergent ones (see Santilli *(loc. cit.)* and footnote[21], p. 172). We should finally indicate that the proposed isograndunification needs no sources for the gravitational fields (even in interior conditions), because they are provided by the identification of the gravitational field with the electromagnetic, weak and strong fields themselves responsible for the structure of matter, according to Theorem V.3.1.

APPENDIX V.A: GASPERINI'S ISOTOPIES OF EINSTEINS GRAVITATION

In three pioneering papers, Gasperini (1984a, b, c) presented the first Lie-isotopic lifting of Einstein's gravitatiion with a locally Lorentz-isotopic structure following Santilli's ((1978a), (1981a), (1982a). (1983a)) submission of the foundations of the Galilei-isotopic and Lorentz-isotopic relativities.

In this appendix we shall review Gasperini's results, not only to compare them with the results of the main text of this paper, but also because, being based on the gauge formulation of gravitation (see, e.g., Trautman (1972) or Ivanenko (1973)), constitute an important complement to the treatment presented in the main text.

It should be indicated that Gasperini presented his isotopies of Einstein's gravitation following preceding works (Gasperini (1983a, b)) on the isotopies of gauge theories (see also gauge isotopies of Santilli (1979b)), which will not be reviewed here for brevity.

Consider a *conventional* Riemannian space $R(x,g,\mathfrak{R})$ in (3+1)-dimension. Let small Greek indeces μ, ν, ... denote conventional Lorentz indeces, and small Latin indeces a, b, ... denote anholonomic Lorentz indeces. Let P_a and M_{ab} be the conventional generators of the Poincaré symmetry P(3.1). Then, in the gauge language of gravity(*loc. cit.*) the *frame one-form* is given by

$$V^a = V^a{}_\mu \, dx^\mu, \tag{A.1}$$

and the *connection one-form* can be written

$$\omega^{ab} = \omega^{ab}{}_\mu \, dx^\mu. \tag{A.2}$$

The *standard potential of Einstein's gravitation* is then given by

$$h = h^A X_A = V^a P_a + \omega^{ab} M_{ab}, \tag{A.3a}$$

$$h^A = \{ V^a, \omega^{ab} \}, \qquad X_A = \{P_a, M_{ab}\}, \tag{A.3b}$$

where the capital Latin indeces run over all values of a, ab, ...

Following Santili's (1983a) construction of the Lorentz-isotopic symmetry, Gasperini (1984a) introduced the following isotopy of "potential" (A.3)

$$\hat{h} = h^A T_A{}^B X_A$$

$$= V^a T_a{}^b P_b + V^a T_a{}^{ab} M_{ab} + \omega^{ab} T_{ab}{}^b P_b$$

$$+ \omega^{ab} T_{ab}{}^{cd} M_{cd}. \tag{A.4}$$

which can be written

$$\hat{h} = \hat{h}^A X_A, \tag{A.5}$$

where

$$\hat{h}^A = \{ \hat{V}^a, \hat{\omega}^{ab} \}, \qquad X_A = \{P_a, M_{ab}\}. \tag{A.6a}$$

$$\hat{V}^a = h^A T_A{}^a = V^b T_b{}^a + \omega^{bc} T_{bc}{}^a, \tag{A.6b}$$

$$\hat{\omega}^{ab} = h^B T_b{}^{ab} = V^c T_c{}^{ab} + \omega^{cd} T_{cd}{}^{ab}. \tag{A.6c}$$

Note the correct preservation of the conventional generators of the Poincaré symmetry under the isotopy, which is a central requirement of the Lie-isotopy (Santilli (1978a)) , and which is also at the basis of Theorem V.2.3.

Now, in the formulation of gravitation under consideration, torsion is represented by the one-form

$$R^a = dV^a + \omega^a{}_b \wedge V^b, \tag{A.7}$$

and curvature by the two-form

$$R^{ab} = d\omega^{ab} + \omega^a{}_c \wedge \omega^{cb}. \tag{A.8}$$

Gasperini (*loc. cit.*) then showed that, in the particular case of constant isotopic elements, the torsion (A.7) and curvature (A.8) admit the isotopic liftings

$$\hat{R}^a = d\hat{V}^a + \hat{\omega}^a{}_b \wedge \hat{V}^b, \tag{A.9a}$$

$$\hat{R}^{ab} = d\omega^{ab} + \omega^a{}_c \wedge \omega^{cb}. \tag{A.9b}$$

For the standard case, one imposes the torsion R^a to be null, resulting in the standard Einstein's action for the case without source

$$A = \frac{1}{4K} \int \epsilon_{abcd} R^{ab}(\omega) \wedge V^c \wedge V^d, \tag{A.10a}$$

$$K = 16\pi G/c_o^4, \tag{A.10b}$$

which admits the immediate isotopic liftings

$$\hat{A} = \frac{1}{4K} \int \epsilon_{abcd} \hat{R}^{ab}(\hat{\omega}) \wedge \hat{V}^c \wedge \hat{V}^d. \tag{A.11}$$

Gasperini (*loc. cit.*) then considers the particular case of symmetric constant isotopic elements T in the particular form

$$T_{ab}{}^c = T_c{}^{ab} = 0, \quad T_{ab} = \eta_{ac} T^c{}_b \neq \eta_{ab}, T_{ab}{}^{cd} = \delta^c{}_a \delta^d{}_b \tag{A.12}$$

where the Minkowski metric is

$$\eta = \text{diag.}\ (1, 1, 1, -1) \tag{A.13}$$

Then, isoaction (A.11) can be explicitly written

$$\hat{A} = \frac{1}{4K} \int d^4x\, \Delta^{\frac{1}{2}}\{\tfrac{1}{2}R\phi^2 - \tfrac{1}{2} R\, T^{\alpha}{}_{\beta}\, T^{\beta}{}_{\alpha} - 2\, R_{\nu}{}^{\alpha}\, T_{\alpha}{}^{\nu}\, \phi$$

$$+ 2R_{\nu}{}^{\alpha}\, T_{\alpha}{}^{\beta}\, T_{\beta}{}^{\nu} + R_{\mu\nu}{}^{\alpha\beta}\, T_{\alpha}{}^{\mu}\, T_{\beta}{}^{\nu}\}, \tag{A.14}$$

where

$$R_{\mu\nu}{}^{\alpha\beta} = 2V_a{}^{\alpha}\, V_b{}^{\beta}(\partial_{[\mu}\, \omega_{\nu]}{}^{ab} + \omega_{[\mu}{}^{ac}\, \omega_{\nu]\,c}{}^{b} \tag{A.15}$$

is the curvature tensor,

$$R_{\mu\nu} = R_{\mu\alpha\nu}{}^{\alpha}, \tag{A.16}$$

is the Ricci tensor,

$$F = g^{\mu\nu} T_{\mu\nu} \equiv \eta^{ab} T_{ab}, \tag{A.17}$$

and the world metric tensor is given by

$$g_{\mu\nu} = V_{\mu}{}^{a} V_{\nu}{}^{b} \eta_{ab}. \tag{A.18}$$

It is easy to see that, starting with a torsionless theory, $R^a = 0$, the isotopic theory has a generally non-null torsion, i.e.,

$$R^a = 0, \Rightarrow \hat{R}^a \neq 0. \tag{4.19}$$

LEMMA V.A.1 (Gasperini's (1984a, b) Lemma): The Lie-isotopic liftings of Einstein's gravitation induce, even in the absence of a source term, a Riemann-Cartan geometrical structure , with the isotopic element acting as a source of torsion.

As an explicit example of the modification of Einstein's equation induced by the lifting, Gasperini considers the particular case

$$T_a{}^b = \sigma \delta_a{}^b + t_a{}^b, \tag{A.20}$$

where 4σ is the trace of $T_a{}^b$. Then, via some algebra, Gasperini derived the following modified Einstein's equations

$$G_\alpha{}^\beta = R_\alpha{}^\beta - \tfrac{1}{2} \delta_\alpha{}^\beta R = \sigma^{-1}(F_\alpha{}^\beta - t_\alpha{}^\nu G_\alpha{}^\beta) + \sigma^{-2} t_\alpha{}^\nu F_\nu{}^\beta \tag{A.21}$$

where

$$F_\alpha{}^\beta = R_\alpha{}^\nu t_\nu{}^\beta + t_\alpha{}^\nu R_\nu{}^\beta - \tfrac{1}{2}Rt_\alpha{}^\beta - R_\nu{}^\mu t_\mu{}^\nu \delta_\alpha{}^\beta + R_{\mu\alpha}{}^{\nu\beta} t_\nu{}^\mu \tag{A.22}$$

which can be solved, e.g., for small isotopic elements (see (*loc. cit.*) for details).

The emerging theory is a gravitational theory with two metrics. In fact, in addition to the conventional Riemannian metric

$$g_{\mu\nu} = V_\mu{}^a V_\nu{}^b \eta_{ab}, \tag{A.23}$$

we have the isotopic metric for the tangent space

$$\hat{\eta}_{ab} = g^{\mu\nu} \hat{V}_{\mu}{}^{a} \hat{V}_{\nu}{}^{b}, \tag{A.24}$$

which is different than the conventional Minkowski metric η. This allowed Gasperini to introduce the Poincaré-isotopic symmetry $\hat{P}(3.1)$ in the tangent space via a restriction which is simply given by $T > 0$.

LEMMA V.A.2 (Santilli (1983a(), (1988d)): Under the condition that the isotopic elements are positive-definite, the local Poincaré symmetry is not lost in the isotopies of Einstein's gravitation, but reconstructed as exact at the isotopic level.

Needless to say, the reaching of two metrics g and $\hat{g}$ is not new, because inherent in the very structure of Einstein's gravitation (where the two metrics are g and η). Also, generalized metrics were reached in a number of cases (see, e.g., Papapetrou (1951), Rosen (1980), and quoted literature). What is new is the reconstruction of the exact Poincare' symmetry at the isotopic level of the tangent space.

The above results also allowed the formulation of the following

ISOTOPIC PRINCIPLE OF EQUIVALENCE (Gasperini (1984)) : Gravitational effects on a Riemannian space $R(x,g,\Re)$ can be made to disappear locally when the metric g is transformed into the metric $\hat{\eta}$ of the tangent Poincaré-isotopic space.

The above principle was generalized in Section V.2 to isoriemannian spaces.

In the remaining paper (1984b), Gasperini worked-out the following additional results:

1) The extension of the lifting to the case with a matter source, with results structurally equivalent to the preceding ones;

2) An explicit example of isotopic gravitation for the particular case when the tangent metric is the characterized by the low energy values of Nielsen-Picek (1983) for the interior of pions or kaons, i.e.

$$g = ((1 - \alpha/3), (1 - \alpha/3), (1 - \alpha/3), -(1 + \alpha)) \tag{A.25}$$

where the "Lorentz asymmetry parameter" α has the *negative* value for pions

$$\alpha \approx -3.8 \times 10^{-3}, \tag{A.26}$$

and the *positive* value for kaons

$$\alpha \approx +0.6 \times 10^{-3}, \tag{A.28}$$

which is particularly significant for this analysis, inasmuch as they allow a first approximation of the gravitational isotopies;

3) The detailed derivation of the gravitational equations for tangent metric (A.25) which are particularly useful and instructive;

4) The deviation of the "orbital" motion from the conventional one also for the case of tangent metric (A.25). The primary intent was the identification of upper limits in the isotopy for our Solar system (see below for comments);

5) An approximate generalization of the Schwarzschild metric also for tangent metric (A.25);

and other results.

We are now in a position to identify the most salient differences between Gasperini's isotopies of Einstein's gravitation, and the isotopies studied in this volume.

I: Gasperini formulated his isotopies for the entire space-time, while our isotopies are solely restricted to the interior gravitational problem. In particular, in his limits for admissible isotopies at our planetary level Gasperini (1984c) obtained very small possible values [8]. In our treatment, the liftings are restricted to the interior gravitational problem, while recovering the conventional theories identically in the exterior problem (Sect. V.2). As a result, no upper limit exist for the value of the isotopic elements in our approach. In fact, the total conserved quantities are the conventional ones, Eq.s (V.2.14), which are unaffected by the non-potential interior terms, because the latter represent local, internal, energy-exchanges which are compatible, by construction, with the total conserved quantities. As a result, no *exterior* experimental information can provide numerical limits for our *interior* isotopies.

II: Gasperini formulated his isotopies on a conventional Riemannian space, while our isotopies are formulated in a generalized Riemannian space. This implies that Gasperini's gravitational equations are based on the conventional Riemannian geometry, while our equations are based on a structural generalization of the Riemannian geometry. The latter generalization was needed for several reasons, including the need to identify the most general possible theory of gravitation, model (V.3.33), which is permitted by the current axioms. In different terms, Gasperini generalized only the tangent space in an isotopic form by preserving the conventional Einstein's equations, while we generalize both the tangent and the Riemannian space in an isotopic manner by reaching certain isotopic forms of Einstein's equations.

III: Gasperini treatment remains local-differential, while our treatment is intrinsically nonlocal-integral. This is a direct consequence of our generalized isoriemannian geometry and, more particularly, of its isounits $\hat{1} = T^{-1}$, which permit the most general possible, nonlinear, nonlocal, nonlagrangian and non-Newtonian interior trajectories. By comparison, the gauge language of Gasperini's (*loc. cit.*) gravity remains strictly local-differential.

The above comparative comments allow a deeper understanding of several aspects of our analysis. For instance, the non-null character of torsion under isotopic liftings is explicitly expressed in Gasperini's approach via Eq.s (A.9a), while our approach implies a further generalization into the isoriemannian spaces in which the isotopic torsion is identically null, Eq.s (II.11.90), although the torsion in the physical space-time remains non-null, Eq.s (II.11.91).

Gasperini's studies remain also useful for numerous other aspects, e.g., the study of the local Lorentz-isotopic space in the interior of pions and kaons, the understanding of the modification to the Schwatzschild metric expected from a true representation of the interior physical media, and other aspects (see also Sect. V.4).

CHAPTER VI: MUTUAL COMPATIBILITY OF THE ISOTOPIC RELATIVITIES

VI.1: STATEMENT OF THE PROBLEM

Important element for appraising the isotopic liftings of Galilei's, Einstein's special and Einstein's general relativities studied in these volumes are given by their individual consistency and experimental verification, as well as their mutual compatibility.

In fact, the conventional relativities are deeply inter-related and mutually compatible, as well known, and these properties must evidently persist for all possible generalized relativities.

These issues will be studied in this chapter by, first, re-inspecting the mutual compatibility of conventional relativities (Sect. VI.2), and then passing to the compatibility of their isotopic coverings (Sect. VI..3).

As we shall see, the isogalilean, isospecial and isogeneral relativities result to be as deeply inter-related and mutually compatible as the conventional relativities to such an extent, that any possible disproofs, to be consistent, must be shown to be compatible with the physical systems of our environment, such as spinning tops with decaying spin, trajectories with decaying angular momenta, etc.

Moreover, the study of compatibility yields a most important theoretical result of these volumes: the unification of all known, linear and nonlinear, local and nonlocal, Hamiltonian and nonhamiltonian, relativistic and gravitational, exterior and interior systems via one single notion, the isopoincaré symmetry.

VI:2: MUTUAL COMPATIBILITY OF CONVENTIONAL RELATI-VITIES.

The problem under consideration is multi-fold: first we have the compatibility of the special with the Galilean relativity, then that of the general with the special relativity, and finally the general mutual compatibility of all relativities.

As we shall see, our isotopic techniques permit a rather intriguing unified formulation of these mutual compatibilities via the conventional Poincaré synmmetry, only realized in an isotopic way (Santilli (1988c, d))

As well known, the compatibility of the special with the Galilean relativity cann be based on the property that the nonrelativistic limit of the Poincaré algebra **P**(3.1) yields precisely the Galilei algebra **G**(3.1) via the so-called *Inonü-Wigner contraction*.

This property is well presented in a number of books, e.g.,Gilmore (1974). For the reader's convenience, let us recall the basic ideas.

Consider our Minkowski spaces $M^{I}(x,\eta,\mathfrak{R})$ and write the fundamental invariant in the form

$$\frac{1}{c_o^2} r^k r^k - tt = -\frac{R^2}{c_o^2}, \qquad (2.1)$$

then it is easy to see that at the limit

$$\epsilon = \frac{R}{c_o} \Rightarrow 0, \qquad (2.2)$$

we have the contraction

$$M(x,\mathfrak{R})_{|c_o/R \Rightarrow \infty} \Rightarrow \mathfrak{R}_t \times E(r,\delta,\mathfrak{R}), \qquad (2.3)$$

Namely, the Minkowski space contracts, for the space coordinates, into the Euclidean space $E(r,\delta,\Re)$ with separation $r^2 = r^t \delta r$, $\delta = \text{diag.}(1, 1, 1)$, multiplied the field representing time $\Re_t$.

The next step is the recovering of the Galilei's symmetry as a contraction of the Poincaré one via the Inonü-Wigner contraction (Gilmore (*loc. cit.*)). Consider basis (IV.6.5) of the Poincaré algebra P(3.1) and decomposit it in the form

$$\mathbf{P}(3.1) = \mathbf{g}_0 \oplus \mathbf{g}_1 = \{J_{ij} + P_k + P_4\} \oplus J_{k4}; \tag{2.4}$$

Redefine it in the vicinity of the "north pole" (0,R) (see Gilmore (*loc. cit.*) p. 451), and perform the contractions

$$J_k = \mathrm{Lim}_{c_o/R \Rightarrow \infty} \epsilon_{kij} J_{ij} = \epsilon_{kij}(r_i p_j - r_j p_i), \tag{2.5a}$$

$$P_k = \mathrm{Lim}_{c_o/R \Rightarrow \infty} p_k = p_k, \quad H = \mathrm{Lim}_{c_o/R \Rightarrow \infty} p_4 = p_4 = E, \tag{2.6b}$$

$$G_k = \mathrm{Lim}_{c_o/R \Rightarrow \infty} J_{k4}/R = \mathrm{Lim}_{R \Rightarrow 0}(x_k p^4 - x^4 p_k)/R, \tag{2.7c}$$

$$i, j, k = 1, 2, 3$$

where we have asumed the new expression for the energy. Then we have the contraction

$$\mathbf{P}(3.1)_{c_o/R \Rightarrow \infty} = \mathbf{G}(3.1) \tag{2.8}$$

which is amply sufficient to establish the compatibility between the special and Galilean relativities.

The compatibility between the general and the special relativity is generally presented via the original Riemann (1868) formulation of normal coordinartes (Sect. II.11) and related tangent planes, in which the general relativity recovers the special in its entirety (see, e.g., Pauli (1958), Lovelock and Rund (1975)).

Despite the clear consistency of these results, there is a considerable methodologiocal discrepancy between the compatibility of the special and Galilean, with that between the general and the special relativities.

In fact, the first is centrally dependent on Lie symmetries, while these symmetry are manifestly absent in the latter, evidently because the general isometry of a conventional Riemannian manifold is unknown in the conventional luiterature.

This discrepancy is resolved by the isotopic techniques via

Theorem V.2.1, which establishes that the isopoincaré symmetry is the general group of isometries of a conventional Riemannian space.

The result is based, first, on the reformulation of the Riemannian space as an isotope of the Minkowski space

$$R(x,g,\Re) \approx \hat{M}(x,\hat{\eta},\hat{\Re}), \quad g = T_g\eta \equiv \eta, \quad \hat{\Re} = \Re\,\hat{1}, \quad \hat{1} = T_g^{-1} \quad (2.9)$$

and then the construction of the isopoincaré symmetry $\hat{P}_g(3.1)$ with respect to the isounit $\hat{1} = T_g^{-1}$.

In conclusion, the isotopioc techniques permit a new unified, mutual compatibility of Einstein's general, Einstein's special and Galilei's relativities in term of one single, abstract notion, the isopoincaré symmetry, according to the sequence

$$\hat{\mathbf{P}}_g(3.1) \Rightarrow \mathbf{P}_\eta(3.1) = \hat{\mathbf{P}}_\eta(3.1)\,|_{T_g = I} \Rightarrow \mathbf{G}(3,1) = \mathbf{P}_\eta(3.1)\,|_{c_o/R\Rightarrow\infty} \quad (2.10)$$

Note that, as a consequence of the above chain, the isopoincaré symmetry $\hat{P}_g(3.1)$ unifies all local-differential, Lagrangian-Hamiltonian, exterior systems.

VI.3: MUTUAL COMPATIBILITY OF ISOTOPIC RELATIVITIES.

The preceding analysis of mutual compatibility of the conventional relativities admits a consistent isotopic generalization, thus establishing the mutual compatibility of the isogalilean, isospecial and isogeneral relativities. This is not surprising, for the isorelativities themselves were built to verify such mutual compatibility.

To begin, the conventional theory of *contraction of Lie groups* (see, e.g., Gilmore (1974), Ch. 10) admits a rather simple and consistent isotopic lifting.

Consider our Minkowski isotopic spaces $\hat{M}^I(x,\hat{g},\hat{\Re})$ and write the fundamental isoinvariant (IV.3.4) in the form

$$\frac{1}{c_o^2}\, r^k\, b_k^{\,2} r^k - t b_4^{\,2} t = -\frac{R^2}{c_o^2}, \quad (3.1)$$

then it is easy to see that at the limit

$$\epsilon = \frac{R}{c_o} \Rightarrow 0, \tag{3.2}$$

we have the contraction

$$\hat{M}^I(x,\hat{g},\hat{\mathfrak{R}})_{|c_o/R\Rightarrow\infty} \Rightarrow \hat{\mathfrak{R}}_t \times \hat{E}(r,\hat{G},\hat{\mathfrak{R}}), \tag{3.3}$$

Namely, the Minkowski-isotopic spaces contract into the Euclidean-isotopic spaces $\hat{E}(r,\hat{G},\hat{\mathfrak{R}})$ for the space coordinates,

$$\hat{E}(r,\hat{G},\hat{\mathfrak{R}}) : \quad r^{\hat{2}} = r^t \, \hat{G} \, r, \tag{3.4a}$$

$$\hat{G} = \text{diag.} \, (\hat{b}_1{}^2, \hat{b}_2{}^2, \hat{b}_3{}^2) > 0, \tag{3.4b}$$

$$\hat{\mathfrak{R}} = \mathfrak{R}\hat{1}_2, \; \hat{1}_2 = \text{diag.} \, (\hat{G}^{-1}, \hat{G}^{-1}), \tag{3.4c}$$

multiplied the isotime field

$$t^{\hat{2}} = t \, \hat{b}_4{}^2 \, t \in \hat{\mathfrak{R}}_t = \mathfrak{R}\hat{1}_t, \; \hat{1}_t = \hat{b}_4^{-2}, \tag{3.5}$$

This justifies the statement of Sect. III.5 to the effect that the isounit of time in nonrelativistic Hamilton-isotopic mechanics can be best seen as a contraction from the relativistic formulations.

Needless to say, the preceding limit parallels the conventional contraction, Eq.s (VI.2.3).

It is a simple exercise for the interested reader to prove (with the use of Sect.s II.5 and II.6) the following generalization of the contraction theorem of Gilmore (*loc. cit.*), p. 449.

THEOREM VI.3.1 (Isotopic Inönü-Wigner contractions, Santilli (1988c)):: Let $\hat{\mathbf{g}}$ *be a (finite-dimensional) Lie-isotopic algebra defined on an isofield* $\hat{F}$ *of real or complex numbers, and consider its direct-sum decomposition as isovector space*

$$\hat{\mathbf{g}} = \hat{\mathbf{g}}_o \oplus \hat{\mathbf{g}}_1, \tag{3.6}$$

Let $\hat{U}(\epsilon)$ *be an isotransformation on* $\hat{\mathbf{g}}$ *which becomes singular at the limit* $\epsilon \Rightarrow 0$*, and which is such that*

$$\hat{U}(0) * \hat{\mathbf{g}}_o = \mathbf{g}_o, \tag{3.7a}$$

$$\hat{U}(0) * \hat{g}_1 = 0. \tag{3.7b}$$

Then $\hat{g}$ can be contracted with respect to $\hat{g}_0$ into a new isoalgebra $\hat{g}'$ iff $\hat{g}_0$ is a closed subgroup of $\hat{g}$, in which case:

1) $\hat{g}_0$ is a subalgebra of both $\hat{g}$ and $\hat{g}'$;

2) $\hat{g}'_1$ becomes an isobelian invariant subalgebra of $\hat{g}'$; and

3) $\hat{g}$ is non-semisimple.

The application of the above theorem to the isopoincaré algebra $\hat{P}(3.1)$ is straightforward. Consider the basis (II.6.5a) of $\hat{P}(3.1)$; decompose it as an isovector space in the form

$$\hat{P}(3.1) = \hat{g}_0 \oplus \hat{g}_1 = (J_{ij} + P_k + P_4\} \oplus J_{k4}; \tag{3.8}$$

redefine it in the vicinity of the "north pole" (0,R) (see Gilmore (*loc. cit.*) p. 451), and perform the contractions (virtually identical to the conventional ones because the generators do not change under isotopy, Sect. II.3)

$$J_k = \mathrm{Lim}_{c_o/R \Rightarrow \infty} \epsilon_{kij} J_{ij} = \epsilon_{kij}(r_i p_j - r_j p_i), \tag{3.9a}$$

$$P_k = \mathrm{Lim}_{c_o/R \Rightarrow \infty} p_k = p_k, \quad H = \mathrm{Lim}_{c_o/R \Rightarrow 0} p_4 = p_4 = E, \tag{3.9b}$$

$$G_k = \mathrm{Lim}_{c_o/R \Rightarrow \infty} J_{k4}/R = \mathrm{Lim}_{R \Rightarrow 0}(x_k p^4 - x^4 p_k)/R, \tag{3.9c}$$

$$i, j, k = 1, 2, 3$$

where we have used definition (IV.9.34), and assumed the new nonrelativistic expression for the energy.

Then, it is easy to see that the isocommutation rules of $\hat{P}(3.1)$, Eq.s (IV.6.6), are contracted to the isocommutation rules of the isogalileian algebras $\hat{G}(3.1)$, Eq.s (III.5.19),

$$\hat{P}(3.1)\,|_{c_o/R \Rightarrow \infty} = \hat{G}(3.1) \tag{3.10}$$

in the same way as the commutation rules of **P**(3.1) contract into those of **G**(3.1).

The contraction of Lie-isotopic groups and algebras is, however, richer than the conventional one. As an example, we have used in the above considerations the contraction on isospaces of Class I. We leave it to the interested reader the study of the same contraction, but in the more general isospaces of Class III. Since the abstract **O[4]** isotopes on $\hat{M}^{III}(x,\hat{g},\hat{\mathfrak{R}})$ unify all simple six-dimensional Lie groups (Sect. IV.5), we expect the possibility of unifying all possible contractions of the Lie groups of a given dimension.

Equally recommendable for the interested reader is the study of the *expansion of isotopic groups and algebras* , which we cannot possibly consider here for brevity.

The compatibility between the isogeneral and isospecial relativities is two-fold. First, it is provided by the existence of a consistent isotopies of Riemannian's normal coordinates (Sect. II.11). As a result, our isogravitations admit the isospecial relativities in their local, internal, tangent isoplanes (which must be so, at any rate, by construction).

Second, the general isosymmetries of isogravitation on isospaces $\hat{R}(x,\hat{g},\hat{\mathfrak{R}})$ are given precisely by the most general known realization of the isopoincaré symmetry $\hat{P}_{\hat{g}}(3.1)$, according to Theorem V.2.2. This allows the consistent isotopic lifting of property (2.10)

$$\hat{\mathbf{P}}_{\hat{g}}(3.1) \Rightarrow \hat{\mathbf{P}}_{\hat{\eta}}(3.1) = \hat{\mathbf{P}}_{\hat{g}}(3.1)\,|_{T_{\hat{g}}\,=\,I} \Rightarrow \hat{\mathbf{G}}_{\hat{\delta}}(3,1) \;=\; \hat{\mathbf{P}}_{\hat{\eta}}(3.1)\,|_{c_o/R\Rightarrow\infty} \quad (3.11)$$

The above mutual compatibilities are not a pure mathematical occurrence, because they carry considerable implications in the physical application of the isotopic relativities.

Recall that the isogalilean relativities need no experimental verification because they are constructed from the equations of motion of the physical reality in our clasical environment. The compatibility of the isospecial relativities with the isogalilean therefore implies considerable credibility for the predictions of novel relativistic effects in interior physical media, to such an extent that any possible future disproof of the isospecial relativities, for consistency, must be proved to be with our classical nonrealitivistic physical reality.

Similarly, the compatibility of our isogravitations with the isogalilean relativities implies that any possible disprof of the formers, to be final, must be proved to be compatible with the physical reality of our interior environment, that is, with continuously decaying spin, trajectories in our atmospheres with continuously decaying angular momenta, etc. For further comments, see Figure VI.3.1.

GEOMETRIC UNIFICATION OF ALL EXTERIOR AND INTERIOR RELATIVITIES

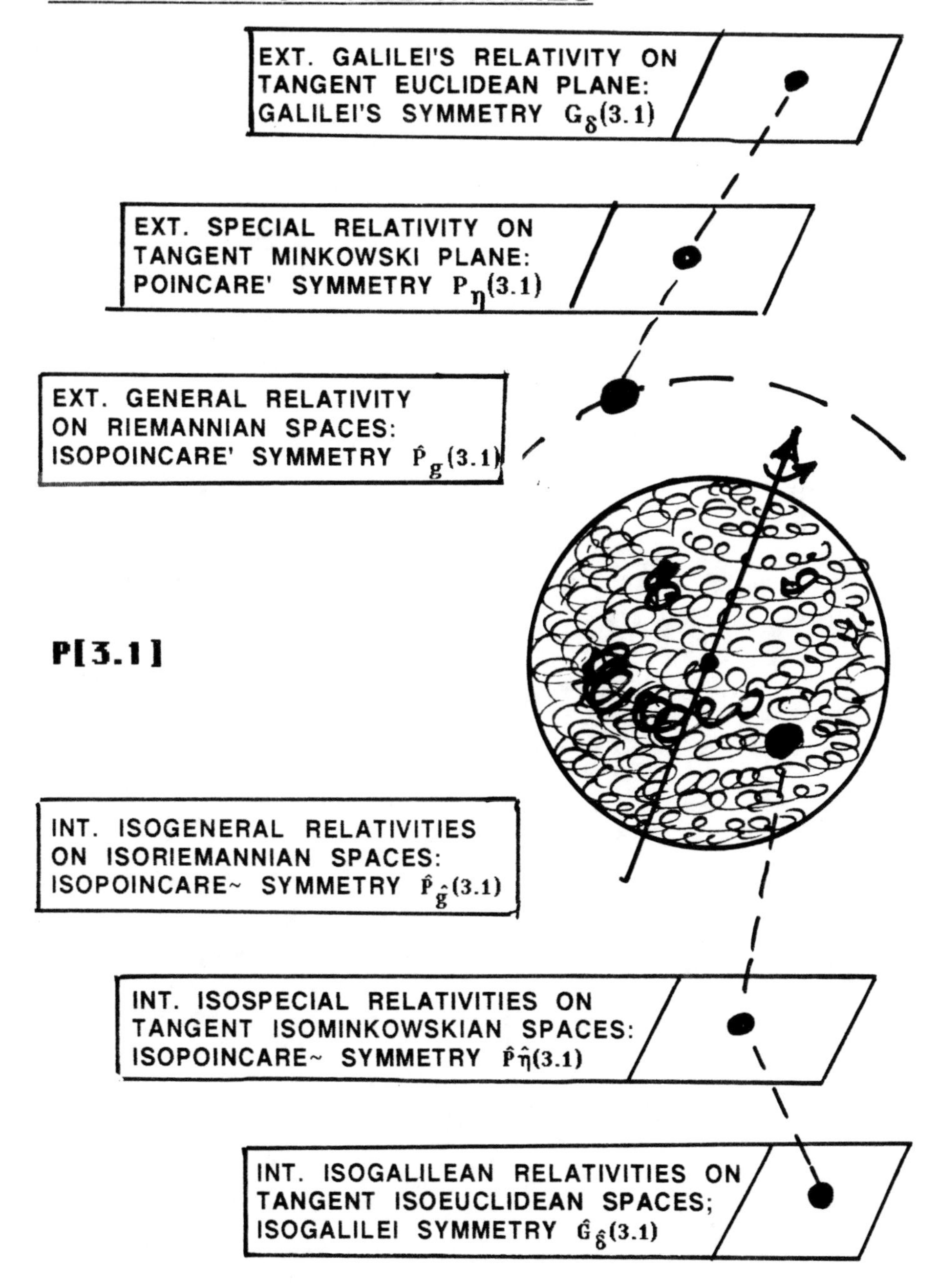

FIGURE VI.3.1: A schematic view of the unification of all interior and exterior relativities proposed in Santilli (1988d), (1991b). It is based on one, ultimate, abstract isosymmetry, the *Poincaré-isotopic symmetry*, say, **P[3.1],** which is realized in isogravitations (V.2.12) or (V.3.26) in a multiple variety of ways of increasing complexity and methodological needs:

A) As the exterior, general, differential and nonlinear isosymmetry $\hat{P}_g(3.1)$ of the conventional general relativity;

B) As the exterior, local, differential and linear conventional Poincaré symmetry $P_\eta(3.1)$ of the special relativity in the Minkowskian tangent plane;

C) As the exterior local, differential and linear conventional Galilei's symmetry $G_\delta(3.1)$ under group contration of the special relativity in the Euclidean tangent plane;

D) As the interior, global, integral and nonlinear isopoincaré symmetries $\hat{P}_{\hat{g}}(3.1)$ of the isogeneral relativities;

E) As the interior, local, integral and nonlinear isopoincaré symmetries $\hat{P}_\eta(3.1)$ of the isospecial relativities for the isominkowskian tangent plane;

F) As the interior, local, integral and nonlinear isogalilean symmetries $\hat{G}_{\hat{\delta}}(3.1)$ under isogroup contraction for the isogalilean relativities in the tangent isoeuclidean space;

under the condition, automatically verified by positive-definite isotopic elements T, that all isotopic symmetries are locally isomorphic to the corresponding conventional ones,

$$\hat{P}_{\hat{g}}(3.1) \approx \hat{P}_{\hat{\eta}}(3.1) \approx P_g(3.1) \approx P_\eta(3.1),$$

$$\hat{G}_{\hat{\delta}}(3.1) \approx G_\delta(3.1).$$

The studies reviewed in these volumes therefore allow the reduction of all possible linear and nonlinear, differential and integral, Lagrangian and nonlagrangian, exterior and interior, relativistic andgravitational systems to one, single, unique geometric notion: the isopoincaré symmetry **P[3.1].**

CHAPTER VII: EXPERIMENTAL VERIFICATION OF THE ISOTOPIC RELATIVITIES

VI.1: STATEMENT OF THE PROBLEM

In the Preface we stated that physics is a science with an absolute standard of values: the experimental verification. It is therefore time to confront the problem of ***the experimental verification of the isogalilean, isospecial and isogeneral relativities in the physical conditions of their conception, motion of extended and deformable particles or electromagnetic waves within inhomogeneous and anisotropic physical media.***

The mental attitude necessary for the study of the issue is therefore that of abandoning all Einsteinian tests available in the literature (see, e.g., the historical review in Pauli (1958)), because conducted in the homogeneous and isotropic vacuum, and therefore inapplicable to the new relativities.

On the contrary, in order to truly test the new relativities, it is necessary that the environment has structural geometric differences with that of conventional relativities, namely, as it is the case for an inhomogeneous and anisotropic physical medium.

As elaborated in more details in the concluding remarks, the mental attitude needed is that of ***conducting tests under the***

necessary condition of maximizing the departures from the physical conditions of conventional Einsteinian tests . If one minimizes them or, worst, keeps the same conditions, no departure from the Einsteinian laws is expected, and no test of the new relativities has actually occurred.

As pointed out earlier, the isogalilean relativities are verified by construction in our *classical* environment, and need no additional classical test. Nevertheless the implications in *particle physics* are far reaching, and in need of numerous direct tests.

The isospecial relativities, instead, needs a number of independent, direct tests, in both classical and particle physics, because they predict numerical variations from all basic postulates of the special relativity, depending on the physical media considered, according to Postulates I–V of Sect. IV.9.

The isogeneral relativities also need direct tests, although those for the isospecial are local, internal tests on tangent isoplanes of the isogeneral and, as such, sufficient for this preliminary study.

Since the treatment of these volumes is purely classical, we shall provide primary attention to ***classical tests*** . Particle tests will be merely mentioned for completeness without a detailed treatment at this time.

By keeping in mind the verification of the isogalilean relativities by construction, ***the primary objective is to identify classical tests of the isospecial relativities on isominkowski spaces that are currently feasible .***

For this purpose, let us recall the *isominkowski spaces* (IV.9.2), i.e.,

$$\hat{M}^I(x,\hat{g},\hat{\mathfrak{R}}): \quad x^{\hat{2}} = x^\mu \hat{g}_{\mu\nu} x^\nu =$$

$$= x^1 \hat{b}_1^2 x^1 + x^2 \hat{b}_2^2 x^2 + x^3 \hat{b}_3^2 x^3 - x^4 \hat{b}_4^2 x^4,$$

$$= x^1 \frac{1}{\hat{n}_1^2} x^1 + x^2 \frac{1}{\hat{n}_2^2} x^2 + x^3 \frac{1}{\hat{n}_3^2} x^3 - t \frac{c_o^2}{\hat{n}_4^2} t, \tag{1.1a}$$

$$x = (r, x^4) = (r, c_o t), \quad r \in \hat{E}_2(r,\hat{G},\hat{\mathfrak{R}}) \tag{1.1b}$$

$$\hat{g} = T_2\, \eta, \tag{1.1c}$$

$$\eta = \text{diag.}\,(1,1,1,-1) \in M(x,\eta,\mathfrak{R}), \tag{1.1d}$$

$$T_2 = \text{diag.}\,(\hat{b}_1^2, \hat{b}_2^2, \hat{b}_3^2, \hat{b}_4^2) > 0, \quad \hat{\mathfrak{R}} = \mathfrak{R}\,\hat{1}_2, \quad \hat{1}_2 = T_2^{-1}, \tag{1.1e}$$

$$\hat{b}_\alpha = 1 / \hat{n}_\alpha = \hat{b}_\alpha(s, x, u, a, \mu, \tau, n, ...) > 0, \ \alpha = 1, 2, 3, 4, \quad (1.1f)$$

$$\hat{b}_k = 1/\hat{n}_k = \hat{b}_3 = 1/\hat{n}_3, \ k = 1, 2, 3 \quad (1.1g)$$

where the last conditions are imposed to avoid any isogalilean contribution, and restrict the tests to pure isospecial settings.

The fundamental symmetries to be tested are the *isolorentz symmetries* with *isotransformations*

$$x'^1 = x^1, \ x'^2 = x^2, \ x'^3 = \hat{\gamma}\,(x^3 - \beta\, x^4), \ x'^4 = \hat{\gamma}\,(x^4 - \hat{\beta}\, x) \quad (1.2)$$

where

$$\hat{\beta} = \frac{\hat{n}_4}{\hat{n}_3}\,\beta; \qquad \hat{\gamma} = \left| 1 - \frac{\hat{n}_4^{\,2}}{\hat{n}_3^{\,2}}\,\hat{\beta}^2 \right|^{-\frac{1}{2}} \quad (1.3)$$

The isospecial relativities and related isospaces $\hat{M}^I(x,\hat{g},\hat{\Re})$ can be characterized by their primary implications, the values of the *maximal causal speed* V_{Max} and the *fourth component of the isometric* (Sect. IV.9) which are given by

$$V_{Max} = c_o\frac{\hat{b}_4}{\hat{b}_3} = c_o\,\frac{\hat{n}_4}{\hat{n}_3} \ \substack{<\\=\\>}\ c_o \quad (1.4a)$$

$$c = c_o\hat{b}_4 = c_o/\hat{n}_4 \ \substack{<\\=\\<}\ c_o. \quad (1.4b)$$

The above values are deeply linked to the *geometrization of physical media* provided by the isospaces $\hat{M}^I(x,\hat{g},\hat{\Re})$ which, from predictions (1.4), implies the existence of the following *nine isorelativistic media*

TYPE 1: $n_3 = n_4$, $n_4 = 1$; $\hat{\beta} \equiv \beta$, $\hat{\gamma} \equiv \gamma$; $V_{Max} = c_o$, $c \equiv c_o$., (1.5a)

TYPE 2: $n_3 = n_4$, $n_4 > 1$; $\beta \equiv \beta$, $\hat{\gamma} \equiv \gamma$; $V_{Max} \equiv c_o$, $c < c_o$, (1.5b)

TYPE 3: $n_3 = n_4$, $n_4 < 1$; $\hat{\beta} \equiv \beta$, $\hat{\gamma} \equiv \gamma$; $V_{Max} \equiv c_o$, $c > c_o$, (1.5c)

TYPE 4: $n_3 < n_4$, $n_4 > 1$; $\hat{\beta} > \beta$, $\hat{\gamma} < \gamma$; $V_{Max} < c_o$, $< c_o$, (1.5d)

TYPE 5: $n_3 < n_4$, $n_4 = 1$; $\hat{\beta} > \beta$, $\hat{\gamma} < \gamma$; $V_{Max} < c_o$, $c = c_o$, (1.5e)

TYPE 6: $n_3 < n_4$, $n_4 < 1$; $\hat{\beta} > \beta$, $\hat{\gamma} < \gamma$; $V_{Max} < c_o$, $c > c_o$, (1.5f)

TYPE 7: $n_3 > n_4,\ n_4 > 1;\quad \hat{\beta} < \beta,\ \hat{\gamma} > \gamma;\quad V_{Max}\ c_o, < c_o,$ (1.5g)

TYPE 8: $n_3 > n_4,\ n_4 = 1;\quad \hat{\beta} < \beta,\ \hat{\gamma} > \gamma;\quad V_{Max} > c_o,\ c = c_o,$ (1.5h)

TYPE 9: $n_3 > n_4,\ n_4 < 1;\quad \hat{\beta} < \beta,\ \hat{\gamma} > \gamma;\quad V_{Max} > c_o,\ c > c_o.$ (1.5i)

In essence, the infinite variety of different physical media existing in the Universe, such as fluids, gases, conductors, superconductors, nuclei, hadrons, stars, etc., is reduced by our isominkowski spaces to nine essential classes which are geometrically significant and, as such, suitable for tests.

A central pre-requisite for a true test of the new relativities is therefore ***to identify physical media in classification (1.5) which are most suitable for experimental tests .***

We assume the reader is familiar with the main properties of the isospecial relativities, such as the fact that the basic local invariant is the maximal causal speed V_{Max} and not the speed of light (recall that the relativistic addition of two speeds of light in water *does not* yield the same speed, but the isorelativistic addition of two $V_{Max} > c$ does yield V_{Max}); the quantity $c = c_o\ /\ \hat{n}_4$ represents the local speed of light only for transparent media, otherwise it represents a geometric quantity, as in general relativity; etc.

The general experimental objective is therefore ***to test the isospecial relativities for each essential physical media via the deviations from Einsteinian values predicted by the isotopic quantities $\hat{\beta}$ and $\hat{\gamma}$.***

Needless to say, we are not in a position to reach any conclusion at this time, whether in favor or against the isospecial relativities. Nevertheless, the currently available data are sufficiently encouraging to warrant additional tests.

To facilitate the experimental task, let us briefly review the above physical media each with a representative example:

Media of Type 1 coincide with the conventional Minkowski space for the exterior problem in vacuum and, as such, they will be ignored.

Media of Type 2 represent homogeneous and isotropic fluids such as water, in which the isospecial relativities can represent:

1) the actual speed of light $c = c_o/n_4 < c_o$, where n_4 is the index of refraction[34];

2) the speed of massive particles (such as electrons) which is bigger than the local speed of light, and can attain maximal values c_o, as established by the Cerenkiov light;

3) the correct invariance of the maximal causal speed (which cannot be the speed of light according to the special relativity because of the manifest violation of the relativistic addition of the speeds, but V_{Max});

4) the correct isorelativistic addition of speeds;

5) the homogenuity and isotropy of the media via the scalar isotopy (IV.10.6); and other aspects.

Nevertheless, additional, independent ispections are needed before claiming that the isospecial relativities are exactly valid in media of Type 2.

Media of Type 3 represent superconductors, with $V_{Max} = c_o$ describing the speed of the electrons, and $c > c_o$ representing a purely geometrical quantity (the deviation from the geometry of empty space caused by matter). The isospecial relativities can therefore provide potentially new possibilities in superconductivity which, as such, deserve an inspection (see below for preliminary results).

Media of Type 4 represent ordinary atmospheres and, as such, they are particularly suited for experiments. Since, for these media, $\hat{\beta} > \beta$ and $\hat{\gamma} < \gamma$, the isospecial relativities predict a natural redshift for light propagating within atmospheres which appears to be measurable with contemporary technology, as we shall see in Sect. VII.4. These media will therefore be of primary relevance for the experiments proposed in this chapter.

Media of Types 5 and 6 represent the interior of nuclei and, as such, they will be considered in future works, jointly with the operator formulation of the isotopic relativities.

Media of Types 7, 8 and 9 represent the interior of hadrons and of stars . They will be studied in detail in future works because they need operator formulations for genuine predictions of the isorelativities. Nevertheless, we shall present in Sect. 4 certain classical interpretations of experimental data which appear to confirm that hadrons are precisely media of this type.

It is evident from the above considerations that a virtually endless variety of experiments can be formulated in most branches of physics,

[34] We here continue to use our notation whereby the $\hat{b}$ and $\hat{n}$ quantities are functions while the corresponding averages b and n are constants.

as expected, because relativities are truly at the foundations of our contemporary knowledge, with endless implications. In a situation of this type, we can evidently present in this chapter only a few representative experiments.

To begin, ***all historical experiments on Einstein's special relativity can be repeated in physical media of progressively increasing geometrical complexity, e.g., first in homogeneous, isotropic, and transparent fluids (Type 2), then in inhomogeneous and anisotropic gases (Type 4), etc.***

The deviations from the Einsteinian predictions are readily computable with the isospecial relativities. The study of these tests will therefore be left to the interested experimenter.

In this chapter we shall study a few basically new experiments, as an indication of the possibilities of isotopic techniques.

The reader should be aware that the most intriguing and promising experimental results are recently emerging from the operator formulations of the isotopic relativities in particle and other branches of physics, which we cannot possible review here.

We only mention, as an example, the studies by Animalu (1991b) on the ***application of isotopic theories to superconductivity***, in particular, via the representation of the *Cooper's pair* as a generalized bound state of two extended wavepackets in conditions of total mutual immersion, with short-range nonlocal interactions precisely of the type studied these volumes. As one can see in Animalu's paper (*loc. cit.*), there are considerable phenomenological data supporting the isotopic formulation of superconductivity. The undestanding is that a number of additional tests are needed before reaching final conclusions.

Further studies worth a mention are those by Santilli (1992) on the ***application of isotopic theories to Bose-Einstein correlations***, e.g., of particles emerging from $p-\bar{p}$ high energy reactions. As well known, such correlation is basically unresolved and can be interpreted in ordinary quantum mechanics only via semiphenomenological models, without any axiomatic background.

The studies here considered assume that the origin of correlation is precisely the historical open legacy of the ultimate nonlocality of the strong interactions due to deep mutual wave overlappings (Sect. I.1), with consequential short-range nonlocal interactions which, being intrinsically nonpotential-nonhamiltonianm, are represented via the isotopy of quantum mechanics into hadronic mechanics, that is, via the generalization of the conventional unit $\hbar = 1$, into an operator isounit

$\hat{1}$[35]. The capabilities of the isotopic theory of representing available experimental data on Bose-Einstein correlations, as compared to the manifest insufficiencies of conventional theories, has been rather rewarding (see the phenomenological fits of UA1 data in Santilli (1992)). These results evidently support other applications where nonlocal internal effects are expected.

While considering the experiments outlined in the subsequent sections, the reader should keep in mind the following fundamental law:

No mathematical or physical theory can predict the value of its own unit.

More specifically, the isosymplectic, isoaffine and isoriemannian geometries provide general characteristics to be verified by their own isounits, but they cannot possibly predict the numerical value of the characteristic b-quantities in isoseparation (1.1a).

To be even more specific in this point of fundamental experimental character, the expectation that the isospecial relativities can predict the numerical values of the characteristics b-quantities, say, for our atmosphere is equivalent to the expectation that the special relativity should predict the numerical value of the Hamiltonian for each given

[35] For the reader's convenience we briefly touched on hadronic mechanics in Eq.s (II.6.23-26), and footnotes[10, 32] of pages 94 and 281, respectively. It may be useful to outline here the main idea of the *isotopic representation of Bose-Einstein correlations*. A central point is that *the notion of correlation itself is outside the axiom of expectation value of quantum mechanics* because, for a state of two particles $|a_1, a_2>$, the expectation value is given by the familiar expression

$$< a_1, a_2 | a_1, a_2 > = < a_1 | a_1 > + < a_2 | a_2 > \in \Re .$$

For the covering hadronic mechanics, the corresponding axiom of *isoexpectation value* is given instead by

$$< a_1, a_2 \lceil a_1, a_2 > = < a_1, a_2 | T | a_1 a_2 > \hat{1} = (\textstyle\sum_{ij} < a_i | T_{ij} | a_j >) \hat{1} =$$

$$= (K_1 < a_1 | a_1 > + K_2 < a_2 | a_2 > + K_{12} < a_1 | T_{12} | a_2 >) \hat{1} \in \hat{\Re}, \ K\text{'s} \in \Re,$$

where $\hat{1} = T^{-1}$, $T = T^{\dagger} > 0$, is the isounit of the theory. As a consequence, *the correlation term $C_{12} = K_{12} < a_1 | T_{12} | a_2 >$ emerges from the very axioms of the covering isotopic theory.* The direct and quantitative interperetation of the experimental data is then reduced to the appropriate selection of the isotopic elements (Santilli (1992)). The current experiments on Bose-Einstein correlation, such as the UA1 experiments at CERN, therefore appear to have a truly fundamental chartacter for all of the studies presented in these volumes because, if treated in a scientifically constructive environment, can provide experimental evidence for the existence of nonlocal nonhamiltonian effects in the ultimate structure of matter.

phenomena and, as such, has no physical value[36].

The isorelativities provide a geometrization of media of Type 4, that is, of the atmosphere of: Earth, Jupiter, Venus, Sun, as well as other different atmsopheres existing in the Universe.

Within such a setting, the characteristic b-quantities of each of them must be computed via experiments because ***the numerical value of n_4 is the average index of refraction for the propagation of light through the atmosphere considered, and a similar situation happens for the space term n_3***.

It is evident that: these numerical values vary from atmosphere to atmosphere; they are expected to exist in a large variety of different numerical values; they cannot possibly be individually predicted by any theory; and must be measured via experiments.

VII.2: CLASSICAL TREATMENT OF RAUCH'S EXPERIMENTS ON THE ROTATIONAL SYMMETRY

We begin our experimental studies with an application of the fundamental isosymmetry to particle physics, the isorotations $\hat{O}(3)$ of Sect. III.3, as the central part of the isogalilean relativities.

Of all the existing experiments in particle physics, those most significant for the isoprotational symmetry in particular, and of the isogalilean relativities in general, are the ***interferometric tests of the rotational symmetry of thermal neutrons conducted by Rauch and his associates*** (see the review by Rauch (1981), (1983) and quoted experimental papers).

These experiments have been studied via the *operator* formulation of isotopic theories by Santilli (1981), (1989c), and (1991d)), Eder (1981) and (1983)), and others. However, it is important to show that the *classical and nonrelativistic* methods presented so far can already provide an approximate, yet quantitative and physically meaningful representation of the experimental data.

[36] The author would like to express his surprise at the fact that the most common "objection" to the isospecial relativities is that they cannot predict the numerical values of the characteristics b-quantities and, as a consequence, they are not significant because "any number fits the theory". Comments of this type are equivalent to comments such as: "Einstein's gravitation is not relevant because any gravitational mass fits the theory", or "Einstein's special relativity is not relevant because any value of the charge fits the theory", and, as such, they have no physical value.

In essence, neutrons are not massive points as represented by Galilei's (and Einstein's) relativity, but possess an extended charge distribution of the order of 1F in radius.

Such charge distribution cannot be perfectly rigid, and it is therefore expected to admit deformations under sufficiently intense external fields.

When the neutrons are represented with their actual shape characterized by the isogalilean relativities, e.g., an oblate spheroidal ellispoid (Fig. III.3.2), they become the isoparticles of Definition III.7.1, and we shall call them *isoneutrons*. All possible deformations of shape, which are also directly represented by the isogalilean relativities, then become the simplest possible *mutations* of the intrinsic characteristics.

A physical consequence is that these expected deformations imply the necessary mutation of the intrinsic magnetic moment of the particle, with fundamental mathematical, theoretical and experimental relevance for this analysis (See Fig. VII.2.1 for more details).

EXPECTED DEFORMABILITY OF HADRONS / MUTATION OF INTRINSIC MAGNETIC MOMENTS

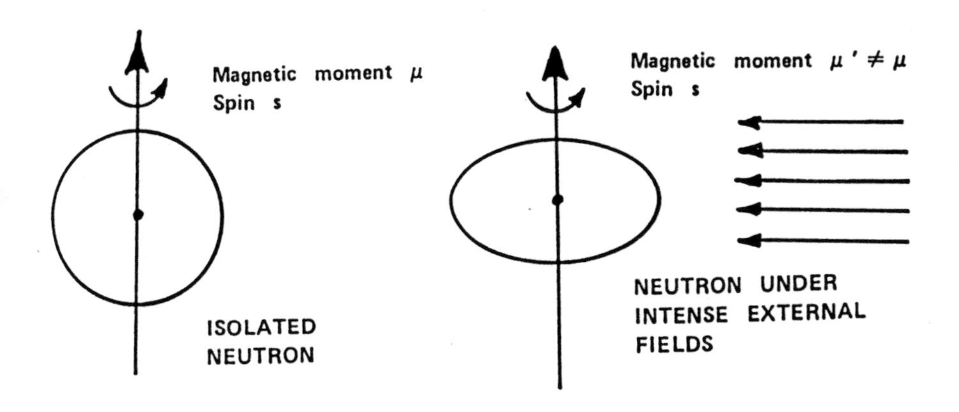

FIGURE VII.2.1. A schematic view of a fundamental application of the isogalilean relativity: the direct, quantitative, characterization of the deformation of shape of extended charge distributions under sufficiently intense external fields, and the consequential alteration of their intrinsic magnegic moments. Consider, as a simple example, a classical, charged and spinning sphere with a given magnetic moment μ. Suppose now that the shape of the sphere is deformed as in the figure. Then, this implies the *necessary* alteration of the magnetic moment μ into a new value μ' depending on the physical conditions considered (original value μ, original size, value of the angular momentum, intensity of the external forces on the sphere, etc.). This physical occurrence can be easily proved via the use of the ordinary, classical, Maxwell's electrodynamics. The same occurrence evidently persists in quantum mechanics, as proved in atomic and nuclear physics. In fact, the deformation of shape of an atom or of a nucleus causes a necessary alteration of their magnetic moments, as experimentally established. The same occurrence has been experimentally tested for elementary particles by Rauch and his collaborators ((1981), (1983)) for neutrons under external nuclear fields, resulting in an apparent measurement of deformation of shape-alteration of the magnetic moment which is however only preliminary at this writing. In this section we provide a first, rudimentary, classical and nonrelativistic description of Rauch's fundamental tests, and in the final section we propose specific additional experiments for the final resolution of the issue.

With reference to Figure VII.2..2, a neutron interferometer is essentially constituted by a neutron beam which is first subjected to a coherent splitting into two branches via a perfect crystal, and then their recombined. The neutron beam is generally monochromatic, unpolarized and with high flux. The perfect crystal is generally given by a Si crystal with extremely low impurities which allows the achievement of angles of separation of the two branches sufficiently wide to permit experiments in one branch or in both.

In his experiments, Rauch used: a thermal neutron beam with a cross section of about 2×1.4 mm^2; a characteristic wavelength of the crystal of 1.83 A°; about 1 cm of electromagnetic gap; and a magnetic field of the intensity of 7,496 G which is calibrated to produce two, complete and exact spin-flips, say, around the third axis ($\theta_3 = 720°$), for neutrons with their conventional magnetic moment

$$\mu_n = - 1.91304211 \pm 0.0000011 \quad 2\hbar/2m_p c_o. \qquad (2.1)$$

The experimenters filled up the electromagnet gap with Mu-metal sheets for the primary purpose of reducing stray fields. It is this latter, rather accidental, feature that renders the experiments truly fundamental, inasmuch as Rauch's measures test the rotationals ymmetry of the neutrons under external magnetic and nuclear interactions.

RAUCH'S NEUTRON INTERFEROMETRIC EXPERIMENTS ON THE ROTATIONAL SYMMERY

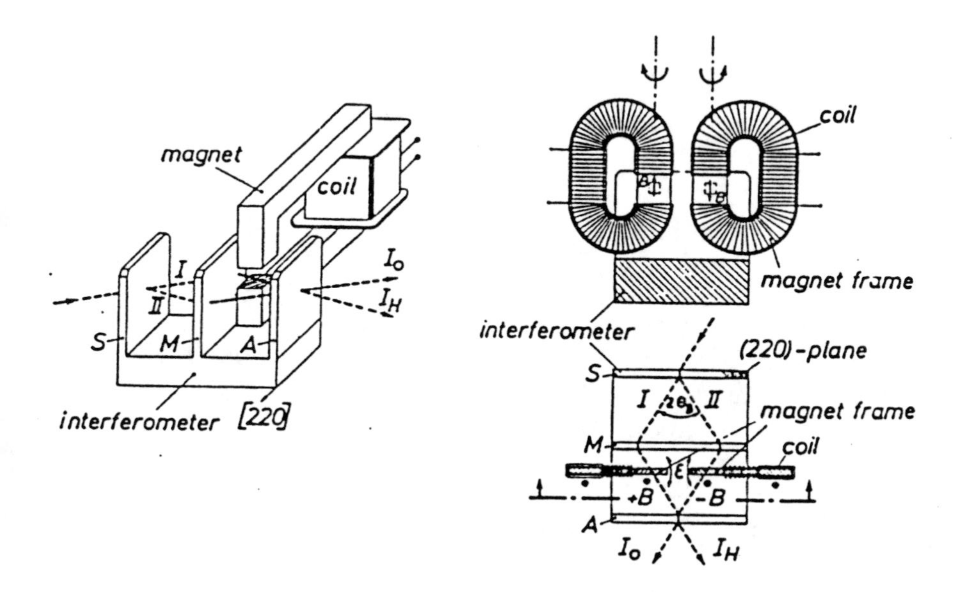

FIGURE VII.2.2: A schematic view of Rauch's fundamental experiments on the rotational symmetry of neutrons via an interferometer, showing the coherent splitting and recombination of the thermal neutron beam via a perfect crystal. In essence, Rauch applied to one branch of the beam a magnetic field originating from an electromagnet in a preset intensity (see the text) which should have produced two complete and exact spin flips for a rotation of 720°. An air gap is shown in the figure, although the experimenters filled up the magnet gap with Mu-metal sheets to reduce stray fields. This effectively produced a joint interaction of the neutron beam with the magnetic field of the electromagnet as well as with the nuclear fields of the Mu-metal sheets. In all the tests, rather than finding the expecteed 720°, Rauch found instead a median angle consistently *lower* than 720°, with the last best measures (2.2) not including the 720° in the minimal and maximal experimental values (2.2b). As well known, contemporary experiments are generally done on the center-of-mass system (as for inelastic scattering experiments) and result to be consistently in favor of conventional quantum mechanics. This is readily predictable via the notion of closed nonhamiltonian system (III.2.3) for which no detection of the generalized internal structure can beachieved via outside, center-of-mass tests. Actual measures of physical quantities (e.g., magnetic moments) are solely done via *external, long range electromagnetic fields* . Rauch's experiments are of fundamental character inasmuch as they are the firsts to conduct measures under *external, short range, nuclearinteractions* , namely of a configuration exactly as needed to test interior dynamical conditions.

The best available measures are given by (Rauch (*loc. cit.*))

$$\hat{\theta}_3 = 715.87^\circ \pm 3.8^\circ, \tag{2.2a}$$

$$\hat{\theta}_3^{max} = 719.67^\circ, \quad \hat{\theta}_3^{min} = 712.07^\circ, \tag{2.2b}$$

It should be immediately indicated that *the above measures do not establish the violation of the rotational symmetry because the deviation should be of the order of four to five times the statistical error to achieve a sufficient degree of confidence.* Thus, to establish the value 715.87°, the error should be of the order of $\pm$ 1° to $\pm$0.7°.

Despiote this unsettled nature, the implications of the above measures are intriguing because:

1) Measures (2.2) do not confirm the exact rotational symmetry for the neutrons in the open conditions considered, by indicating a conceivable violation of about 1%.

2) None of the median angles measured by Rauch coincide with 720°. On the contrary, all experiments show a median angle consistently lower than 720°, an occurrence called "angle slow-down effect" (Santilli (1981)).

3) The measurements of the intensity modulation for measures (2.2) do not confirm the exact rotational symmetry because the modulation curve is not an exact co-sinusoid, as well as for other reasons.

Measures (2.2) are therefore valuable for the following reasons. First, the sole possible origin for an angle $\hat{\theta}_3$ different than the expected 720° is the alteration of the magnetic moment (2.1), i.e., the mutation of Fig. VII.2.1

$$\mu_n \Rightarrow \hat{\mu}_n, \tag{2.3}$$

In turn, such an alteration of the magnetic moment can best occur under the deformation of shape indicated earlier.

Furthermore, Rauch's measures have an important mathematical and theoretical value. As stressed earlier, the *amount* of the deviation under Rauch's conditions is scientifically open at this time. Nevertheless, the *deformability* of the charge distribution of the neutron remains out of any scientific doubt and so is its consequential mutation of the magnetic moment.

It is therefore important to achieve a quantitative representation of Rauch's experiments by leaving open the finalization of the numerical amount of mutation for given conditions to future experiments.

We therefore pass to a representation of Rauch's experiments via the isogalilean relativities which, even though only classical and nonrelativist (and therefore necessarily approximated) it is nevertheless direct and quantitative.

Recall that there are no appreciable "contact interactions" between the neutron beam and the Mu-metal nuclei. Thus we can effectively use the case of extended and deformable isoparticles under external *potential* forces only (see Sect. III.7).

Suppose that $T^*E(r,\delta,\Re)$ is the conventional phase space of the neutron beams. Then, the isogalilean relativity uniquely follows by assuming the isospace

$$T^*\hat{E}(r,\hat{\delta},\hat{\Re}): \ \hat{\Re} = \Re\, \hat{1}, \quad \hat{1}_1 \equiv \hat{1}_2 \equiv \hat{1} = \mathrm{diag.}\,(\hat{\delta}^{-1}, \hat{\delta}^{-1}), \tag{2.4a}$$

$$\hat{\delta} = \mathrm{diag.}\,(b_1^{\,2}.\ b_2^{\,2},\ b_3^{\,2}), \quad b_k = \mathrm{constants} > 0. \tag{2.4b}$$

and the representation via the Hamilton-isotopic equations (III.2.9)

$$\dot{a}^{\mu} = \omega^{\mu\alpha}\, \hat{1}_{2\alpha}{}^{\nu}(a)\, \frac{\partial H(a)}{\partial a^{\nu}} = \begin{cases} \dot{r}_i = b_i^{-2}\, \dfrac{\partial H(r,p)}{\partial p_i}, & (2.5a) \\[2ex] \dot{p}_i = -\, b_i^{-2}\, \dfrac{\partial H(r,p)}{\partial r_i}, & (2.5b) \end{cases}$$

with Hamiltonian

$$H = \hat{p}^2 / 2m_n + V = p_k b_k{}^2 p_k / 2m_n + V(r), \tag{2.6a}$$

$$r = | r_i b_i{}^2 r_j |^{\frac{1}{2}}, \tag{2.6b}$$

where V(r) represents the external electromagnet and nuclear fields (see Eder (1981) and (1983) for their detailed study).

Our treatment is independent from the explicit form of V(r) and essentially based on the assumption that the application of the external field implies a deformation of the shape representable via our isotopies of the Euclidean space

$$\{ \delta = \text{diag.}\ (1, 1, 1),\ V = 0 \} \Rightarrow \{ \hat{\delta}' = \text{diag.}\ (b'_1{}^2, b'_2{}^2, b'_3{}^2),\ V \neq 0 \}, \tag{2.7}$$

under the evident condition of being volume preserving,

$$b_1{}^2\, b_2{}^2\, b_3{}^2 = b'_1{}^2 b'_2{}^2 b'_3{}^2, \tag{2.8}$$

Since we have at best a small deviation, it is reasonable to assume that the mutation of shape is also small. In first approximation, we have from data (2.2) that the deviation is of the order of

$$716°/720° \cong 0.9944, \tag{2.9}$$

which can be assumed to be of the order of magnitude of the oblateness caused by the external nuclear fields (Figure VII.2.1).

Then, our purely classical nonrelativistic treatment implies that mutation (2.7) for values (2.9) under condition (2.8), assumes the explicit form

$$\{ \delta = \text{diag.}\ (1,1,1),\ V = 0\} \Rightarrow \{\hat{\delta}' = \text{diag.}\ (1.0028,\ 1.0028,\ 0.9944),\ V \neq 0 \}. \tag{2.10}$$

We then have a consequential mutation of the magnetic moments of the order of 6×10^{-3}, i.e.,

$$\mu_n \approx -1.913\ e\hbar/2m_p c_o \Rightarrow \hat{\mu}_n \approx -1.902\ e\hbar/2m_p c_o, \tag{2.11}$$

which does indeed provide a first, approximate, but quantitative interpretation of Rauch's data (2.2).

Intriguingly, we are not only in a position to represent measures (2.2), but also the "angle slow-down effect" (Santilli (1981)), namely the fact that the median angles measured by the experimenters during the several years of the conduction of the tests have been consistently lower than the needed 720°.

In fact, mutated value (2.11) is *lower* than the original value (2.1), thus implying angles of spin-flips necessarily lower than 720° for a magnetic field of 7,946 G.

Moreover, the conventional rotational symmetry is evidently broken for values (2.2). Nevertheless, *our isogalilean relativities reconstruct the exact rotational symmetry for the deformed neutrons.* This is another aspect that we believe needs an identification, first, at the primitive Newtonian level, and then at the operator counterpart.

Furthermore, the true symmetry tested by Rauch at the particle level is the spinorial SU(2) symmetry, rather than the O(3) symmetry. Nevertheless, we believe that the issue deserves an analysis, first, within the context of the rotational symmetry O(3), and prior to a study within the covering SU(2) extension, in order to separate the rotational from the spinorial contribution.

For this purpose, consider the subgroup of $\hat{G}(3.1)$ given by our covering isorotational symmetries $\hat{O}(3)$ of Sect. III.3. As now familiar, the isotopes $\hat{O}(3)$ provide the form-invariance of all possible ellipsoidical deformations of the sphere, while being locally isomorphic to the conventional rotational symmetry O(3). This establishes the reconstruction of the exact rotational symmetry for deformed charge distributions (2.10), of course, at our isotopic level.

However, the mechanism of such reconstruction deserves a deeper inspection because important for Rauch's experiments.

Consider our isorotation around the third axis, i.e.,

$$\hat{O}(3): \quad r' = \hat{R}(\theta) * r = \{\exp \theta_3 \, \omega^{\mu\sigma} \, \hat{1}_\sigma{}^\nu \, (\partial_\nu J_3) \, (\partial_\mu)\} \, r, \tag{2.12a}$$

$$\hat{1} = \text{diag.} \, (\hat{\delta}'^{-1}, \hat{\delta}'^{-1}), \tag{2.12b}$$

where $\hat{\delta}'$ is that of Eq.s (2.4), which is explicitly given by Eq. (III.3.42), i.e.,

$$r' = \hat{R}(\theta_3) * r = S_\delta(\theta_3) \, r = \tag{2.13}$$

$$\begin{pmatrix} r'_1 \\ \\ r'_2 \\ \\ r'_3 \end{pmatrix} = \begin{pmatrix} r_1 \cos(\theta_3 b_1 b_2) - r_2 \frac{b_2}{b_1} \sin(\theta_3 b_1 b_2) \\ \\ r_1 \frac{b_1}{b_2} \sin(\theta_3 b_1 b_2) + r_2 \cos(\theta_3 b_1 b_2) \\ \\ r_3 \end{pmatrix}.$$

The reconstruction of the exact rotational symmetry is then based on mechanism (II.3.43) originating from the values b_1 and b_2 of Eq. (2.10) and Rauch's median angle (2.2), i.e.,

$$\hat{\theta}'_3 = b_1 \, b_2 \, \hat{\theta}_3 \Big|_{\substack{\theta_3 = 716^\circ \\ b_1 = b_2 = 1.0028}} \cong 720^\circ, \tag{2.14}$$

namely, *our geometric isospace $\hat{E}(r,\hat{\delta},\hat{\Re})$ reconstructs the angle θ'_3 = 720° needed for the exact symmetry from an actual rotation of $\hat{\theta}_3$ = 716° in our physical space $E(r,\delta,\Re)$.*

In conclusion, irrespective of whether Rauch's measures (2.2) are confirmed or adjusted by future experiments, in this section we have shown that the covering isogalilean relativities, at its primitive classical level, can:

a) directly represent the actual shape of the neutron;

b) directly represent all possible deformations of said shape caused by sufficiently intense external fields and/or collisions;

c) directly represent the consequential mutation of the intrinsic magnetic moment of the particle;

d) directly represent the apparent "angle slow-down effect" because of the decreased value of the magnetic moment, and

e) reconstruct the exact rotational and Galilei symmetries at the more general isotopic level.

Not surprisingly, the operator and relativistic treatments (see Santilli (1989c) and (1991d)) confirm in full the above rudimentary, classical and nonrelativistic results.

The reader should keep in mind the societal, let alone physical impocations. In fact, the large efforts on the controlled fusion have been conducted until now via the use of the *magnetic confinement* which, in turn, is based on the assumption that the Galilean-Einsteinian neutrons and protons preserve their intrinsic magnetic moment in vacuum under the fusion conditions.

But the large public expenditures done on controlled fusion until now have not produced the desired results, decade after decade, thus warranting a reinspection of the ultimate theoretical structure in the field: the Galilean-Einsteinian notion of particle.

If Rauch's experiments on the rotational asymmetry of neutrons are confirmed, they would establish that protons and neutrons experience a mutation of their intrinsic magnetic moments at the time of initiation of the fusion process which, in turn, would invalidate the engineering design of magnetic confinement, let alone its practical realization .

VII.3: ISOTOPIC BEHAVIOR OF THE MEANLIFE OF UNSTABLE HADRONS WITH SPEED

As reviewed in Sect. IV.3, the nonlocal and nonhamiltonian effects expected in the *interior* of hadrons from the historical legacy of Fermi, Bogolubov and others, are expected to have no manifestation in the *exterior* . In particular, *the center-of-mass of a hadron in a particle accelerator must strictly obey Einstein's special relativity, irrespective of whether the particle is stable or unstable, as experimentally established*

Nevertheless, a number of phenomenological studies, including those by Blockhintsev (1964), Redei (1966), Kim (1978), Nielsen and Picek (1983) and others, provide serious arguments on ***the possibility that an internal inapplicability of Einstein's special relativity could manifest itself in the exterior of unstable hadrons via a departure from the Einsteinian behavior of the meanlife with speed***

$$\tau = \eta \tau_0 , \quad \gamma = (1 - \beta^2)^{-\frac{1}{2}}, \quad \beta = v / c_0, \qquad (3.1)$$

and produced several quantitative predictions of expected deviations.

As a result of these efforts, two experiments were subsequently performed on the K°-system, one by Aronson *et al.* (1983), and the second by Grossman *et al.* (1987). The experimental situation at this writing (Fall 1991) is the following):

A) ***There is no significant deviation in the behavior of the K°_S meanlife in the low energy range 0 to 30 GeV;***

B) ***There is an anomalous behavior of the meanlife from 35 and 100 Gev*** (Aronson *et al.* (1883)); and

C) ***The meanlife of the K°_S returns to behave conventionally between 100 and 350 GeV*** (Grossman *et al.* (1987)).[37]

As pointed out in Sect. IV.4 and IV.10, the isominkowski spaces (1.1) and isospecial relativities appear to be ideally suited for a representation of the interior of hadrons. In fact, they imply the *isotopic meanlife* (Postulate IV of Sect. IV.9)

$$\hat{\tau} = \hat{\gamma}\,\tau_o, \quad \hat{\gamma} = \left|\, 1 - \frac{\hat{b}_3^{\,2}}{\hat{b}_4^{\,2}}\,\beta^2 \right|^{-\frac{1}{2}}, \quad \beta = v \,/\, c_o, \tag{3.2}$$

which has been proved to be "directly universal" (Aringazin (1989)), that is, capabble of recovering all individual generalized meanlives suggested by the above quoted authors via different power series expansions, different truncations and differemnt coefficients (see Sect. IV.3).

In particular, isotopic law (3.2) can be used either with characteristics b-constants or, depending on the desired approach, with characteristic $\hat{b}$-functions, in which case the most significant dependence is expected to be that in the velocity or, equivalently, in the energy.

This isorelativistic representation of unstable hadrons was studied by Cardone *et al.* in two subsequent papers (1992a) and (1992b) with rather encouraging results. The first study deals with the isotopic

[37] The author has been surprised by several authoritative claims that the behavior of the meanlife is Einsteinian because it is so between 100 to 350 GeV. But then, one could equally state that the behaviour is noneinstenian because it is so between 35 to 100 GeV. The only true scientific claim which can be stated today is that the final behaviour of the meanlife is unknown, and must be resolved by future experiments.

representation of the experimental data by Aronson *et al.* (1983) and it is summarized in Figures VII.3.1, VII.3.2 and VII.3.

LINEAR EINSTEINIAN FIT OF THE DATA BY ARONSON ET AL. ON THE $K^o{}_S$ MEALIFE

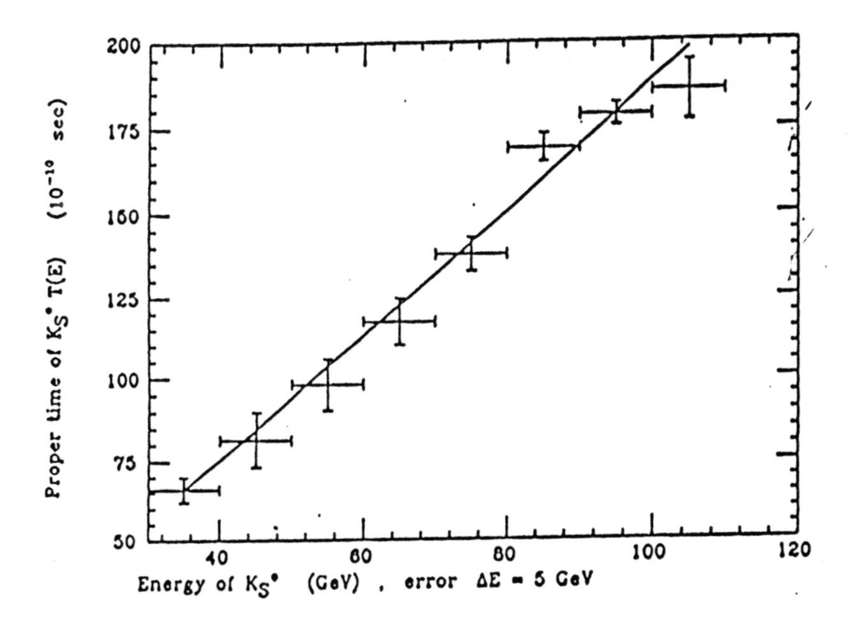

FIGURE VII.3.1: The diagram presents a linear fit on the experimental data by Aronson *et al.* (*loc. cit.*) on the behavior of the meanlife with speed via the Einsteinian law (3.1), done by Cardone *et al.* (1992a).. The fit parameter is a = τ_0/m_0; the reduced chi-square is χ_n^2 = 0.9. The value found for τ_0 is (0.9375 ± 0.0021) x 10^{-10} s, which is sensibly different than the value of the same meanlife given by the Particle Data (0.8922 ± 0.0021) x 10^{-10} s; the confidence level is 0.39 giving a probability of 61% that the meanlife τ_0 at rest is greater than the true value. It is therefore concluded that a linear fit with the Einsteinian law (3.1) is not satisfactory. In particular, a nonlinear dependence of the meanlife on the speed (or, more precisely, on the energy E) is needed.

FIT OF THE DATA BY ARONSON ET AL. ON THE MEANLIFE OF THE $K^o{}_s$ WITH A POWER LAW IN THE ENERGY

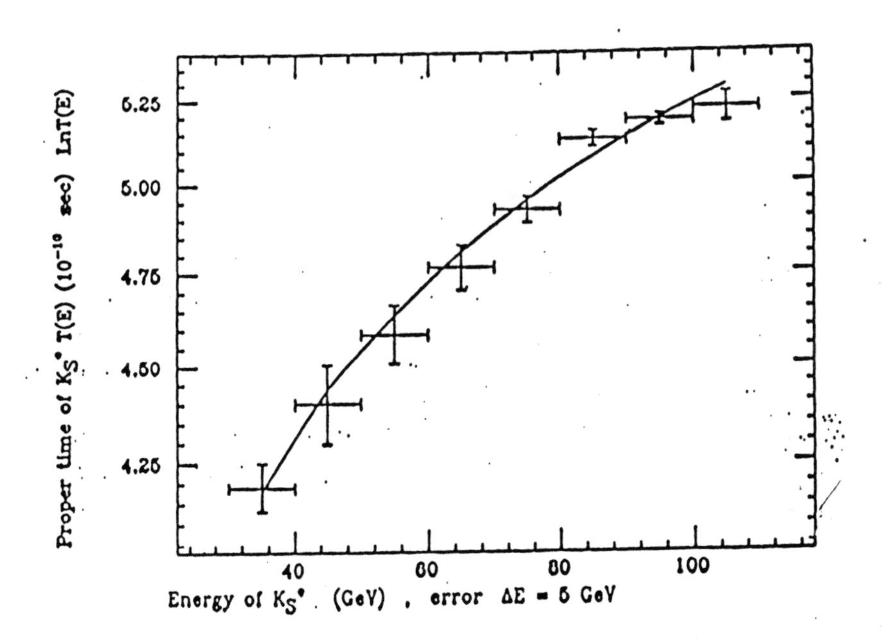

FIGURE VII.7.2: Logarithmic fit of the experimental data by Aronson *et al.* (*loc. cit.*) on the $K^o{}_s$ meanlife via the power law

$$\tau = \frac{\tau_o}{m_o} E^n \qquad (3.3)$$

done by Cardone *et al.* (1992a). The parameters are $a = \tau_o/m_o$ and n. In this second case $\chi_n{}^2 = 0.85$, n derived from the data is $n = 1.013 \pm 0.003$, and the deviation from linearity is 1.3% at 100 GeV. The fit is still insufficient.

ISOTOPIC FIT OF THE DATA BY ARONSON ET AL. ON THE $K^o{}_S$ MEANLIFE

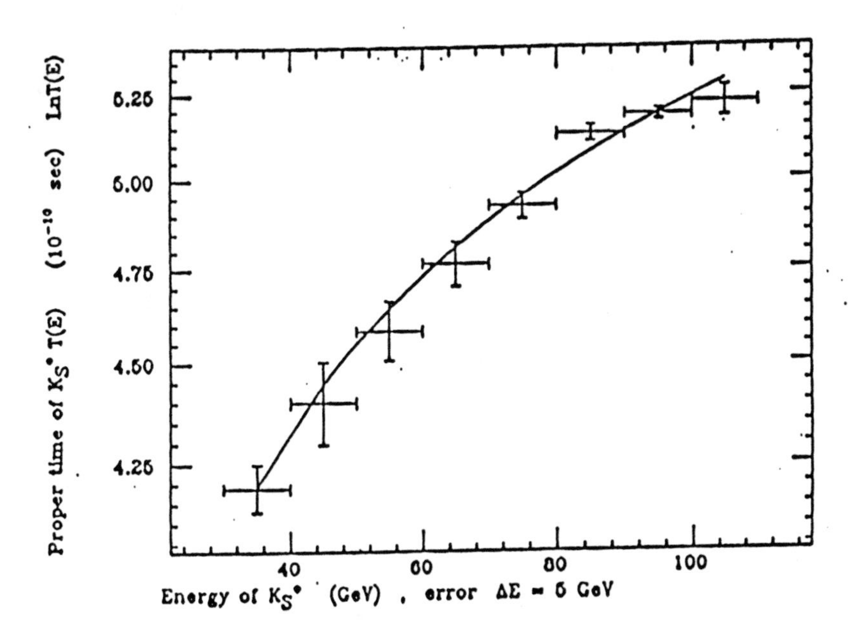

FIGURE VII.3.3: Logarithmic fit of the experimental data by Aronson *et al.* (*loc. cit.*) on the behavior of the $K^O{}_S$ meanlife between 30 and 100 GeV via the use of isotopic law (3.2), done by Cardone *et al.* (1992a). The fit parameter is

$$a = b_3^2 / b_4^2 = 0.898 \pm 0.021 \tag{3.4}$$

with $\chi^2 = 0..86$. Additional fits compatible with the above ratio yield

$$b_3^2 = 0.9023 \pm 0.0004, \quad b_4^2 = 1.003 \pm 0.0021. \tag{3.5}$$

The improvement of the fit over the preceding ones is evident. The fit is however, still an approximation because of the expected energy-dependence of the characteristic b-quantities, thus requiring the use of the full isotopic law (3.2) with functional dependence on the b-quantities.

We now pass to the representation of the data on the K°_S meanlife by joining those from Aronson *et al.* (1983) and by Grossman *et al.* (1987), as done by Cardone et al. (1992b). The first point regards the capability of law (3.1) to represent the correct *meanlife of the particle at rest* as given by the Particle data book.

As well known, *meanlifes are not measured at rest, but at various energies, and then scaled at rest via the use of the Lorentz transformations*. The possible lack of exact character of the Lorentz transform,ation then directly implies the possibility that the meanlifes at rest as current given need inspection.

With reference to the diagram of Fig. VII.3.4, the use of the Einsteinian law (3.1) yields the meanlife at rest of the K^{o}_S $\tau_o = (0.91444 \pm 0.00193) \times 10^{-10}$ s, with $\chi 2 = 1.23$. The confidence level is 19% with a probability of 81% that the actual value of τ_o is greatere than that given by the fit.

Moreover, by denoting with $\sigma^f_{\tau_o}$ the error of the fitted vaue and with σ_{τ_o} that of the Particle Data , Cardone *et al.* (1992b) obtain

$$\sigma^f_{\tau_o} = 0.00193, \quad \sigma_{\tau_o} = 0.0020, \tag{3.7a}$$

$$\Delta \tau_o = \tau_o^f - \tau_o = 0.022 = 11.1\, \sigma^f_{\tau_o}, \tag{3.7b}$$

namely, ***the einsteinian law fails to predict the correct value of the meanlife of the K^o_S at rest***.

The results for the case of the covering isotopic law are given in Figure VII.3.4.

Note that in the transition from the fit of the data by Aronson *et al.* (Figure VII.3.3) to that for the data by Grossman *et al.* (Figure VII.3.4) the values of the characteristic b-quantities changes minimally, Eq.s (3.5) and (3.9).

An inspection of the data by Cardone *et al.* (*loc. cit.*) indicates that ***the K^o_S is a medium of (Clas I) Type 9*** for which

$$\text{TYPE 9:} \quad n_3 > n_4,\ n_4 < 1; \quad \hat{\beta} < \beta,\ \hat{\gamma} > \gamma; \quad V_{Max} > c_o,\ c > c_o, \tag{3.8}$$

In turn, this is in agreement with the isotopic interpretation of all preceding phenomenological studies, such as those by Nielsen and

ISOTOPIC FIT OF THE DATA BY ARONSON ET AL. AND GROSSMAN ET AL. ON THE $K^o{}_S$ MEANLIFE

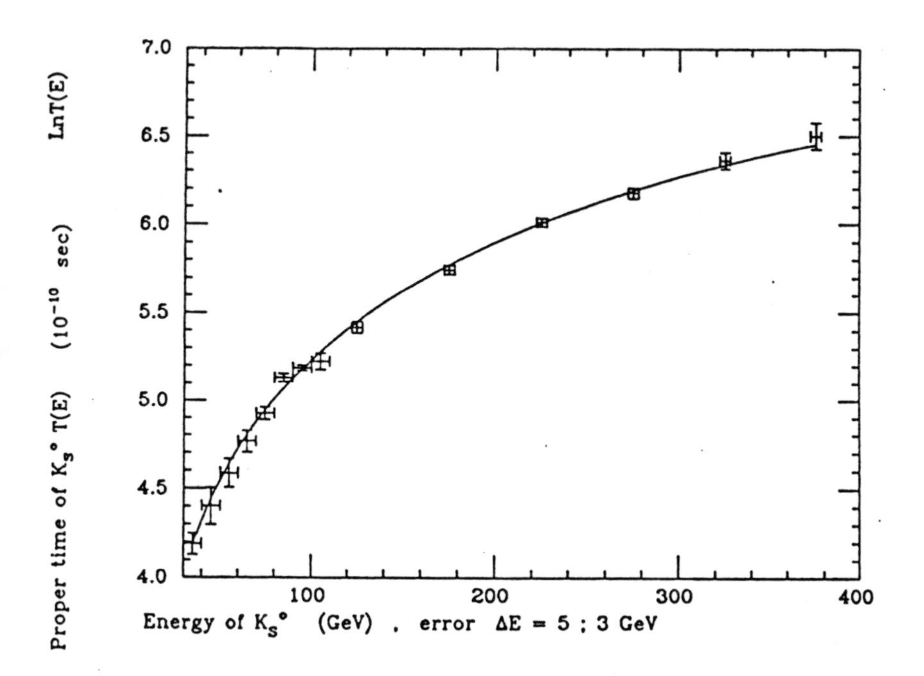

FIGURE VII.3.4: The fit of the experimental data by Aronson *et al.* (loc. cit.) and by Grossman *et al.* (loc. cit.) on the $K^\circ{}_S$ meanlife via isotopic law (3.2), as done by Cardone *et al.* (1992b). The objective is to see whether the two data are compatible under a suitable statistical weight. The parameters are assumed to be b_3, b_4 and the difference $\Delta = b_4^2 - b_3^2$. The results of the fit are the following:

$$\chi^2 / n = 0.71, \quad \Delta = b_4^2 - b_3^2 = (3.926 \pm 0.002) \times 10^{-7}, \qquad (3.9a)$$

$$b_3^2 = 0.009080 \pm 0.00004, \quad b_4^2 = 1.002 \pm 0.007. \qquad (3.9b)$$

The diagram therefore shows the remarkable capability by the isospecial relativities of achieving full compatibility between the seemingly contrasting data on the meanlife behavior with energy. Moreover, the fits indicate that

$$\Delta b_3^2 = 0.007 = 170\, \sigma_{|b_3|}, \qquad (3.10a)$$

$$\Delta b_4^2 = 0.001 = 0.50\,\sigma_{|b_4|}, \qquad (3.10b)$$

which provide additional support for the Lie-isotopic theory in the range 30-400 GeV, because b_4 changes very littel with energy, while b_3 varies considerably. This is also predicted by the isospecial relativities, because b_4 (n_4) geometrizes the medium inside the K^O_S, while the dependence on the energy is expected to be assumed by the space part b_3 (or n_3). Finally note that ***numerical values (3.9) are valid specifically for the K^O_S, and that different numerical values are expected for different hadrons***. As indicated in Chapter IV, this is expected from the fact that hadrons have approximately the same size, thus resulting in different densities for different hadrons. In turn,different densities necessarily imply different values of the characteristic b-quantities. Despite these differenmces and with the sole known exception of the pions, ***the characteristic b-quantities of all hadrons are expected to belong to the same geometrization of Class I, Type 9*** (see text).

Picek (1983) (see Eq.s (10.9)), De Sabbata and Gasperini (1982) (see Eq.s (9.11)) and others, to the effect that ***all hadrons beginning from the kaons on are expected to be constituted by the most general possible medium predicted by the isospecial relativities, those of Type 9.***

In conclusion, as anticipated in Sect. VII.1, the most significant experimental data provided by the phenomenological studies by Cardone *et al.* (1992a, b) are the numerical values of the characteristic b-quantities for the K^O_S, and the consequential identification of the type of physical media from classification (VII.1.5).

In fact, a possible conformation of these data, with the consequential finalization of the medium indicated, would permit truly fundamental, novel possibilities, such as: the representation of hadrons as generalized bound states (closed nonhamiltonian systems verifying the isogalilean or the isopoincaré symmetry); the possible achievement of a true quark confinement with an identically null probability of tunnel effect for free particles with fractional charges; the possible identification of quarks with conventional physical particles whose characteristics are mutated because of the internal nonlocal and nonhamiltonian effects; the achievement of an unambiguous isotopic quantization of gravity; the study of iso-grand-unification; and other possibilities indicated earlier.

VII.4: ISODOPPLER REDSHIFT FOR QUASARS AND PLANETARY ATMOSPHERES

As recalled in Sect. IV.9, Postulate IV and isodoppler's law (IV.9.48) were submitted in Santilli (1988c) for the intent of avoiding the violation of Esteinian laws by quasars' under Einsteinian conditions (motion in vacuum at speeds higher than c_o).

Recall that quasars are very massive and bright bodies which, as such, are expected to have an atmosphere as any other similar astrophysical body; light is emitted in the quasars structure; it propagates first within the quasars' inhomogenous and anisotropic atmospheres; and then propagates through intergalactic distances to reach us.

The hypothesis sumbitted in Santilli (*loc. cit.*) is that the currently measured quasars' redshift is first caused by propagation of light within the quasars' atmosphere according to the *isotopic redshift law*

$$\hat{\omega} = \hat{\gamma}\,(1 - \hat{\beta}\cos\alpha), \quad \hat{\beta} = \frac{\hat{b}_3}{\hat{b}_4}\,\beta, \quad \hat{\gamma} = |\,1 - \frac{\hat{b}_3{}^2}{\hat{b}_4{}^2}\,\beta^2\,|^{-\frac{1}{2}}, \qquad (4.1)$$

and then by propagation of light through intergalactic distances according to the *Einsteinian redshift law*

$$\omega' = \eta\,(1 - \beta\,\beta\cos\alpha), \quad \beta = v\,/\,c_{ot}, \quad \gamma = (1 - \beta^2)^{-\frac{1}{2}}, \qquad (4.2)$$

due to the speed of the quasars.

The latter one can be decomposed into two contributions, one due to the expansion of the galaxy associated to the quasar, and the second to the expulsion of the quasars from the said galaxy (see Fig. VII.4.1 for details).

Mignani (1992) conducted explicit calculations along these proposals by computing for the first time numerical estimates for the characteristic b-quantities of quasars' atmospheres.

Mignani essentially assumed, as a first approximation, that quasars are at rest with respect to the associated galaxy, in which case the difference between their neasured redshift and that of the associated galaxy is entirely of isotopic origin.

Under this assumption, he identified the following expression for the ratio b_3/b_4 , where the $\hat{b}$'s are now assumed to be averaged to constant b's,

$$B = \frac{b_3}{b_4} = \frac{(\omega'_1 + 1)^2 - 1}{(\omega'_1 + 1)^2 + 1} \times \frac{(\hat{\omega}_2 + 1)^2 - 1}{(\hat{\omega}_2 + 1)^2 + 1}, \quad (4.3)$$

where ω'_1 represents the measured Einstenian redshift for galaxies according to law (4.2), and $\hat{\omega}_2$ represents the isotopic redshift for quasars according to law (4.1).

ISOTOPIC ORIGIN OF QUASARS' REDSHIFTS

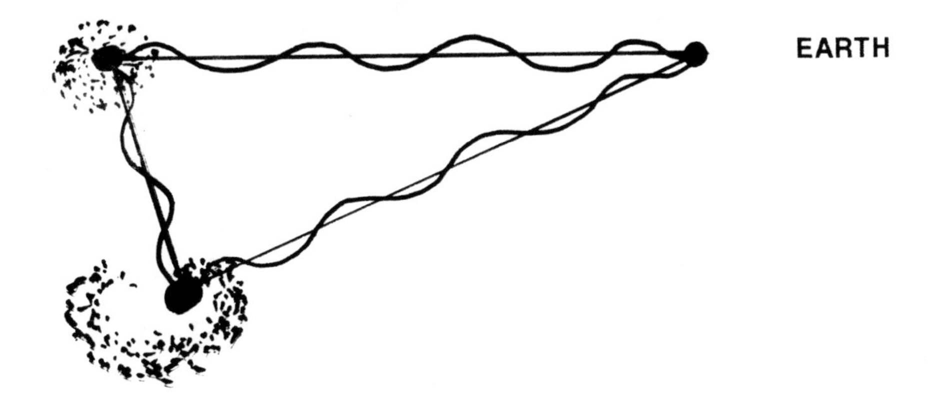

Figure VII.4.1:The Einsteinian interpretation of recent quasars' redshifts implies speeds in excess of the speed of light in vacuum, and actually of the order of 10 c_0 and more. In particular, these speeds originate in the expulsion of quasars from their associated galaxies, as computed via the Einsteinian interpretation of the difference between the redshift of the quasars and that of the associated galaxies. The isodoppler law (4.1) was submitted by Santilli (1988c) to avoid the violation of Einsteinian laws under Einsteinian conditions (motion in vacuum at speeds in excess of c_0). The main argument is essentially the following. Quasars are extrmely massive and bright objects. As such, they are expected to possess an atmosphere as any other similar astrophysical body. But all atmospheres are inhomogeneous and anisotropic, as well known and, as such, at variance with the geometrical structure at the foundations of the Einsteinian Doppler's law. The amount of deviation from the Einsteinian law (4.2) in the conditons considered is evidently debatable at this time. But ***the lack of exact applicability of law (4.2) within inhomogeneous and anysotropic media should be out of scientific doubts***. The submitted hypothesis assumed that the ***quasars' light experiences a redshift during the propagation within such atmospheres and prior to reaching empty space*** . This would decrease the relative speed between quasars and associated galaxies down to such values to avoid the violation of Einsteinian laws under Einsteinian conditions. The above use of the isodoppler's law was submitted as a small correction to the expectedly primary Einsteinian redshift due to the expansion of the Universe. Studies subsequent to Santilli (1988c) have shown, see later on Eq.s (4.9) and (4.10), that the isodoppler contribution can be numerically higher than that of the basic expansion of the galaxies, thus rendering unsettled a number of aspects of current cosmology (e.g., the distance of the quasars). The original proposal in Santilli (1988c) also suggested ***an additional isotopic correction due to propagation of light in intergalactic distances*** . In essence, space can be considered "empty" for distances of the order of our Solar system. At intergalactic distances, space becomes a medium filled up with radiation, particles, dark matter, etc. This renders conceivable a further isotopic interpretation of the redshift of both quasars and galaxies. This latter possibility shall not be considered here for brevity, but it is hoped that experts in the field will consider it because, if experimentally confirmed, can evidently imply corrections on our current views on the distance of the galaxies themselves and the dimension of the known Universe.

From known astrophysical data, Mignani (*loc. cit.*) then computed the following numerical values

GAL.	ω'_1	QUASAR	B	$\hat{\omega}_2$
NGC	0.018	UB1	31.91	0.91
		BSOI	20.25	1.46
NGC 470	0.009	68	87.98	1.88
		68D	67.21	1.53
NGC 1073	0.004	BSO1	198.94	1.94
		BSO2	109.98	0.60
		RSO	176.73	1.40
NGC 3842	0.020	QSO1	14.51	0.34
		QSO2	29.75	0.95
		QSO3	41.85	2.20
NGC 4319	0.0056	MARK205	12.14	0.07
NGC 3067	0.0049	3C232	82.17	0.53

(4.4)

In summary, according to the above model, the Einsteinian expansion of the Universe is unchanged, because it is that of the Galaxies, but the violation of Einstein's special relativity by speeds higher than c_o in vacuum is eliminated.

Needless to say, values (4.4) are only preliminary and in need of additional studies because of the possibility that the quasars are indeed expelled by the galaxies but at Einstenian speeds.

However, these latter possibilities can at best imply small corrections of the characteristic B-values and, as such, they have no major impact on the analysis of this section.

Despite these unsettled aspects, values (4.4) are significant because they constitute the first numerical values for the characteristic B-quantity of interior physical media reached in the literature. As such, they have particular referral value, e.g., for the experiments proposed below in this section.

As an incidental note, the isotopic interpretation of the quasars' redshift is not unique, and other attempts have been made to avoid noneinsteinian quasars' speeds (see, e.g., Arp *et al.* (1990) and quoted papers). These latter attempts, however, have to retort to rather unusual quantum mechanical and other assumptions (such as quantum mechanical polarization of the vacuum and the like). By comparison, our hypothesis is based on macroscopic physical characteristics (the inhomogenuity and anisotropy of the quasars' atmospheres), and provides a more direct, quantitative and plausible interpretation, as the reader is encouraged to verify.

Also, the "direct universality" of isodoppler's law (4.1) (Aringazin (1989)) should be kept in mind because it implies, as particular cases via different expansions, all existing or otherwise possible alterations of the Einsteinian laws based on, or equivalent to a topology peserving alteration of the Minkowski metric.

An inspection of Mignani's (1992) results points out that ***the fundamental meaning of values (4.4) is the identification of the quasars' atmospheres as physical media of Class I, Type 4***,

$$\text{Type 4:}\ n_3 < n_4,\ n_4 > 1;\ \ \hat{\beta} > \beta,\ \ \hat{\gamma} < \gamma;\ \ V_{Max} < c_o,\ \ c < c_o. \quad (4.5)$$

In fact, all Mignani's values are positive, bigger than one, and imply a shift toward the red, thus characterizing media (4.5).

These results are not trivial because the isotopic geometrization of physical media is so broad, that there was no *a priori* reason to expect a necessary "redshift", because a "blueshift" was equally expectable.

As an example, *the media of Type 9 identified in the preceding section as those existing in the interior of hadrons do imply a shift toward the blue*, as the reader is encouraged to verify.

A further meaning of Mignani's values is that of permitting numerical predictions for the ***direct tests of the isotopic law (4.1) in planetary atmospheres***, as submitted in Santilli (1988c).

The proposals essentially consist in measuring the possible redshift of light within the inhomogeneous and anisotropic asmospheres of the planets of our solar system, and are based on the approximate proportion

$$\frac{<\hat{\omega}_{Quasars}>}{\hat{\omega}_{Atmosphere.}} \approx \frac{<M_{Quasars}>}{M_{Planet}} \tag{4.6}$$

where: $<\hat{\omega}_{Quasars}>$ is the expected average part of of the quasars redshift entirely of isotopic character (that is, due to their atmospheres); $\hat{\omega}_{Atmosphere}$ is the expected isoredshift in planetary atmospheres; $< M_{Quasar} >$ is the average mass of the quasars; and M_{Planet} is the mass of the planet considered.

Needless to say, proportion (4.6) is merely intended to give an order of magnitude of the expected redshift, while its exact counterpart can be reached only after a number of future refinements.

Mignani's values (4.4) now render proportion (4.6) suitable for explicit computation. In fact the average of Mignani's values is readily computed giving the value

$$<|B|> \approx 72.78, \tag{4.7}$$

with corresponding *average redshift of the quasars*

$$<|\hat{\omega}_2|> \approx 1.15, \tag{4.8}$$

while the *average redshift of the associated galaxies* is

$$<|\omega'_1|> \approx 0.01. \tag{4.9}$$

The isodoppler's redshift law, under the indicated limit assumption, therefore implies the following *average value of the redshift expected from the inhomogenuity and anisotropy of the quasars' atmospheres*

$$< \omega_{Quasars} > \; = \; <|\hat{\omega}'_2|> - <|\omega'_1|> \approx 1.14. \qquad (4.10)$$

From astrophysical information we can assume that the quasars' atmospheres are of the order of 10^5 densier than the atmosphere of the desired planet, say, Jupiter. If, in first approximation, the isotopic deviation from the conventional redshift is assumed to be proportional to the density of the atmosphere (and in fact it is absent for null densities), we have the following *expected numerical value of the characteristic B-ratio of Jupiter's atmospheres*

$$< B_{Jupiter} > \; = \; < b_3 \, / \, b_{4 \; Jupiter} > \approx 7.3 \times 10^{-4}, \qquad (4.11)$$

with *isotopic redshift expected from the inhomogenuity and anisotropy of Jupiter's atmosphere*

$$<|\hat{\omega}_{|\, Jupiter}|> \approx 1.14 \times 10^{-5}, \qquad (4.12)$$

which is fully within current experimental capabilities. The understanding is that its possible final value must be identified by experiments (which would then yield the possible final value of ratio (4.11)).

Similar orders of magnitude are obtained by using the atmospheres of other planets, e.g., that of our Earth, or of the Sun.

If the quasars are not at rest with respect to the associated galaxies but expelled from them at speeds compatible with the special relativity, the corrections to values (4.4) can at best be small, thus leaving value (4.12) still within current experimental capabilities.

VII.5: PROPOSED EXPERIMENTS

Not only physics is a science with an absolute standard of value, the experimental verification, but experiments themeselves have their own standard of value, with priority for the test of fundamental laws, because of their evident scientific and societal implications, over the test of lesser basic aspects.

In these volumes we have studied the apparent inapplicability of the Lorentz symmetry and related special relativity for the classical description of relativistic dynamics within inhomogeneous and anisotropic physical media in favor of covering isotopic formulations, and presented preliminary tests which, even though inconclusive, are encouraging. This is a clearly fundamental, open aspect of contemporary physics which, as such, deserves an experimental resolution one way or the other.

The tests here recommended must: A) be classical; B) imply a sufficient departure from the geometry of the special relativity to warrant the tests; and C) be feasible with current technology.

The best experiments along these lines appear to be those proposed in Santilli (1988c) on the test of Postulate IV (Sect. IV.9) and related isodoppler's redshift (4.1) within inhomogeneous and anisotropic atmospheres which, as indicated in the preceding section, verify all conditions A, B and C above.

More specifically, the test submitted are the following:

EXPERIMENT I: Measure in our laboratories the possible isoredshift of light from a quasar (before and) after going through the atmosphere of a member of our solar system, such as Jupiter or the Sun .

The main idea is to consider first a light which is known to be redshifted, namely, which is known to be moving at a finite speed v away from us, according to the Einsteinian law

$$\omega' = \gamma\,\omega, \quad \gamma = (1 - \beta^2)^{-\frac{1}{2}}, \quad \beta = v / c_o, \quad \alpha = 0, \qquad (5.1)$$

and then see whether such a light experiences an additional redshift when propagating within an inhomogeneous and anisotropic atmosphere according to the isotopic law

$$\hat{\omega}' = \hat{\gamma}\,\omega', \quad \hat{\gamma} = \left| 1 - \frac{b_3^{\,2}}{b_4^{\,2}}\beta^2 \right|^{-\frac{1}{2}}, \quad \beta = v / c_o, \; \alpha = 0, \qquad (5.2)$$

The experiment is readily feasible via the use of exactly the same apparata used for the measure of the quasars redshift, and requires the measure of light just before and then through Jupiter's atmosphere to minimize the decrease of the distance from the center of the planet. In turn,. this renders ignorable gravitational corrections to the expected isoredshift.

The predictions of the isospecial relativities reviewed in the preceding sections are of the order of

$$\hat{\Delta}_\omega = \hat{\omega}' - \omega' \cong 1.14 \times 10^{-5} \tag{5.3}$$

and, therefore, fully within current experimental capabilities.

If successful, these measures will first produce the actual value of the characteristic ratio $B = b_3/b_4$ of Jupiter's atmosphere according to the simple espression

$$B = \frac{b_3}{b_4} = \frac{[1 + (\gamma + \hat{\Delta}_\omega)^2]^{\frac{1}{2}}}{\beta} \tag{5.4}$$

But the average speed of light through Jupiter's atmosphere is given by

$$c = c_o b_4 = c_o / n_4 \tag{5.5}$$

where n_4 is the average index of refraction. As such, the numerical value of $b_3 = n_4^{-1}$ can be reached via a number of estimates, e.g., via comparative values on Earth' s atmosphere adjusted for the different Jovian conditions of density, etc. The knowledge of ratio (5.4) will then permit the identification of the characteristics b_3 quantity of the Jovian atmosphere, by reaching in this way values for both $b_3 = n_3^{-1}$ and $b_4 = n_4^{-1}$.

EXPERIMENT II: Measure in a satellite the possible isoredshift of light from a quasar (before and) after going through the entire Earth's atmosphere.

If successful, Experiment I will produce the numerical value of the characteristics b_3 and b_4 quantities, specifically, for Jupiter. Experiment II is suggested so as to achieve the numerical value of the same characteristic quantities, this time, for our atmosphere (which are evidently expected to be different than the Jovian ones due to different physical characteristics of density, size, etc.).

In turn, such a knowledge is essential for a variety of other studies of the interior dynamics of our Earthly environment.

EXPERIMENT III: measure in our laboratories the possible isoredshift of sun light in the transition from the zenit to the equator.

The main idea is essentially to see whether the natural redshift of light we have all visually observed at sunset has an (expectedly small) component due to the inhomogenuity and anisotropy of our atmosphere.

The theoretical implications of this third experiment are however deeper, and complementary to the preceding ones. In fact, the experiment can be conceived in such conditions to minimize the relative speed v between the Sun and Earth. As a result, if successful, the experiment can provide information on the functional dependence of the characteristic b-quantities.

Specifically, for the Einsteinian case the value $v = 0$ implies that $\gamma = 1$ and $\omega' = \omega$ resulting in no redshift. In the transition to isominkowskian geometries, the situation is not that simple, because the $\hat{b}_3$-quantity has a primary dependence on speed (see Sect. VII.3, in particular, the diagram of Figure VII.3.2). Isodoppler's law (5.2) therefore reads explicitly

$$\hat{\omega}' = \omega' \left| 1 - \frac{v\, \hat{b}_3(v)^2\, v}{c_o\, b_4^2\, c_o} \right|^{-\frac{1}{2}}, \quad \alpha = 0, \tag{5.2}$$

where b_4 is assumed to be constant as in Fig. VII.3.4.

As a result, there is a possibility that an isodoppler redshift occurs also for sources at rest with respect to each other, and the issue evidently calls for a separate experimental study.

To state it in different terms, the experimental test of the isodoppler redshift predicted by the isospecial relativities is multifold. First there is the problem whether a distant light which is already redshifted admits an additional redshift when passing through an atmosphere (Experiments I and II). Then there is the different problem whether a nonredshifted light can experience a redshift when passing through an inhomogeneous and anisotropic medium, which is a significance of Experiment III.

Note that Experiments I and II can be successful even under negative results for experiment III. Note also that the characteristic b-quantities of our atmosphere can be measured via Experiment II and not III.

Needless to say, Experiments I, II and III are intended to be merely representative of a virtually endless variety of classical experiments which can be conceived via the repetition of the historical tests of the special relativity within physical media of progressive geometric complexity (Sect. VII.1), and other means.

Since the isospecial relativities have a *classical* applicability, this author insists in their experimental resolution at the *classical* level, and in a way independent from any possible, separate test in particle and/or nuclear physics. This is not merely due to epistemological aspects, but also to technical reasons related to the appearance of additional degrees of freedom in the operator formulation of the isospecial relativities, in addition to the isounit $\hat{1}$ (or isotopic element T).[38]

Nevertheless there are particle experiments which are semiclassical and therefore suitable for additional classical tests. This is typically the case for the data elaboration of experiments on the behaviour of the meanlife with speed, as outlined in Sect. VII.3. This leads us to suggest the following

EXPERIMENT IV: Achieve final experimental resolution of the behaviour of the meanlife of unstable hadrons with speed, not only for the $K^0{}_s$ particle, but also for other unstable hadrons.

In essence, as stressed in Sect. VII.3, both data by Aronsoin *et al.* (1983) showing deviations for 30 – 100 GeV, and by Grossman *et al.* (1987) showing verification of the Einsteinian law between 100 to 350 GeV, are preliminary and in need of final resolution, one way or the other.

[38] The interested experimenter should be aware of these aspects to avoid possible erroneous elaborations of experimental data. In essence, the isospecial relativities are characterized, at the classical level, by only one degree of freedom, the isotopic element T of the universdal enveloping associative algebra $\hat{\xi}$ with basic isoassociative product $A*B = ATB$, which also characterizes the underlying isofield $\hat{\Re} = \Re\,\hat{1}$, $\hat{1} = T^{-1}$. In the transition to operator formulations, isospecial relativities are characterized by two independent degrees of freedom: the isotopic, Hermitean and positive-definite operator T of the enveloping isoassociative operator algebra $\hat{\xi}$, which also characterizes the isofields of reals $\hat{\Re} = \Re\hat{1}$ and of complex numbers $\hat{C} = C\hat{1}$, plus a second, independent isotopic operator $G = G^\dagger > 0$ characterizing the underlying *isohilbert space* $\hat{\mathcal{H}}$ with *isoinner product* which, in its most general possible form, can be written $<\phi \hat{|} \psi> = <\phi | G | \psi> \hat{1} \in \hat{C}$. The evident result is that possible experimental data in particle physics on isosotpic relativities do not necessarily characterize the operator counterpart of the classical element T because of the general presence of the additional isotopic element G. It is true that in practical applications one usually assumes T = G (in which case the generalized operation of *isohermiticity*, $H^{\hat{\dagger}} = T^{-1}G\, H^\dagger\, G^{-1}T$ coincides with the conventional Hermiticity, $H^{\hat{\dagger}} \equiv H^\dagger$, and all conventional observables remain observables under isotopies). The point nevertheless persists that the possible differences between T and G must be investigated with care before claming final experimental conclusions. For more details, the interested reader may consult Santilli (1989).

Owing to the evidently basic character of the issue for all of theoretical physics, it is therefore essential to finalize the experimental data, by either confirming or denying the current experimental status.

The experiments should also be conducted for particles other than the K^{o}_{S}, because of *the indirect nature of the measure of time for such a particle owing to its neutral character*, as compared to other decays, e.g. that of charged pions, in which the behavior of the meanlife with speed can be reached in a more direct way with less theoretical elaborations of the data.[39]

The experimenter is however warned, that, ***to achieve acceptance of the results by the physics community at large, the theoretical assumptions in the data elaboration must be clearly such not to suppress any possible deviation.***

We are here referring to the unfortunate theoretical assumption by Grossman *et al.* (1987) of a frame in which the CP violation is null, because it is nowaday well know that the origin of CP violation in the K^{o}-system may be exactly that triggering the anomalous behaviour of the meanlife, as discussed in detail by Kim (1978) and others.

In different terms, the verification of the Einsteinian behaviour of the meanlife of the data by Grossman *et al.* may be due precisely to the theoretical assumption in the data elaboration of the frame with null CP violation (as well as other theoretical assumptions). As a resultm a re-elaboration of exactly the same data by Grossman *et al.* in different frames ensuring a nontrivial violation of CP-symmetry may produce well anomalous data similar to those by Aronson *et al.*

Finally, the experimenter should keep in mind that ***a possible, future experimental verification of the Einsteinian behaviour of the meanlife for the entire range of values from 30 to 400 GeV has no implication whatever on the validity or invalidity of the isospecial relativities .***

In fact, as stressed in Sect. IV.3, isospecial relativities have been conceived and constructed for the ***interior of hadrons***, and not for their ***exterior behavior*** in a particle accelerator. As a matter of fact, ***the exterior, center-of-mass behavior of closed, nonhamiltonian and isopoincaré invariant systems verifies, in their most general form, the totality of the***

[39] ***The measure of the behavior of the meanlife of the muon with speed is strongly discouraged as a first step, and recommended only after the, or jointly with the basic tests on unstable "hadrons".*** This is due to the realistic possibility that the muon is indeed an elementary particle, only in an excited state, in which case no internal nonlocal effects are conceivable, and no deviation from the Einsteinian meanlife is possible.

Einsteinian laws by construction, including the behavior of the meanlife (see Sect. IV.2).

In fact, the verification is such that the admission of the noneinsteinian behavior of the meanlife requires theoretical justifications via the removal of one or more subsidiary constraint.

The behavior of the meanlife of unstable hadrons is one of those intriguing special cases to deserve serious experimental studies.

All the above tests are specifically referred to the boosts component of the isolorentz symmetries. Our presentation would be grossly deficient without an experimental proposal for the ultimate, fundamental isosymmetry from which all others can be derived: the isorotational symmetry $\hat{O}(3)$ (Sect.s III.3 and VII.2). This leads us to the following final proposal:

EXPERIMENT V: ***Repeat Rauch's experiments on the apparent rotational asymmetry of netrons (Rauch (1981), (1983)) via: 1) a better accuracy; 2) with multiple spin flips; and 3) with nuclear interactions in the electromagnet gap of increased mass.***

The last experimental run by Rauch and his associates dates back to 1978. In the meantime, the technology and accuracy of neutron interferometers have improved considerably. The repetition of the experiments with the reduction of the error to about 20% of current values would evidently resolve the issue, and this clarifies Condition 1.

Also, as recalled in Sect. VII.2, Rauch's tests were done for only two spin-flips, while neutron interferometry can today achieve up to fifty spin-flips and more. The repetition of the tests with multple spin-flips is important because the effects is eminently nonlinear, and therefore expected to have a nonlinear increase with the number of spin-flips, and this clarifies Condition 2.

Finally, the deviation is expected to be due, in a generally nonlinear fashion,, to the intense fields in the vicinity of Mu-metal nuclei (Santilli (1981), Eder (1983)). The use of heavier nuclei is then recommendable to *maximize* (rather then minimize) the deviations.

Moreover, in each of the preceding runs, the experimenter should

a) Measure the basic angle of spin-flip and related error;

b) Verify or disprove that the median angle is always lower than the amount expected ("angle slow-down effect"); and

c) verify that the behaviors of the intensity and polarization modulations are indeed sinusoidal and with the appropriate phase (because angles a and b could be conventional, yet a rotational asymmetry may emerge from data c).

In the final analysis, the experimenter should keep in mind the central concept at the foundation of all isotopic relativities:

> ***the direct representation of the actual size and shape of the considered charge distribution, as well as of all its infinitely possible deformations, with consequential, necessary alteration of the intrinsic magnetic moment.***

In a scientific setting of this nature, the ***amount of deformation*** for given external conditions is unknown at this writing, and can only be resolved via experimental measures. But the ***existence of the deformation*** is beyond any scientific doubt, thus directly activating the isotopic rotational symmetries with consequential isotopic relativities.

APPENDIX VII.A: NECESSARY CONDITIONS FOR TRUE TESTS OF QUANTUM MECHANICS

It is appropriate to close this chapter by recalling that, *with the sole exception of Rauch's measures, all experiments conducted until now for the test of quantum mechanics have provide no deviation whatever. This includes the negative results of the experiments testing a possible nonlinearity of quantum mechanics* (see Zeilinger *et al.* (1983), Ellis *et al.* (1984), Ignat'ev *et al* (1987) and quoted experimental papers).

These results are generally assumed as "evidence" on the lack of need of isotopic symmetries and relativities in particle physics. It is therefore important to indicate the reasons why ***the latter tests are fundamentally insufficient to test hadronic mechanics, that is, to test possible nonlinear and nonlocal effects in the interior of hadrons*** on a number of independent counts.

The mean reasons is that the latter experiments are conceived to approach the conditions of applicability of quantum mechanics as close as possible, thus resulting in predictable lack of deviations. Another reason is that, most of the tests considered are of atomic type (e.g., searching for a possible nonlinearity in the atomic structure) and, as such, strictly inapplicable to the interior of strong interactions. Also, the latter experiments search for a very special type of nonlinearity, that in the wavefunction ψ , while our studies indicate that the most important nonlinearity is that in the derivatives of the wavefunctions $\partial\psi$ (because all drag effects are dependent on the velociitiy)[40]; and other reasons.

[40] The interested reader may inspect the basic equations of motion of hadronic mechanics, Eq.s (II.6.23) and (II.6.24), with related emphasis on the "direct universality" for the most general conceivable linear and nonlonear, local and nonlocal equations

The experimental guidelines provided by these volumes are that ***the possible experimental detection of deviations from quantum mechanics demands the maximization of the departures of the physical conditions of the experiments from those of the original conception of the theory.***

To begin, none of the experiments conducted until now (with the sole exception of Rauch's experiments, see below) focus the attention on the true, ultimate limitations of (relativistic and nonrelativistic) quantum mechanics, the representation of particles as being point-like.

We reach in this way the following:

CONDITION VII.A.1: A first condition to truly test possible deviations from quantum mechanics, is that the tests must be fundamentally dependent on the limitations of the theory caused by its point-like approximation of particles, and be essentially depedent on physical conditions in which the extended and therefore deformable character of the charged distributions and/or wavepackets of the particles is expected to produce measurable effects.

The limitations of quantum mechanics caused by its point-like approximation of particles are so evident, that the studies of its generalizations to represent particles in their actual, finite, and deformable, character (e.g., via hadronic mechanics) have physical value even prior to, and independently from the experimental resolution of the issue.

Despite these manifest limitations, the experimental detection of possible deviations from quantum mechanics has continued to be elusive, again, because of the lack of identification of the appropriate physical conditions under which deviations are admissible.

The most visible case is that of experiments on inelastic scatterings of hadrons. These are particles which possess a fully established, extended character. Nevertheless, all tests conducted until now have shown no appreciable deviation[41].

in all variables.

[41] There are few exceptions we can only indicate here without treatment, such as the difference in the cross section for inelastic scatterings of polarized hadrons, depending on whether the particles have spon parrallel or antiparallel. These differences could be evidently due to the extended character of the particles which, when colliding, render singlet and triplet states inequivalent, as discussed in Appendix III.A with the "gear model". This is a typical topic currently under study

The point is that all such experiments have been conducted in their center-of-mass frame, or in equivalent frames. Extended-deformable hadrons in their center-of-mass are precisely the closed nonhamiltonian (nonselfadjoint) systems studied in these volumes. We have stressed throughout this analysis that these generalized systems, when seen from the outside in their closed conditions, show no deviation from the conventional atomic setting.

The very idea of the isotopic relativities is that of expressing the complete identity of a closed nonhamiltonian with a closed Hamiltonian system when studied from the outside.

Thus, the studies of this volume establish that experiments conducted in the center-of-mass systems, and/or in any of their equivalent frames, are substantially insensitive to the internal structure of the systems. In the final analysis, nonlinear, nonlocal and nonhamiltonian forces are strictly internal effects without potential energy , and without measurable effect in the outside center-of-mass system (with the sole possible exclusion of the meanlife behavior, see Sect. VII.4)..

We reach in this way the following:

CONDITION VII.A.2: A second condition to truly test possible deviations from quantum mechanics is that measures are conducted under the maximization of nonconservative conditions due to external interactions.

To understand this point, the reader should always recall that conventional quantum mechanics provides a form-invariant description of stable systems via the stability of each individual orbit, as requested by its fundamental rotational symmetry. If the experimenter ***maximizes the compliance*** with these conditions of stability, it is evident that no deviation is possible. On the contrary, if the experimenter ***maximizes the departures*** from stable orbits, measurable deviations are conceivable, and actually expected on a number of counts.

One way to illustrate the above occurrence, is by considering the current experimental beliefs on the validity of the *time-reversal symmetry in strong interactions*

As well known the experiments in the field have been systematically conducted in the center-of-mas frame or in equivalent frames. Now, a study of the issue reveals that there is no real need to

with the operator formulation of the isotopic relativities within the context of hadronic mechanics.

conduct experiments to test the exact T-symmetry in the center-of-mass conditions. In fact, by observing Jupiter (see Fig. I.1.1 and the appendices of Chapter II), one can easily see that its center-of-mass trajectory in the solar system is strictly time-reversal invariant. Nevertheles, its internal structure is irreconcilably irreversible.

Thus, the scientifically correct statement is that *the T-reversal symmetry is exactly valid for strong interactions* ***in their center-of-mass system***. Scientific caution requires the statement that *the validity or invalidity of the T-reversal symmetry for strong interactions* ***in their open nonconservative conditions****, e.g., for one hadronic constituent while considering the remaining constituents as external, is fundamentally unresolved at this writing, theretically and experimentally.*

We reach in this way our

CONDITIONS VII.A.3: A third and final condition to truly test possible deviations from quantum mechanics is that the physical conditions VII.A.1 and VII.A.2 should be elaborated with a theory strictly independent from the law to be tested.

The T-reversal symmetry is again a good example to illustrate this latter occurrence. Suppose that experimenters do indeed finally pass, from center-of-mass measures, to *bona fide* open conditions under strong interactions. However, if the data from these latter conditions are elaborated via conventional quantum mechanics with a Hermitean Hamiltonian, the tests acquires a scientifically misleading character because the theory assumed in the data elaboration (unitary time evolution) does verify the theorem of detailed balancing in its entirety. An exact T-symmetry under open conditions is therefore expected to result from the theoretical assumption, and we have no true "experimental results".

In order to conduct true experiments on the T-symmetry that will resist the test of time, the experimenters first need physical conditions as open-nonconservative as possible (Condition VII.A.2), e.g., a beam of hadrons interacting with an *external*, fixed, sufficiently heavy target., with a measurable *loss of energy or of other physical quantities* of the beam in favor of the external target (with the undersdtanding that the total quantities, includive those of the external target are evidentlhy conserved[42]). In addition, the

[42] Eder (1981) interpreted Rauch's data (1981 and 1983) on the rotational asymmetry of neutrons as a form of *spin fluctuation*. The evident understanding is that the Mu-metal nuclei experience a complementary fluctuation in such a way that the total

experimenters should elaborate these nonconservative conditions with a nonunitary theory.

If these open-nonconservative conditions are truly represented, e.g., with a nonunitary time evolution, then there is no need to conduct tests to establish the *deviation* from the T-symmetry on qualitative grounds, and the test are needed only to establish its quantitative amount for given conditions, because *nonunitary time evolutions violate the theorem of detailed balancing* (see Santilli (1983b, c)).

This is the dichotomy of exact verification of conventional quantum mechanical laws for the center-of-mass treatment of strongly interacting systems, in a way fully compatible with possible deviations for interior conditions. After all, internal nonconservative trajectories imply internal exchanges of energy with no impact in the conservation of the total energy for an isolated composite system, and a similar situation occurs for all other physical quantities.

It is hoped in this way the experimenter can begin to see a new horizon of basically new experiments, which we plan to study in future works after reviewing the operator formulation of the isotopic relativities.

The above comments also illustrate the reasons why, in disagreement with all other tests, Rauch's experiments do indeed reach preliminary deviations from, the prediction of quantum mechanics.

In fact, Rauch's experiments:

1: are centrally dependent on the extended and therefore deformable character of neutrons, by therefore verifying Condition VII.A.1;

2: conduct measures under open conditions (thermal neutron beams interacting with the external Mu-metal target), by therefore verifying Condition VII.A.2; and, last but not least,

3: the measures of the angle of spin-flip is basical independent from the theory to be tested, the rotational symmetry, by therefore verifying Condition VII.A.3.

We therefore close these volumes suggesting scientific caution prior to claiming lack of deviation from the predictions of quantum mechanics via tests that are fundamentally based on physical

---

angular momentum of the neutron beam and of the external Mu-metal sheets is exactly conserved and fully conventional according to quantum mechanics.

conditions and theoretical elaborations of data that are strictly quantum mechanical.

This occurrence has been illustrated with the fact that there is no need to conduct experiments on the T-symmetry to see the existence of its violation by nonconservative conditions when truly represented by nonunitary time evolutions.

Similarly, there is no need to conduct experiments for the existence the deformability of neutrons under sufficiently intense external fields, with the consequential alteration of its intrinsic magnetic moment.

A number of additiona experimental insights for the interior particle problem are proviued by Animalu's (1991b) studies on superconductivity, via the representation of the Cooper pairs with internal nonlocal effects which are structurally outside the capabilities of quantum mechanics.

But, above all, the most significant experiments are those directly testing the foundations of the studies of these volumes:

> ***the legacy of Fermi, Bogoliubov, and others on the ultimate internal nonlocality of the strong interactions due to deep mutual overlappings of the wavepackets of hadrons or of their constituents.***

The test based on the Bose-Einstein correlation are particularly significant for this purpose. In fact, as pointed out in footnote[35], p. 305, the very notion of correlation is outside the quantum mechanical axiom of expectation values, while it is fully predicted by the covering isotopic mechanics and relativities (Santilli (1992).

REFERENCES

1638. Galilei G., *Dialogus de Systemate Mundi*, translated and reprinted by Mc Millan, New York, 1918.

1687. Newton I., *Philosophiae Naturalis Principia Mathematica*, translated and reprinted by Cambridge Univ. Press, Cambridge, England, 1934.

1788. Lagrange J. L., *Méchanique Analytique*, reprinted by Gauthiers-Villars, Paris, 1988.

1834. Hamilton W. H., contribution reprinted in *Hamilton's Collected Papers*, Cambridge Univer. Press, Cambridge 1940.

1837. Jacobi C. G., *Zur Theorie der Variationensrechnung und der Differentualgleichungen*, reprinted by Springer, München 1890.

1868. Riemann B., Gött. Nachr. 13, 133.

1887. Helmholtz H., J. Reine Angew. Math. 100, 137.

1893. Lie S., *Theorie der Transformationsgruppen*, Teubner, Leipzig.

1904. Lorentz H. A., Amst. Proc. 6, 809.

1905. Einstein A., Ann. Phys. 17, 891.
Poincaré H., Compte Rendues, Paris 140, 1504.

1913. Minkowski H., *Das Relativitätsprinzip* , Lipsia.

1916. Einstein A., Ann. Phys. 49, 769.

1920. Rutherford, Proc. Roy. Soc. A97, 374.

1921. Pauli W., *Relativitätstheorie*, Teubner, Lipsia.

1927. Birkhoff G. D., *Dynamical Systems*, A.M.S., Providence, RI.

1939. Freud P., Ann. Math. (Princeton) 40, 417.

1942. Bergmann P. G., *Introduction to the Theory of Relativity*, Dover, New York.

1946. Bliss G. A., *Lectures on the Calculus of Variations*, Univ. of Chicago Press.

1948. Albert A. A., Trans. Amer. Math. Soc. 64, 552.

1949. Fermi E., *Nuclear Physics*, Univ. of Chicago Press, Chicago.

1950. Dirac P. A. M., Can. J. Math. 2, 129.

Goldstein H., *Classical Mechanics*, Addison-Wesley, Reading MA.

Schrödinger E., *Space-Time Structure*, Cambridge University Press, Cambridge, England.

1951. Papapetrou A., Proc. Roy. Soc. 209A, 284.

1952. Blatt J.M. and V.F.Weisskopf, *Theoretical Nuclear Physics*, Wiley, New York (7-th princing of 1963).

1956. Pauli W., *Wellenmechanik*, Univ. of Zurig.

1958. Bruck H. R., *A Survey of Binary Systems*, Springer-Verklag, Berlin.

Dirac P. A. M., Proc. Roy. Soc. A246, 326.

Pauli W., *Theory of Relativity*, Pergamon Press, Londion.

Yilmaz Y., Phys. Rev. 111, 1417.

1959. Martin J. L., Proc. Roy. Soc. A251, 536 and 543.

1960. Bogoliubov N.N. and D.V.Chirkov, *Introduction a la Théorie Quantique des Champs*, Dunod, Paris (1960)

Pond R.V. and G.A.Rebka, Phys. Rev. Letters 4, 337.

1961. Hughes J. B., Suppl. Nuovo Cimento 20, 89.

1962. Schweber S, *An Introduction to Relativistic Quantum Field Theory*, Harper and Row, New York.

Jacobson N., *Lie Algebras*, Wiley, N.Y.

Prigogine I., *Nonequilibrium Statistical Mechanics*, J. Wiley, NY.

1963. Albert A. A., Editor, *Studies in Modern Algebra*, Math. Ass. of Amer. ,Prentice-Hall, Englewood Cliff, N.J.

Edwards W. F., Amer. J. Phys. 31, 432.

1964. Blockhintsev D. I., Physics Letters 12, 272.

Dirac P. A. M., *Lectures in Quantum Mechanics*, Yeshiva University, New York.

1966. Redei L. B., Phys. Rev. 145, 999.

Schafer R. D., *An Introduction to Nonassociative Algebras*, Academic Press, N.Y.

1967. Abraham R. and J.E.Marsden, *Foundations of Mechanics*, Benjamin, New York.

Hill R. N., J. Math. Phys. 8, 1756.

Santilli R. M., Lettere Nuovo Cimento 51A, 570.

1968. Prigogine I., Cl. George and F. Henin, Physica 45, 418.

Santilli, R. M., Suppl. Nuovo Cimento 6, 1225.

1969. Bloom E. D., D.H.Coward, H.De-Staebler, J.Drees, G.Miller, L.Mo, R.E.Taylor, M.Breindenbach, J.I.Friedman, G.C.Hartmann, H.W.Kendall, Phys. Rev. Letters 23, 935.

Santilli, Meccanica 1, 3.

1970. Hagihara Y., *Celestial Mechanics* ,Vol. I, The MIT Press, Cambridge MA.

Hofstadter R., *Electron Scattering and Nuclear and Nucleon Structure*, Noth Holland, NY.

Fujimura K., T. Kobayashi and M. Namiki, Progr. Theor. Phys. 43, 73.

1971. Dirac P. A. M., Proc. Roy. Soc. A322, 435.

Lévy-Leblond J. M., contributed paper in *Group Theory and Its Applications*, E. M. Loebl Editor, Academic Press, N.Y.

Trautman A.R., Bul. Acad. Pol., Series Scie., Math., Astr., Phys. 20, 185, 503 and 895

Yilmaz H., Phys. Rev. Letters 27, 1399.

1972. Dirac P.A.M., Proc. Roy. Soc. A328, 1.

Recami E. and R.Mignani, Letyt. Nuovo Cimento 4, 144.

1973. Ivanenko D. and I. Sardanashvily, Phys. Rep. 94, 1

1974. Behnke H., F.Bachmann, K.Fladt, and W.Suss, Editors *Fundamentals of Mathematics*, Vol. I, MIT Press, Cambridge, MA.

Gilmore, *Lie Groups, Lie Algebras, and Some of Their Applications*, Wiley, New York.

Mann R.A., *The Classical Dynamics of Particles*, Academic Peress, New York.

Santilli R. M., Ann. Phys. 83, 108.

Sudarshan G. and N. Mukunda, *Classical Dynamics: A Modern Perspective*, Wiley, N.Y.

Trautman A. R., Bul. Acad. Pol. , Ser. Scie., Math., Astr. Phys. 20, 185, 503 and 895

1975. Kuchowitz B., Acta Cosmologica Z3, 109.

Roman P., *Some Modern Mathematics for Physicists and other Outsiders*, Vol. I, Pergamon Press, N.Y.

Lovelock D. and H. Rund, *Tensors, Differential Forms and Variational Principles*, Wiley, New York.

1976. Kracklauer A. F., J. Math. Phys. 17, 693.

Hehl P. W., P. von der Heyde, G.D.Kelich, and J.M.Nester, Rev. Mod. Phys. 48, 393.

1977. Barut A.O. and R. Racza, *Theory of Group Representations and Applications*, Polish Scient. Publ., Warszawa.

Bogoslovski G. Yu., Nuovo Cimento 40B, 99 and 116.

Edelen D. G. B., *Lagrangian Mechanics of Nonconservative Nonholonomic Systems*, Noordhoff, Leyden.

Yilmaz H., Lettere Nuovo Cimento 20, 681.

1978. Adler S. L., Phys. Rev. 17, 3212.

Bogoslowski G. Yu., Nuovo Cimento 43B, 377.

Kim D. Y., Hadronic J. 1, 1343.

Santilli R. M., Hadronic J. 1, 228 (1978a).

Santilli, Hadronic J. 1, 574 (1978b).

Santilli, R. M., Hadronic J. 1, 1343 (1978c)

Santilli R. M., *Lie-admissible Approach to the Hadronic Structure*, Vol. I, *Nonapplicability of the Galilei and Einstein Relativities ?*, Hadronic Press, Box 0594, Tarpon Spring, FL 34688 USA (1978d).

Santilli R. M., *Foundations of Theoretical Mechanics*, Vol. I, T*he Inverse Problem in Newtonian Mechanics*, Springer-Verlag, Heidelberg/New York (1978e).

1979. Fronteau J., Hadronic J. 2, 727.

Fronteau J., A.Tellez-Arenas and R.M.Santilli, Hadronic J. 3, 130.

Tellez-Arenas A., J. Fronteau and R.M.Santilli, Hadronic J. 3, 177.

Kokussen J. A., Hadronic J. 2, 321 and 578.

Santilli R. M., Hadronic J. 3, 440 (1979a).

Santilli R. M., Phys. Rev. D20, 555 (1979b).

Ziman J. M., *Models of Disorder*, Cambridge Univ. Press, Cambridge, England.

Yilmaz H., Hadronic J. 2, 1186.

1980. Lichtenberg D.B. and S. P. Rosen, Editors, *Development of the Quark Theory of Hadrons*, Hadronic Press, Box 0594, Tarpon Spring, FL 34688 USA

Ktorides C. N., H.C.Myung and R.M.Santilli, Phys. Rev. D22, 892.

Mignani R., Hadronic J. 3, 1313.

Rosen M., contributed paper in *Torsion, Rotation and Supergravity*, P. Bergmann and V. de Sabbata, Editors, Plenum, New Yor

Santilli R. M., invited talk at the *Conference on Differential Geometric Methods in Mathematical Physics*, Chausthal, Germany, 1980, see Santilli (1981b).

Yilmaz H., Hadronic J. 3, 1478.

1981. Eder G., Hadronic J. 4, 634.

Preparata G., Phys. Letters 102B, 327.

Rauch H., invited talk at the *First International Conference on Nonpotential Interactions and their Lie-admissible Treatment*, Orléans, France, printed in Hadronic J. 4, 1280.

Santilli R. M., *Lie-admissible Approach to the Hadronic Structure*, Vol. II, *Coverings of the Galilei and Einstein Relativities ?* Hadronic Press, Box 0594, Tarpon Springs, FL 34688-0594 (1981a).

Santilli, R.M., Hadronic J. 4, 1166 (1981b)

Schoeber A., Hadronic J. 5, 214.

1982. Animalu A. O. E., Hadronic J. 5, 1764.

De Sabbata V. and M.Gasperini, Lett. Nuovo Ciimento 34, 337.
Eder G., Hadronic J. 5, 750.
Fronteau J., Hadronic J. 5, 577.
Jannussis A., G. Brodimas, D. Sourlas, A. Streclas, P. Siafaricas, P. Siafaricas, L. Papaloukas, and N. Tsangas, Hadronic J. 5, 1901
Mignani R., Hadronic J. 5, 1120.
Myung H. C. and R.M.Santilli, Hadronic J. 5, 1277 (1982a).
Myung H. C. and R.M.Santilli, Hadronic J. 5, 1367 (1982b).
Rauch H., Hadronic J. 5, 729.
Santilli R. M., *Foundations of Theoretical Mechanics*, Vol. II, *Birkhoffian Generalization of Hamiltonian Mechanics*, Springer-Verlag, Heidelberg/New York (1982a).
Santilli R. M., Lett. Nuovo Cimento 33, 145 (1982b).
Santilli R. M., Hadronic j. 5, 264 (1982c).
Santilli, R.M., Hadronic J. 5, 1194 (1982d)
Schoeber A., Hadronic J. 5, 1140.
Tellez-Artenas A., Hadronic J. 5, 733.
Trostel R., Hadronic J. 5, 1023 (1982a).
Trostel R., Hadronic J. 5, 1893 (1982b)
Yilmaz H., Physics letters 92A, 377 (1982a).
Yilmaz H., Intern. J. Theor. Phys. 10, 11 (1982b).
1983. Aronson B. H., G.J.Block,H.Y.Cheng, and E.Fishbach, Phys. Rev. D28, 476 and 495.
Gasperini M., Hadronic J. 6, 935 (1983).
Ivanenko D. and I.Sardanashvily, Phys. Rep. 94, 1.
Jannussis A. ,G. Brodimas, V.Papatheou and H.Ioannidou, Hadronic J. 6, 1434.
Jannussis A., G.N.Brodimas, V.Papatheou, and A. Leodaris, Lettere Nuovo Cimento 36, 545.
Jannussis A., G.Brodimas, V.Papatheou, G.Karayannis, and P. Panagopoulos, Lettere Nuovo Cimento 38, 181.
Nishioka M. Hadronic J. 6, 1480.
Rauch H., invited contribution in *Proceedings of the International Symposium on Foundations of Quantum Mechanics*, Phys. Soc. of Japan, Tokyo.
Mignani R., Lettere Nuovo Cimento 38, 169.
Mignani R., H.C.Myung and R.M.Santilli, Hadronic J. 6, 1878 (1983)
Nielsen H. B. and I.Picek, Nucl. Phys. B211, 269.
Santilli R. M., Lettere Nuovo Cimento 37, 545 (1983a).
Santilli, Lettere Nuovo Cimento 37, 337 (1983b).
Santilli R.M., Lettere Nuovo Cimento 38, 509 (1983c).
Zeilinger A., M.A.Horne and C.G.Shull in *Proceedings of the International Symposium on the Foundations of Quantum*

Mechanics, Phys. Soc. odf Japan, Tokyo, p. 389.
1984. Balzer C., H. Coryell, D.M.Norris, J.Ordway, M.Reynolds, T.Terry, M.L.Tomber and K.Trebilcott, Editors, *Bibliography and Index in NonassociativeAlgebras,* Hadronic Press, Box 0594, Tarpon Springs, FL. 34688.
Bogoslovski G. Yu., Hadronic J. 7, 1078.
Ellis J., J.S.Hagelin, D.V.Nanopoulos and M. Srednicki, Nucl. Phys. B241, 381.
Gasperini M., Hadronic J. 7, 650 (1984a).
Gasperini M., Hadronic J. 7, 971 (1984b)
Gasperini M., Nuovo Cimento 83A, 309 (1984c).
Mignani R., Lettere Nuovo Cimento 39, 406 (1984a).
Mignani R., Lettere Nuovo Cimento 39, 413 (1983b).
Nishioka M., Nuovo Cimento 82A, 351 (1984a)
Nishioka M., Hadronic J. 7, 240 and 1636 (1984b)
Nishioka M, Lettere Nuovo Cimento 40, 309 (1984c).
Santilli R. M., Hadronic J. 7, 1680.
Yilmaz H., Hadronic J. 7, 1.
1985. Gasperini, Hadronic J. 8, 52.
Karayannis G., *Lie-isotopic Lifting of Gauge Theories,* Ph.D. Thesis, Univ. of Patras, Greece (1985a).
Karayannis G., Lettere Nuovo Cimento 43, 23 (1985b).
Jannussis A., Nuovo Cimento 90B, 58.
Jannussis A., Lettere Nuovo Cimento 42, 129.
Jannussis A., KL.C.Papaloukas, P.I.Tsilimigras and N.R.C.Democritos, Lettere Nuovo Cimento 42, 83.
Mignani R., Lettere Nuovo Cimento 43, 355.
Nishioka M., Nuovo Cimento 85a, 331.
Nishioka M., Nuovo Cimento 86A, 151.
Nishioka M., Hadronic J. 8, 331.
Santilli R.M., Hadronic J. 8, 25 (1985a).
Santilli R. M., Hadronic J. 8, 36 (1985b).
Santilli R. M., invited talk at the *International Conference on Quantum Statistics and Foundational Problems of Quantum Mechanics,* Calcutta , India, Hadronic J. Suppl. 1, 662 (1985c).
1986. Animalu A. O. E., Hadronic J. 9, 61.
Arp H. C., G.Burbidge, F. Hoyle, V.J.Nardikar, and N.C. Wicramasinghe, Nature 346, 807.
Chatterjee L. and V.P.Gautam, Hadronic J. 9, 95 (1986).
González-Diaz P. F., Hadronic J. 9, 199.
Karayannis G. and A. Jannussis, Hadronic J. 9, 203.
Jannussis A., Hadronic J. Suppl. 2, 458.

Mignani R., Hadronic J. 9, 103.
Nishioka M., Nuovo Cimento 92A, 132.
Nishioka M., Hadronic J. 10, 253.
1987. Animalu A. O. E., Hadronic J. 10, 321.
Grossman N., K.Heller, C.James, M.Shupe, K.Thorne, P.Border, M.J.Longo, A.Beretvas, A.Caracappa, T.Devlin, H.T.Diehl, U.Joshi, K.Krueger, P.C.Petersen, S.Teige and G.B.Thomson, Phys. Rev. Letters 59, 18.
Ignat'ev Yu.A. and V.A.Kuzmin, Sov. J. Nucl. Phys. 46, 444.
Jannussis A., and DS. Vavougios, Hadronic J. 10, 75.
Nishioka M., Hadronic J. 10, 309.
Veljanoski and A. Jannussis, Hadronic J. 10, 53 and 193.
1988. Jannussis A. and R. Mignani, Physica A152, 469.
Jannussis A. and A.Tsohantjis, Hadronic J. 11, 1.
Nishioka M., Hadronic J. 11, 71 (1988a).
Nishioka M., Hadronic J. 11, 97 (1988b).
Nishioka M., Hadronic J. 11, 143 (1988c).
Santilli R.M. Hadronic J. Suppl. 4A, Issue 1 (1988a).
Santilli R.M. Hadronic J. Suppl. 4A, Issue 2 (1988b).
Santilli R.M. Hadronic J. Suppl. 4A, Issue 3 (1988c).
Santilli R.M. Hadronic J. Suppl. 4A, Issue 4 (1988d).
1989. Aringazin A. K., Hadronic J. 12, 71.
Assis A.K.T., Phys. Letters 2, 301.
Bollinger J.J., D.J.Heinzen, W.M.Itano, S.L.Gilbert and D.J.Wineland, Phys. Rev. 63D, 1031
Logunov A. and N. Mestvirshvili, *The Relativistic Theory of Gravitation,* Mir Publ., Moscow.
Mignani R., Hadronic J. 12, 167.
Santilli R.M., Hadronic J. Suppl. 4B, issue 1 (1989a).
Santilli R.M., Hadronic J. Suppl. 4B, issue 2 (1989b).
Santilli R.M., Hadronic J. Suppl. 4B, issue 3 (1989c).
Santilli R.M., Hadronic J. Suppl. 4B, issue 4 (1989d).
Weinberg S. Ann. Phys. 194, 336.
Yilmaz H., contributed paper in *Space-Time Symmetries* (Wigner's Symposium), Y. S. Kim and W.W.Zachary, Editors, North-Holland, N.Y.
1990. Aringazin A. K., Hadronic J. 13, 183.
Aringazin A. K., A. Jannussis, D.F.Lopez, M. Nishioka and B. Veljanoski, Algebras, Groups and Geometries 7, 211, and addendum 8, 77, 1991.
Arp H. C., G.Burbidge, F. Hoyle, V.J.Nardikar, and N.C. Wicramasinghe, Nature 346, 807.

Assis A.K.T., Hadronic J. 13, 441.

Carmeli M., E. Leibowitz and N. Nissani, *Gravitation*, World Scientific Publisher, New York.

Chupp T.F. and R.J.Hoare, Phys. Rev. Letters 64, 2261.

Graneau P., Hadronic J. Suppl. 5, 335.

Logunov A. and N. Mestvirshvili, *The Relativistic Theory of Gravitation*, Mir Publ., Moscow, USSR.

Mijatovic M., Editor, *Hadronic Mechanics and Nonpotential Interactions*, Nova Science, New York.

Prigogine I., *Nobel Symposium*.

Santilli R. M., Hadronic J. 13, 513 and 533.

Strelt'sov V. N., Hadronic J. 13, 299.

Walsworth R.L., J.F.Silvera, E.M.Mattison and R.F.C.Vessot, Phys. Rev. Letters 64, 2599.

Yilmaz H., Hadronic J. 12, 263 (1990a).

Yilmaz H., Hadronic J. 12, 305 (1990b)

1991. Animalu A. O. E., in *Proceedings of the Fifth Workshop on Hadronic Mechanics*, Nova Science, New York (1991a)

Animalu A. O. E., Preprint, Univ. of Calif., Irtvine (1991b)

Aringazin A. K., Hadronic J. 14, in press.

Aringazin A. K., A. Jannussis, D.F.Lopez, M. Nishioka and B. Veljanoski, *Santilli's Lie-Isotopic Generalizations of Galilei's and Einstein's Relativities*, Kostarakis Publisher, 2 Hippokratous St., 10679 Athens, Greece.

Jannussis A., R.Mignani and R.M.Santilli, "Some problematic aspects of Weinberg's nonlinear generalization of quantum mechanics and their possible resolution", Univ. of Rome preprint, submitted for publication.

Jannussis A., M.Mijatovic and B.Veljanoski, Physics Essays 4, 202.

Jannussis A., G. Brodimas and R.Mignani, J. Phys. A24, L775.

Kadeisvili J.V., *Santilli's Isotopies of Contemporary Algebras, Geometries and Relativities*, Hadronic Press, box 0594, Tarpon Springs, FL 34688 USA.

Mignani M. and R.M.Santilli, "Isotopic liftings of SU(3) symmetries", ICTP preprint IC/91/48

Nishioka M. and R.M.Santilli, Physics Essays 5, 44.

Rapoport-Campodonico D. L., Algebras Groups and Geometries 8, 1 (1991a).

Rapoport-Campodonico D. L., Hadronic J., 14, in press (1991b).

Rund H., Algebras, Groups and Geometries 8, 267.

Santilli R.M., Algebras, Groups and Geometries 8, 169 (1991a).

Santilli R.M., Algebras, Groups and Geometries 8, 287 (1991b).

Santilli R.M., "Lie-isotopic generalization of the Poincaré symmetry", ICTP preprint IC/91-45, submitted for publication (1991c).

Santilli R.M., "Theory of mutation of elementary particles", ICTP preprint IC/91-46, submitted for publication (1991d)

Santilli R.M., "Apparent consistency of Rutherford'ds model on the neutron structure within the context of hadronic mechanics", ICTP preprint IC/91-47, submitted for publication (1991e).

Santilli R.M., "Inequivalence of interior and exterior dynamical problems", ICTP preprint IC/91/258, submitted for publication (1991f).

Santilli R.M., "Closed systems with nonhamiltonian internal forces", ICTP preprint IC/91/259, submitted for publication (1991g).

Santilli R.M., "Generalized two-body and three-body systems with nonhamiltonian internal forces ", ICTP preprint IC/91/260, submitted for publication (1991h).

Santilli R.M., "Rotational-isotopic symmetries", ICTP preprint IC/91/261, submitted for publication (1991i).

Santilli R.M., "Euclidean-isotopic symmetries", ICTP preprint IC/91/262, submitted for publication (1991j).

Santilli RM., "Galilei-isotopic symmetries", ICTP preprint IC/91/263, submitted for publication (1991k).

Santilli R.M., "Galilei-isotopic relativities", ICTP preprint IC/91/264, submitted for publication (1991-l).

Santilli R.M., "The notion of nonrelativistic isoparticle", ICTP preprint IC/91/265, submitted for publication (1991m).

Santilli R.M., "Lie-admissible structure of Hamilton's original equations with external terms", ICTP prepriont IC/91/266, submitted for publication (1991).

Santilli R.M. "Proposal to measure a possible light redshift caused by the inhomogenuity and anisotropy of planetary atmospheres", IBR preprint TP/91/05, Palm Harbor, FL, submitted for publication (1991n).

Weiss G.F., *Scientific, Ethical and Accountability Problems in Einstein's Gravitation,* Andromeda Publisher, Via Allende 1, I-40139 Bologna, Italy

1992. Cardone F., R.Mignani and R.M.Santilli, J. Phys. G 18, L61.

Cardone F., R. Mignani and R.M.Santilli, "Lie-isotopic energy dependence of the K°s lifetime", J. Phys. G, in press.

Mignani R., " Quasars redshift in iso-Minkowski space", Phys. Essay, in press.

Santilli R.M., Hadronic J. 15, 1.

INDEX

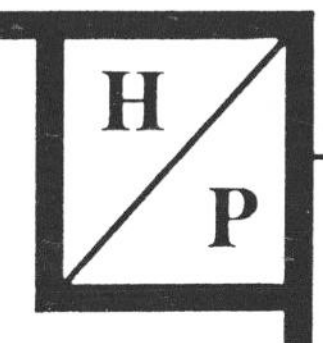

HADRONIC PRESS, INC.

35246 US 19 N. #131, PALM HARBOR, FL 34684, U.S.A.

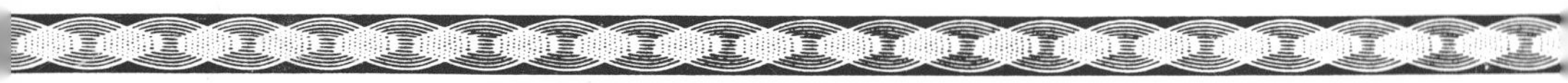

About the Author

Ruggero Maria Santilli obtained his PhD Degree at the Istitute of Theoretical Physics of the University of Turin, Italy, in 1966. He then moved with his family to the USA where he held several faculty positions at various universities, including: the Center for Theoretical Physics of the University of Miami, Coral Gables, Florida (1967–1968); Department of Physics of Boston University, Boston, MA (1968–1974); Center for Theoretical Physics, M.I.T., Cambridge, MA (1975-1977); Lyman Laboratory of Physics, Harvard University (1977–1978); and Department of Mathematics, Harvard University, Cambridge, MA (1978-1981). Santilli is the founder of The Institute for Basic Research, Cambridge, MA, and Palm Harbor, FL, of which he is president and full professor of theoretical physics (1981–). He also is the founder and editor in chief of three Journals, one in pure mathematics *Algebras, Groups and Geometries* (nine years of publication) and two in physics, the *Hadronic Journal* and the *Hadronic Journal Supplement* (fifteen years of publication). Santilli has organized twelve international workshops and conferences in the Lie-admissible theory, Lie-isotopic theory, and hadronic mechanics. As research associate, co-investigator or principal investigator, he has been the recipient of several reseaerch grants from U.S. Governmental Agencies, including: AFOSR, NASA, NSF, ERDA and DOE. In addition, Santilli is the author of seven research monographs in theoretical physics, and over one hundred papers published in various Journals, mostly devoted to mathematical, theoretical and experimental studies of nonlinerar, nonlocal and nonhamiltonian systems in classical and quantum mechanics. As a result of this intense academic activity, Santilli received several honors, including Gold Medals from the Cities of Orléans, France, and Campobasso, Italy. He was recently listed by the Estonian Academy of Sciences among the most illustrious scientists of all times.

ISBN 0–911767–55–X (Volumes I and II)